AF615024

Power System Harmonics

Power System Harmonics

J. ARRILLAGA
Department of Electrical Engineering,
University of Canterbury, New Zealand

D. A. BRADLEY
Department of Engineering,
University of Lancaster, UK

and

P. S. BODGER
Department of Electrical Engineering,
University of Canterbury, New Zealand

A Wiley–Interscience Publication

JOHN WILEY & SONS
Chichester · New York · Brisbane · Toronto · Singapore

Library of Congress Cataloging in Publication Data:
Arrillaga, J.
Power system harmonics.
'A Wiley–Interscience publication.'
Includes bibliographies and index.
1. Electric power systems. 2. Harmonics (Electric waves)
I. Bradley, D. A. II. Bodger, P. S. III. Title
TK3226.A75 1985 621.319′21 84–22097

ISBN 0 471 90640 9

British Library Cataloguing in Publication Data:
Arrillaga, J.
Power system harmonics.
1. Electric apparatus and appliances.
2. Harmonics (Electric waves) 3. Oscillations
I. Title II. Bradley, D. A. III. Bodger, P. S.
621.31′042 · TK454.2

ISBN 0 471 90640 9

Photosetting by Thomson Press (India) Ltd., New Delhi
Printed by Page Brothers, Norwich.

Contents

Preface

While the behaviour of a power system at the main generated frequency is reasonably well understood, the presence of harmonic frequencies means different things to different people and very often doesn't mean anything at all!

Is it then necessary and possible to derive enough information of sufficient general interest and practical application at this stage?

Judging by the present activity of international committees and by the growing number of conferences and technical papers in the field of power system harmonics, a book on this subject is relevant and timely.

The selection of the main underlying concepts and generally accepted techniques would have been a subjective and difficult task for a single author. However, after many individual proposals by the three authors involved, a reasonably unified pattern has eventually emerged which will, we hope, satisfy the present needs of the professional engineers involved in power system planning and operation. The book should also stimulate teachers, researchers and students in the power system and power electronics disciplines.

As well as describing the main harmonic causes and effects, the book covers four areas dealing with analysis, instrumentation, penetration and elimination. The present philosophies used for the setting of harmonic standards and limitations are also discussed.

The vast amount of information used in the preparation of the manuscript makes it impossible to acknowledge them here individually; instead they will be referred to at the end of the chapters. However, we must single out the support and encouragement received from Professor J. K. Bargh, Head of the Department of Electrical and Electronic Engineering (University of Canterbury) and from New Zealand Electricity, especially from K. D. McCool (General Manager), P. S. Barnett, M. C. Underhill, P. J. Morfee, L. A. Wilson, P. R. Hyland and R. J. Simpson.

We would also like to acknowledge the material and ideas derived from many discussions with J. F. Baird (System Software and Instrumentation, Christchurch), N. W. Ross (Central Canterbury Electric Power Board) and two of our colleagues, D. B. Watson and D. J. Byers.

Several graduate students have also taken part in the research programme associated with some of this work and in particular we would like to acknowledge the assistance of B. J. Harker, T. J. Densem, J. F. Eggleston and S. J. Grindrod.

Finally we wish to thank Mrs A. Haughan for her active participation in the preparation of the manuscript.

1

Introduction

1.1 BACKGROUND

In an ideal electrical power system, energy is supplied at a single and constant frequency, and at specified voltage levels of constant magnitudes. However, none of these conditions are fulfilled in practice. The problem of voltage and frequency deviations, and the means of keeping them under control, are the subject matter of conventional power system analysis. The problem of waveform distortion, so far neglected in power system treatises, constitutes the basis of this book.

Power system distortion is not a new phenomenon, and containing it to acceptable proportions has been a concern of power engineers from the early days of alternating current. The recent growing concern for this problem results from the increasing numbers and power ratings of the highly non-linear power electronic devices used in the control of power apparatus and systems.

The deviation from perfect sinusoids is generally expressed in terms of harmonic components. In this introductory chapter we are setting the scene by briefly defining the nature of harmonics and their importance. Subsequent chapters will deal in detail with the harmonic causes, effects, monitoring and penetration. Some consideration will also be given to the setting of limits.

To put the subject in historical perspective, it is necessary to go back to the 18th and 19th centuries when various mathematicians, and in particular J. B. J. Fourier (1768–1830), set up the basis for harmonic calculations.

With reference to power system harmonics, it was largely in Germany in the 1920s and 1930s when the subject of waveform distortion caused by static convertors was developed. The most influential source of convertor theory published during that period in the English language is the book by Rissik.[1] A classical paper on harmonic generation by static convertors was written in 1945 by J.C. Read[2] and is still widely used by designers today.

During the 1950s and 1960s the study of convertor harmonics was advanced in the field of high voltage direct current transmission. During this period a large number of papers were published. These are summarized in a book by Kimbark[3] which contains over 60 references in the field of power system harmonics. In the past few years the subject has been regularly discussed at international meetings and extensive bibliographies are produced from time to time. The latest example is the IEEE Power System Harmonics Working Group Report.[4]

1.2 NATURE OF HARMONICS

The term 'harmonic' originates from acoustics, where it signifies the vibration of a string or column of air at a frequency which is a multiple of the basic repetition (or fundamental) frequency. Similarly with electrical signals, a harmonic is defined as the content of the signal whose frequency is an integer multiple of the actual system frequency, i.e. the main frequency produced by the generators.

When a complex signal is viewed in an oscilloscope its shape is observed in the time domain; that is, for any given instant in time, the amplitude of the waveform is displayed. If the same signal is applied to a hi-fi amplifier, then the ear hears the resultant sound as a mixture of frequencies; that is, it sounds like a full musical chord. The waveform may therefore be described by its time domain or its frequency domain data. The transfer between these two domains forms the basis of Chapter 2.

It must be made clear from the outset that such a transfer is only perfectly applicable when the distorted waveform is maintained for an infinite number of cycles. This is not the case in practice, where variations in loading conditions will alter the system harmonic content. However, this problem presents no difficulty provided that the condition to be analysed persists for a reasonable time. It is thus necessary to distinguish between a harmonic, where the waveshape remains unaltered, and a transient where there is significant cycle to cycle variation in the waveshape.

Finally the phase relationship of the harmonic to the fundamental frequency is significant in determining the waveshape. In acoustics it is generally accepted that the audible effect is not affected by such a phase relationship. This is not the case with electrical signals, where the position of the harmonic and the relative phase of the same harmonic from different sources may alter the overall effect considerably.

1.3 IMPORTANCE OF THE SUBJECT

As with many other forms of pollution the generation of harmonics affects the whole (electrical) environment and probably at much larger distances from its points of origin.

Perhaps the most obvious consequence of power system harmonics is the degradation of telephone communications caused by induced harmonic noise. However, there are other less audible, though often more disastrous, effects such as the maloperation of important control and protection equipment and the overloading of power apparatus and systems. Very often the existence of waveform pollution is only detected following expensive casualties (like the destruction of power factor correction capacitors). Moreover, in the absence of an electrical welfare state, the casualties have to be repaired or replaced, and the equipment protected by filters, at the customers' expense, even though such preventive measures provide a general environmental improvement.

In recent years there have been considerable developments in industrial processes that rely on controlled rectification for their operation, and therefore generate current harmonics. However, the design of such equipment generally assumes the existence of a voltage source free of harmonic distortion, a situation which only occurs if the power system supplying the equipment has a very low harmonic impedance. Consequently the smaller industrial users of electricity are being subjected to increasing difficulties caused by the interaction of their own control equipment with the power supply.

Electricity supply authorities normally abrogate responsibility on harmonic matters by introducing standards or recommendations for the limitation of voltage harmonic levels at the points of common coupling between consumers.

However, determining limits on harmonic levels is not a straightforward exercise. Current knowledge is not sufficiently advanced to ascertain the extent to which any given power system can sustain a level of harmonics and remain viable in terms of the functions that the system is required to perform. As most current knowledge of harmonics has stemmed from a background of events, the standards and limitations so far produced have reflected the results of past practical experience with the aim of preventing similar problems in the future. This subject is discussed in Chapter 8.

Until a reasonable understanding of the power system harmonic phenomena is reached, the electric supply industry will remain in a position of higher than acceptable risk of having to respond to problems after they have occurred.

Two major impediments to such understanding are the ability to make accurate measurements (discussed in Chapters 6 and 7) and the state of power supply models capable of in-depth analysis to the degree of complexity required (discussed in Chapter 9).

Concern for waveform distortion should be shared by all electrical engineers in order to establish the right balance between exercising control by distortion and keeping distortion under control. There is a need for early co-ordination of decisions between the interested parties, in order to achieve acceptable economical solutions. The basis for such acceptability must be discussed at the earliest possible stage between manufacturers, power supply and communications authorities.

In order to educate the profession in this respect the electrical engineering curricula should convey the necessary interdisciplinary message. In particular power systems educators should try to broaden their frequency spectrum and lose respect for the frequency domain. They should also clarify the fact that zero sequence and third harmonic are not necessarily the same thing, and that the power factor is not exclusively concerned with the main power frequency. Fourier analysis, and in particular the fast Fourier transform, should be discussed with relevance to power as well as communication waveforms. Finally the teaching of power electronics should give sufficient coverage to its negative contribution, i.e. the problem of waveform distortion.

1.4 REFERENCES

1. Rissik, H. (1935). *The Mercury Arc Current Convertor*, Pitman, London.
2. Read, J. C. (1945). 'The calculation of rectifier and invertor performance characteristics'. *J. IEE, Pt II*, **92**, 495.
3. Kimbark, E. W. (1971). *Direct Current Transmission,* Vol. I, Wiley–Interscience, New York.
4. 'Bibliography of power system harmonics, Parts I and II'. IEEE papers 84WM 214–3 and 84WM 215–0, presented at the Winter Power Meeting, Dallas, January 1984.

2

Harmonic analysis

2.1 INTRODUCTION

In 1822 the French mathematician Jean Babtiste Joseph Fourier (1768–1830) in his work *Theorie analytique de la chaleur*[1] postulated that any continuous function repetitive in an interval T can be represented by the summation of a fundamental sinusoidal component, with a series of higher order harmonic components at frequencies which are integer multiples of the fundamental frequency.

Harmonic analysis is the process of calculating the magnitudes and phases of the fundamental and higher order harmonics of the periodic waveform. The resulting series is known as the Fourier series and establishes a relationship between a time domain function and that function in the frequency domain.

The Fourier series of a general periodic waveform is derived in the first part of this chapter and its characteristics discussed with reference to simple waveforms.

More generally, the Fourier transform and its inverse are used to map any function in the interval from $-\infty$ to ∞, in either the time or frequency domain, into a continuous function in the inverse domain. The Fourier series therefore represents the special case of the Fourier transform applied to a periodic signal.

In practice, data is often available in the form of a sampled time function, represented by a time series of amplitudes, separated by fixed time intervals of limited duration. When dealing with such data a modification of the Fourier transform, the discrete Fourier transform, is used. The implementation of the discrete Fourier transform by means of the fast Fourier transform algorithms forms the basis of most modern spectral and harmonic analysis systems. The development of the Fourier and discrete Fourier transforms is also examined in this chapter along with the implementation of the fast Fourier transform.

2.2 BASIC CONCEPTS

Periodic functions[2]

A function $x(t)$ is said to be periodic if it is defined for all real t and if there is some positive number T such that

$$x(t+T)=x(t) \quad \text{for} \quad \text{all } t. \tag{2.2.1}$$

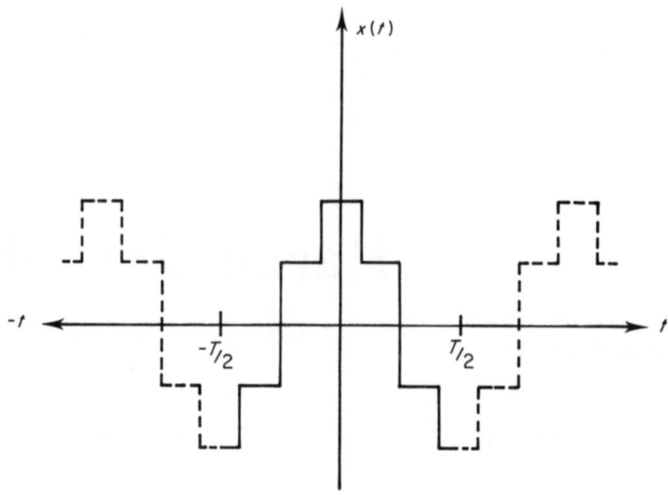

Figure 2.1. Periodic function

T is called the period of the function. Such a function can be represented by the periodic repetition of the waveform at intervals of T, as depicted in Figure 2.1.

If k is any integer then it follows that

$$x(t + kT) = x(t) \quad \text{for} \quad \text{all t.} \tag{2.2.2}$$

If two functions $x_1(t)$ and $x_2(t)$ have the same period T, then the function

$$x_3(t) = ax_1(t) + bx_2(t), \tag{2.2.3}$$

where a and b are constants, also has the period T.

It should be noted that the function

$$x(t) = \text{constant} \tag{2.2.4}$$

is also a periodic function in the sense of the definition, because it satisfies equation (2.2.2) for any positive period T.

Orthogonal functions[2]

Two non-zero functions $x_1(t)$ and $x_2(t)$ are considered orthogonal over an interval $T_1 \rightarrow T_2$ if

$$\int_{T_1}^{T_2} x_1(t)x_2(t)\mathrm{d}t = 0. \tag{2.2.5}$$

Furthermore, a set of r functions

$$\{x_1(t), x_2(t), \ldots, x_r(t)\}$$

form an orthogonal set over the interval $T_1 \rightarrow T_2$ if

$$\int_{T_1}^{T_2} x_i(t)x_j(t)\,dt = 0 \quad \text{for} \quad i=1 \text{ to } r, j=1 \text{ to } r, i \neq j. \tag{2.2.6}$$

2.3 FOURIER ANALYSIS

Fourier series and coefficients[2,3]

The Fourier series of a periodic function $x(t)$ has the expression

$$x(t) = a_0 + \sum_{n=1}^{\infty}\left(a_n \cos\left(\frac{2\pi nt}{T}\right) + b_n \sin\left(\frac{2\pi nt}{T}\right)\right). \tag{2.3.1}$$

This constitutes a frequency domain representation of the periodic function.

In this expression a_0 is the average value of the function $x(t)$, whilst a_n and b_n, the coefficients of the series, are the rectangular components of the nth harmonic. The corresponding nth harmonic vector is

$$A_n\underline{/\phi_n} = a_n + jb_n \tag{2.3.2}$$

with a magnitude

$$A_n = \sqrt{(a_n^2 + b_n^2)}$$

and a phase angle

$$\phi_n = \tan^{-1}\left(\frac{b_n}{a_n}\right).$$

For a given function $x(t)$, the constant coefficient, a_0, can be derived by integrating both sides of equation (2.3.1) from $-T/2$ to $T/2$ (over a period T), i.e.

$$\int_{-T/2}^{T/2} x(t)\mathrm{d}t = \int_{-T/2}^{T/2}\left[a_0 + \sum_{n=1}^{\infty}\left[a_n \cos\left(\frac{2\pi nt}{T}\right) + b_n \sin\left(\frac{2\pi nt}{T}\right)\right]\right]\mathrm{d}t. \tag{2.3.3}$$

The Fourier series of the right hand side can be integrated term by term, giving

$$\int_{-T/2}^{T/2} x(t)\,\mathrm{d}t =$$

$$a_0\int_{-T2}^{T/2}\mathrm{d}t + \sum_{n=1}^{\infty}\left[a_n\int_{-T/2}^{T/2}\cos\left(\frac{2\pi nt}{T}\right)\mathrm{d}t + b_n\int_{-T/2}^{T/2}\sin\left(\frac{2\pi nt}{T}\right)\mathrm{d}t\right]. \tag{2.3.4}$$

The first term on the right hand side equals Ta_0, while the other integrals are zero. Hence the constant coefficient of the Fourier series is given by

$$a_0 = \frac{1}{T}\int_{-T/2}^{T/2} x(t)\mathrm{d}t, \tag{2.3.5}$$

which is the area under the curve of $x(t)$ from $-T/2$ to $T/2$, divided by the period of the waveform, T.

The a_n coefficients can be determined by multiplying equation (2.3.1) by $\cos(2\pi mt/T)$, where m is any fixed positive integer, and integrating between $-T/2$ and $T/2$, as previously, i.e.

$$\int_{-T/2}^{T/2} x(t)\cos\left(\frac{2\pi mt}{T}\right)\mathrm{d}t =$$

$$\int_{-T/2}^{T/2}\left[a_0 + \sum_{n=1}^{\infty}\left[a_n\cos\left(\frac{2\pi nt}{T}\right) + b_n\sin\left(\frac{2\pi nt}{T}\right)\right]\right]\cos\left(\frac{2\pi mt}{T}\right)\mathrm{d}t \tag{2.3.6}$$

$$= a_0\int_{-T/2}^{T/2}\cos\left(\frac{2\pi mt}{T}\right)\mathrm{d}t + \sum_{n=1}^{\infty}\left[a_n\int_{-T/2}^{T/2}\cos\left(\frac{2\pi nt}{T}\right)\cos\left(\frac{2\pi mt}{T}\right)\mathrm{d}t\right.$$
$$\left.+\, b_n\int_{-T/2}^{T/2}\sin\left(\frac{2\pi nt}{T}\right)\cos\left(\frac{2\pi mt}{T}\right)\mathrm{d}t\right]. \tag{2.3.7}$$

The first term on the right hand side is zero, as are all the terms in b_n, since $\sin(2\pi nt/T)$ and $\cos(2\pi mt/T)$ are orthogonal functions for all n and m.

Similarly, the terms in a_n are zero, being orthogonal, unless $m = n$. In this case equation (2.3.7) becomes

$$\int_{-T/2}^{T/2} x(t)\cos\left(\frac{2\pi nt}{T}\right)\mathrm{d}t = a_n\int_{-T/2}^{T/2}\cos^2\left(\frac{2\pi nt}{T}\right)\mathrm{d}t$$
$$= \frac{a_n}{2}\int_{-T/2}^{T/2}\cos\left(\frac{4\pi nt}{T}\right)\mathrm{d}t + \frac{a_n}{2}\int_{-T/2}^{T/2}\mathrm{d}t \tag{2.3.8}$$

The first term on the right hand side is zero while the second term equals $a_nT/2$. Hence the coefficients a_n can be obtained from

$$a_n = \frac{2}{T}\int_{-T/2}^{T/2} x(t)\cos\left(\frac{2\pi nt}{T}\right)\mathrm{d}t \quad \text{for} \quad n = 1 \to \infty. \tag{2.3.9}$$

To determine the coefficients b_n, equation (2.3.1) is multiplied by $\sin(2\pi mt/T)$ and, by a similar argument to the above,

$$b_n = \frac{2}{T}\int_{-T/2}^{T/2} x(t)\sin\left(\frac{2\pi nt}{T}\right)\mathrm{d}t \quad \text{for} \quad n = 1 \to \infty. \tag{2.3.10}$$

It should be noted that because of the periodicity of the integrands in equations (2.3.5), (2.3.9) and (2.3.10) the interval of integration can be taken more generally as t and $t + T$.

If the function $x(t)$ is piecewise continuous (i.e. has a finite number of vertical jumps) in the interval of integration, the integrals exist and Fourier coefficients can be calculated for this function.

Equations (2.3.5), (2.3.9) and (2.3.10) are often expressed in terms of the angular frequency as follows:

$$a_0 = \frac{1}{2\pi}\int_{-\pi}^{\pi} x(\omega t)\mathrm{d}(\omega t), \tag{2.3.11}$$

$$a_n = \frac{1}{\pi}\int_{-\pi}^{\pi} x(\omega t)\cos(n\omega t)\,\mathrm{d}(\omega t), \tag{2.3.12}$$

$$b_n = \frac{1}{\pi}\int_{-\pi}^{\pi} x(\omega t)\sin(n\omega t)\,\mathrm{d}(\omega t). \tag{2.3.13}$$

Simplifications resulting from waveform symmetry[2,3]

Equations (2.3.5), (2.3.9) and (2.3.10), the general formulae for the Fourier coefficients, can be represented as the sum of two separate integrals, i.e.

$$a_n = \frac{2}{T}\int_0^{T/2} x(t)\cos\left(\frac{2\pi nt}{T}\right)\mathrm{d}t + \frac{2}{T}\int_{-T/2}^{0} x(t)\cos\left(\frac{2\pi nt}{T}\right)\mathrm{d}t, \tag{2.3.14}$$

$$b_n = \frac{2}{T}\int_0^{T/2} x(t)\sin\left(\frac{2\pi nt}{T}\right)\mathrm{d}t + \frac{2}{T}\int_{-T/2}^{0} x(t)\sin\left(\frac{2\pi nt}{T}\right)\mathrm{d}t. \tag{2.3.15}$$

Replacing t by $-t$ in the second integral of equation (2.3.14), with limits $(-T/2, 0)$,

$$a_n = \frac{2}{T}\int_0^{T/2} x(t)\cos\left(\frac{2\pi nt}{T}\right)\mathrm{d}t + \frac{2}{T}\int_{T/2}^{0} x(-t)\cos\left(\frac{-2\pi nt}{T}\right)\mathrm{d}(-t)$$

$$= \frac{2}{T}\int_0^{T/2} [x(t) + x(-t)]\cos\left(\frac{2\pi nt}{T}\right)\mathrm{d}t. \tag{2.3.16}$$

Similarly,

$$b_n = \frac{2}{T}\int_0^{T/2} [x(t) - x(-t)]\sin\left(\frac{2\pi nt}{T}\right)\mathrm{d}t. \tag{2.3.17}$$

ODD SYMMETRY

The waveform has odd symmetry if

$$x(t) = -x(-t).$$

Then the a_n terms become zero for all n, while

$$b_n = \frac{4}{T}\int_0^{T/2} x(t)\sin\left(\frac{2\pi nt}{T}\right)\mathrm{d}t. \tag{2.3.18}$$

The Fourier series for an odd function will therefore contain only sine terms.

EVEN SYMMETRY

The waveform has even symmetry if

$$x(t) = x(-t).$$

In this case

$$b_n = 0 \quad \text{for} \quad \text{all } n$$

and

$$a_n = \frac{4}{T}\int_0^{T/2} x(t)\cos\left(\frac{2\pi nt}{T}\right)\mathrm{d}t. \tag{2.3.19}$$

The Fourier series for an even function will therefore contain only cosine terms.

Certain waveforms may be odd or even depending on the time reference position selected. For instance the square wave of Figure 2.2, drawn as an odd

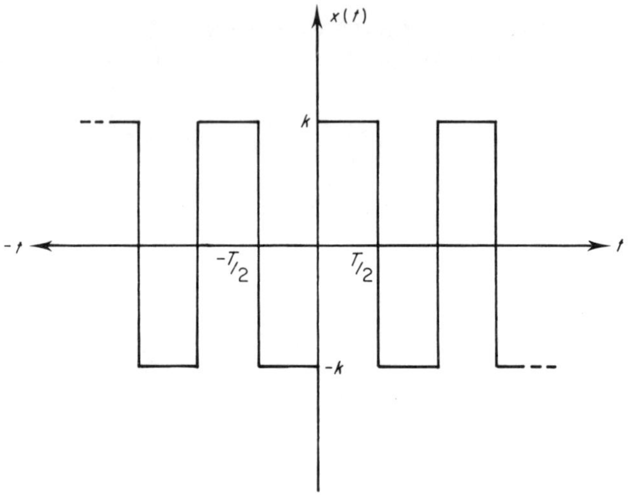

Figure 2.2. Square wave function

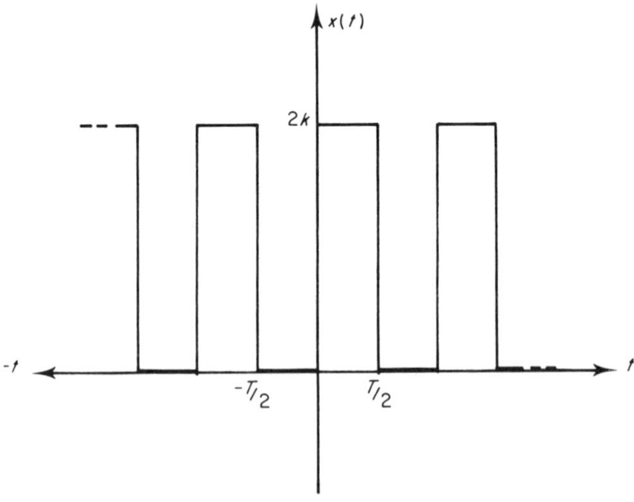

Figure 2.3. Odd function square wave plus constant term

function, can be transformed into an even function simply by shifting the origin (vertical axis) by $T/2$.

It is a matter of analytical convenience and simplicity that a function should be transformed into an odd or even function, provided that the type of waveform makes this possible.

The horizontal axis can also be selected to simplify the Fourier series representation. For example the horizontal axis of the wave function of Figure 2.3 can be relocated at $x(t) = k$ to give an odd function square wave plus a constant term.

HALFWAVE SYMMETRY

A function $x(t)$ has halfwave symmetry if

$$x(t) = -x(t + T/2), \tag{2.3.20}$$

i.e. the shape of the waveform over a period $t + T/2$ to $t + T$ is the negative of the shape of the waveform over the period t to $t + T/2$.

Consequently the square wave function of Figure 2.2 has halfwave symmetry with $t = -T/2$.

Using equation (2.3.9) and replacing (t) by $(t + T/2)$ in the interval $(-T/2, 0)$,

$$\begin{aligned} a_n &= \frac{2}{T}\int_0^{T/2} x(t)\cos\left(\frac{2\pi nt}{T}\right)\mathrm{d}t + \frac{2}{T}\int_{-T/2+T/2}^{0+T/2} x(t+T/2)\cos\left(\frac{2\pi n(t+T/2)}{T}\right)\mathrm{d}t \\ &= \frac{2}{T}\int_0^{T/2} x(t)\left[\cos\left(\frac{2\pi nt}{T}\right) - \cos\left(\frac{2\pi nt}{T} + n\pi\right)\right]\mathrm{d}t \end{aligned} \tag{2.3.21}$$

since by definition $x(t) = -x(t + T/2)$. If n is an odd integer then,

$$\cos\left(\frac{2\pi nt}{T} + n\pi\right) = -\cos\left(\frac{2\pi nt}{T}\right)$$

and

$$a_n = \frac{4}{T}\int_0^{T/2} x(t)\cos\left(\frac{2\pi nt}{T}\right)\mathrm{d}t. \tag{2.3.22}$$

However, if n is an even integer then

$$\cos\left(\frac{2\pi nt}{T} + n\pi\right) = \cos\left(\frac{2\pi nt}{T}\right)$$

and

$$a_n = 0.$$

Similarly,

$$b_n = \begin{cases} \dfrac{4}{T}\displaystyle\int_0^{T/2} x(t)\sin\left(\frac{2\pi nt}{T}\right)\mathrm{d}t & \text{for } n \text{ odd,} \\ 0 & \text{for } n \text{ even.} \end{cases} \tag{2.3.23}$$

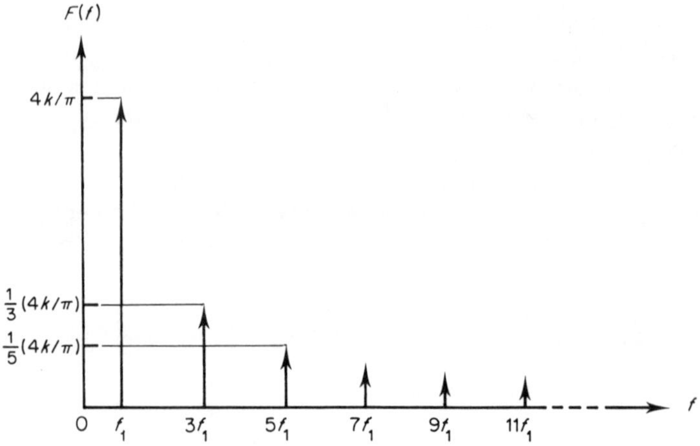

Figure 2.4. Line spectrum representation of a square wave

Thus, waveforms which have halfwave symmetry contain only odd order harmonics.

The square wave of Figure 2.2 is an odd function with halfwave symmetry. Consequently only the b_n coefficients and odd harmonics will exist. The expression for the coefficients, taking into account these conditions, is

$$b_n = \frac{8}{T}\int_0^{T/4} x(t)\sin\left(\frac{2\pi nt}{T}\right)\mathrm{d}t, \tag{2.3.24}$$

which can be represented by a line spectrum of amplitudes inversely proportional to the harmonic order, as shown in Figure 2.4.

2.4 FINITE INTERVAL FUNCTIONS[3]

A function which is only defined for a finite interval [0, T] can be approximated using a Fourier series by assuming that the function is periodic and hence repeats itself in some fashion outside the defined interval. The form of periodicity chosen will give rise to different terms in the Fourier series.

The function, $x(t)$, of Figure 2.5, which is only defined for the interval [0, T], can be considered as forming part of any of the odd, even or halfwave symmetric periodic functions, of period T, shown in Figure 2.6. These have only sine terms,

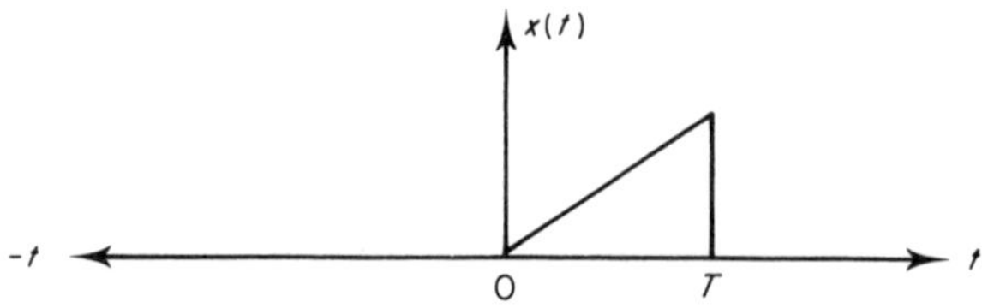

Figure 2.5. Function with finite time definition

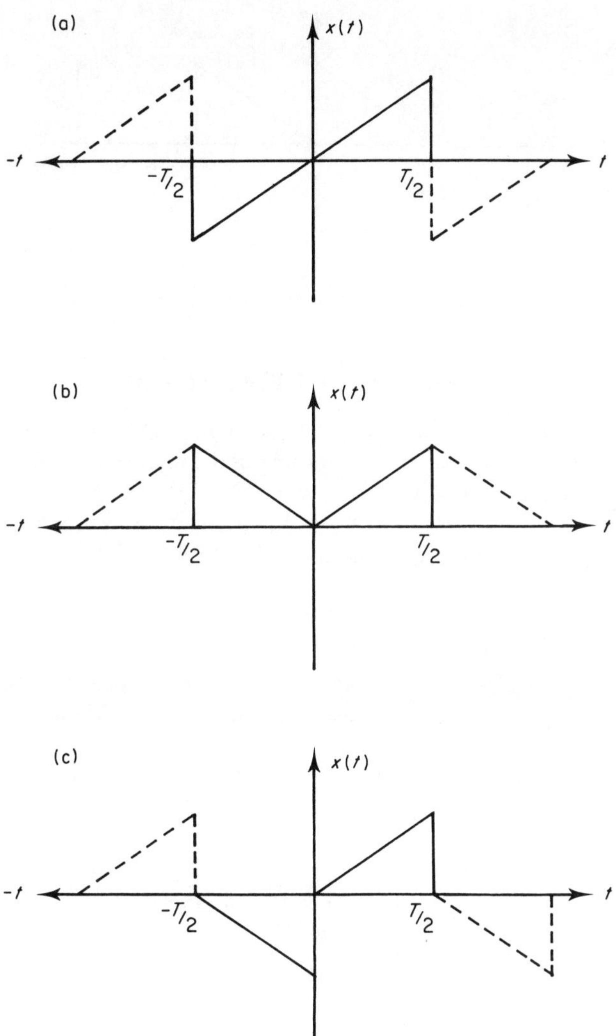

Figure 2.6. Periodic representation of a finite interval function: (a) odd function; (b) even function; (c) half-wave symmetry.

cosine terms or odd harmonic sine and cosine terms respectively in their Fourier series.

Alternatively, the function can be repeated without reflection about either the vertical or horizontal axis. In this case, Figure 2.7, the period of the waveform is half that of the previous representations of Figure 2.6.

This is the form of observation of a waveform on which spectrum analysis instrumentation is based. The techniques and limitations of such analysis are discussed in Sections 2.6–2.12 and Chapter 6.

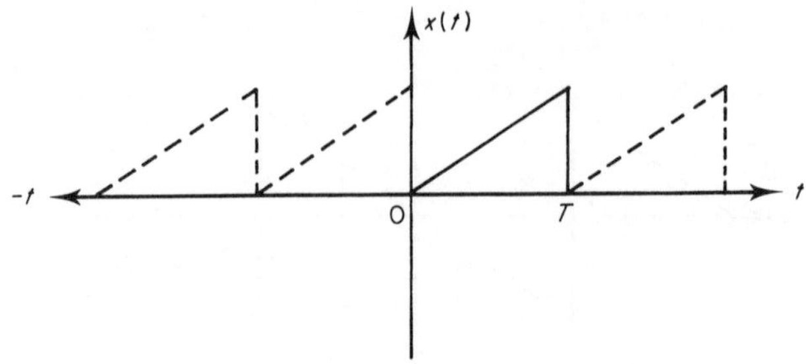

Figure 2.7. Implied repetition of a function

2.5 COMPLEX FORM OF THE FOURIER SERIES

The representation of the frequency components as rotating vectors in the complex plane gives a physical understanding of the relationship of waveforms in the time and frequency domains.[4]

A uniformly rotating vector $(A/2)e^{+j\phi}$ has a constant magnitude $A/2$ and a phase angle ϕ which is time varying according to

$$\phi = 2\pi ft + \theta, \tag{2.5.1}$$

where θ is the initial phase angle when $t = 0$.

A second vector $(A/2)e^{-j\phi}$ will rotate in the opposite direction to $(A/2)e^{+j\phi}$. This negative rate of change of phase angle can be considered as a negative frequency.

The sum of these two vectors will always lie along the real axis, as in Figure 2.8,

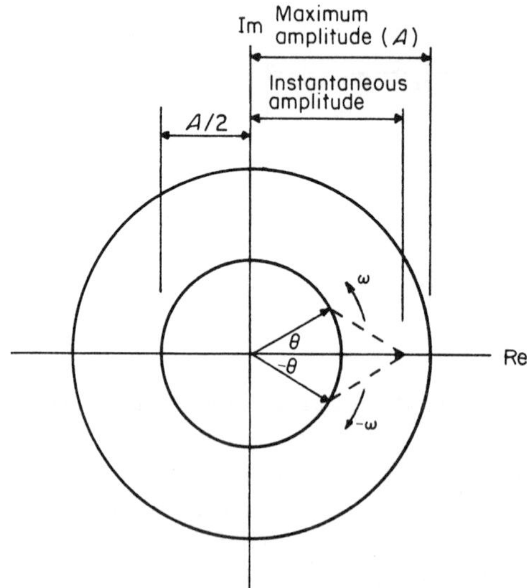

Figure 2.8. Contra-rotating vector pair producing a varying amplitude (pulsating) vector

with the magnitude oscillating between A and $-A$ according to

$$\frac{A}{2}\mathrm{e}^{+j\phi}+\frac{A}{2}\mathrm{e}^{-j\phi}=A\cos\phi. \tag{2.5.2}$$

Rewriting the Fourier series of equation (2.3.1) as

$$x(t)=a_0+A_1\sin(\omega t+\phi_1)+A_2\sin(2\omega t+\phi_2)+\cdots, \tag{2.5.3}$$

where $x(t)$ is periodic with period T and $\omega=2\pi/T=2\pi f$, the nth component of this series, corresponding to the harmonic at a frequency of $f_n=nf$, is given by

$$X(f_n)=\frac{1}{T}\int_{-T/2}^{T/2}x(t)\mathrm{e}^{-j2\pi f_n t}\,\mathrm{d}t, \tag{2.5.4}$$

where $\mathrm{e}^{-j2\pi f_n t}$ is the unit vector and $X(f_n)$ gives amplitude and phase for the harmonic vector (cf. equation (2.3.2)).

For a real-valued signal each component at a frequency f_n will be associated with a component at a frequency $-f_n$ of equal amplitude but opposite phase. Thus, each harmonic component of the real-valued signal can be represented by two half-amplitude contra-rotating vectors of initial values such that

$$X(f_n)=X^*(-f_n), \tag{2.5.5}$$

where $X^*(-f_n)$ is the complex conjugate of $X(-f_n)$.

Expressing equation (2.5.4) in terms of the fundamental frequency f,

$$X(nf)=\frac{1}{T}\int_{-T/2}^{T/2}x(t)\mathrm{e}^{-j2\pi nft}\,\mathrm{d}t. \tag{2.5.6}$$

If the time domain signal $x(t)$ contains a component rotating at a single frequency nf, then multiplication by the unit vector $\mathrm{e}^{-j2\pi nft}$, which rotates at a frequency $-nf$, annuls the rotation of the component such that the integration over time has a finite value.

All components at other frequencies will continue to rotate after multiplication by $\mathrm{e}^{-j2\pi nft}$ and will thus integrate over time to zero.

The original time domain function may then be obtained by multiplying the $X(f_n)$ terms by the unit vector $c^{j2\pi f_n t}$ and summing, i.e.

$$x(t)=\sum_{n=-\infty}^{\infty}X(f_n)\mathrm{e}^{j2\pi f_n t}, \tag{2.5.7}$$

in which $f_{-n}=-f_n$.

2.6 THE FOURIER TRANSFORM[3,4,5]

Fourier analysis, when applied to a continuous, periodic signal in the time domain, yields a series of discrete frequency components in the frequency domain.

By allowing the integration period T to extend to infinity, the spacing between the harmonic frequencies tends to zero and the function $X(f_n)$ of equation (2.5.4) becomes a continuous and infinite function of frequency such that

$$X(f)=\int_{-\infty}^{\infty} x(t)e^{-j2\pi ft}\,dt. \tag{2.6.1}$$

The expression for the time domain function $x(t)$, which is also continuous and infinite, in terms of $X(f)$ is then

$$x(t)=\int_{-\infty}^{\infty} X(f)e^{j2\pi ft}\,df \tag{2.6.2}$$

(cf. equation (2.5.7)). $X(f)$ is known as the spectral density function of $x(t)$.

Equations (2.6.1) and (2.6.2) form the Fourier transform pair. Equation (2.6.1) is referred to as the 'forward transform' and equation (2.6.2) as the 'reverse' or 'inverse transform'.

In general $X(f)$ is complex and can be written as

$$X(f)=\operatorname{Re}X(f)+j\operatorname{Im}X(f). \tag{2.6.3}$$

The real part of $X(f)$ is obtained from

$$\begin{aligned}\operatorname{Re}X(f)&=\tfrac{1}{2}[X(f)+X(-f)] \\ &=\int_{-\infty}^{\infty} x(t)\cos 2\pi ft\,dt.\end{aligned} \tag{2.6.4}$$

Similarly, for the imaginary part of $X(f)$,

$$\begin{aligned}\operatorname{Im}X(f)&=\tfrac{1}{2}j[X(f)-X(-f)] \\ &=-\int_{-\infty}^{\infty} x(t)\sin 2\pi ft\,dt.\end{aligned} \tag{2.6.5}$$

The amplitude spectrum of the frequency signal is obtained from

$$|X(f)|=[(\operatorname{Re}X(f))^2+(\operatorname{Im}X(f))^2]^{1/2}. \tag{2.6.6}$$

The phase spectrum is

$$\phi(f)=\tan^{-1}\left[\frac{\operatorname{Im}X(f)}{\operatorname{Re}X(f)}\right]. \tag{2.6.7}$$

Using equations (2.6.3)–(2.6.7) the inverse Fourier transform can be expressed in terms of the magnitude and phase spectra components:

$$x(t)=\int_{-\infty}^{\infty} |X(f)|\cos[2\pi ft-\phi(f)]\,df. \tag{2.6.8}$$

As an example let us consider a rectangular function such as Figure 2.9, defined by

$$x(t)=\begin{cases}K & \text{for} \quad |t|\leqslant T/2,\\ 0 & \text{for} \quad |t|>T/2;\end{cases}$$

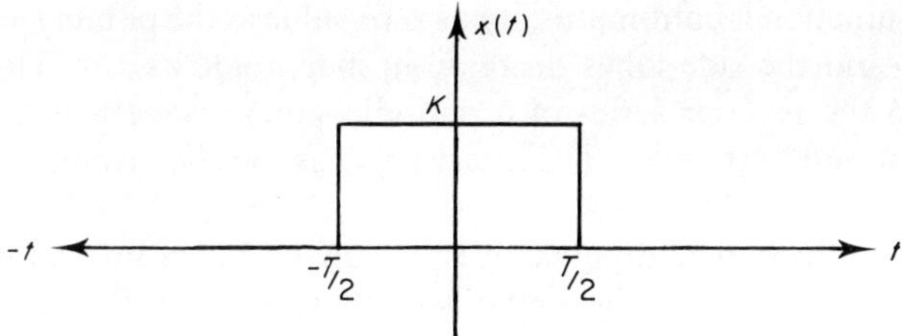

Figure 2.9. Rectangular function

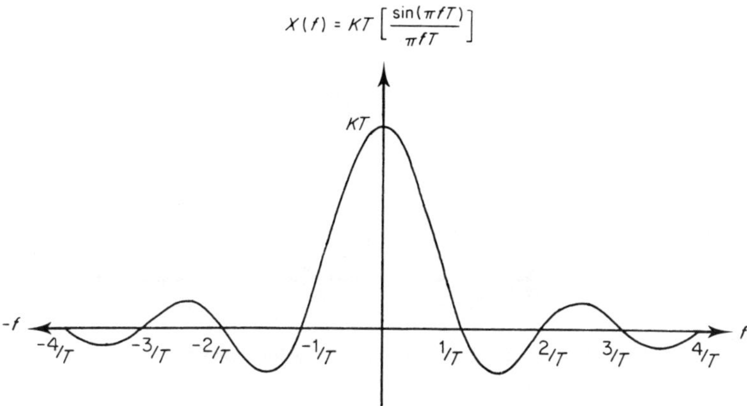

Figure 2.10. The sinc function, $\sin(\pi ft)/\pi ft$.

i.e. the function is continuous over all t but is zero outside the limits $(-T/2, T/2)$.

Its Fourier transform is

$$
\begin{aligned}
X(f) &= \int_{-T/2}^{T/2} x(t)\mathrm{e}^{-j2\pi ft}\,\mathrm{d}t \\
&= \int_{-T/2}^{T/2} K\mathrm{e}^{-j2\pi ft}\,\mathrm{d}t \\
&= \frac{-K}{\pi f}\frac{1}{2j}[\mathrm{e}^{-j\pi fT} - \mathrm{e}^{j\pi fT}],
\end{aligned}
\tag{2.6.9}
$$

and using the identity

$$\sin\phi = \frac{1}{2j}(\mathrm{e}^{j\phi} - \mathrm{e}^{-j\phi})$$

yields the following expression for the Fourier transform:

$$X(f) = \frac{K}{\pi f}\sin(\pi fT) = KT\left[\frac{\sin(\pi fT)}{\pi fT}\right]. \tag{2.6.10}$$

The term in brackets, the $\sin x/x$ or sinc function, is shown in Figure 2.10.

While the function is continuous, it has zero value at the points $f = n/T$ for $n = \pm 1, \pm 2, \cdots$, and the side lobes decrease in magnitude as $1/T$. This should be compared to the Fourier series of a periodic square wave which has discrete frequencies at odd harmonics. The interval $1/T$ is the effective bandwidth of the signal.

The power in the signal, obtained by squaring this function, is shown in Figure 2.11. Most of the power of the signal is contained within the frequency range $-1/T < f < 1/T$. The total power is the area under the curve of Figure 2.11, i.e.

$$P = K^2T^2 \int_{-\infty}^{\infty} \frac{\sin^2(\pi f T)}{(\pi f T)^2} \mathrm{d}f, \tag{2.6.11}$$

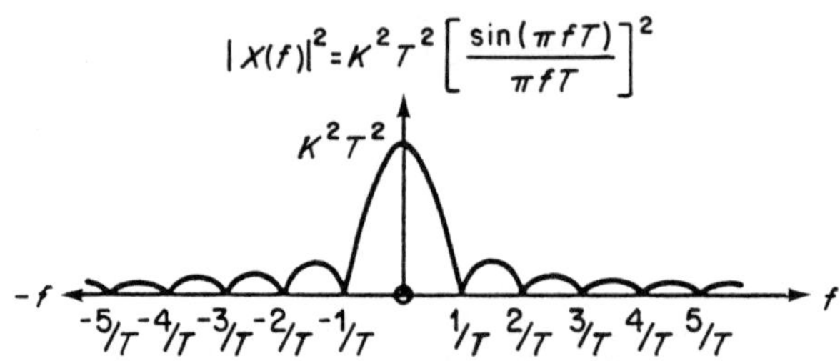

Figure 2.11. Power spectrum of a sinc function

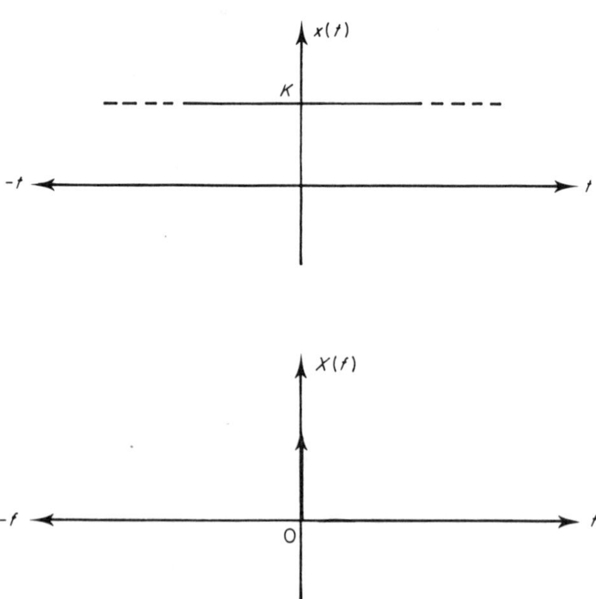

Figure 2.12. Infinite d.c. time domain signal and its frequency spectrum

and since the value of the integral is $1/T$, the total power is

$$P = K^2T^2\left(\frac{1}{T}\right), \tag{2.6.12}$$

i.e. the magnitude of the signal squared multiplied by the effective bandwidth.

It should be observed that the greater the duration of the signal in the time domain (i.e. as T increases) then the narrower the bandwidth becomes. In the limit, the Fourier transform of a square wave function where $T \to \infty$ is an impulse function in the frequency domain at $f = 0$, as shown in Figure 2.12.

The Fourier transform pair is such that if $x(t) \leftrightarrow X(f)$ then $X(t) \leftrightarrow x(-f)$, where $\leftrightarrow$ denotes the Fourier transformation. Thus a rectangular function in the frequency domain, representing a band-limited signal, transforms to a sinc function in the time domain.

Finiteness of energy and power[4]

A signal $x(t)$ is said to be finite in energy, W, if

$$W = \int_{-\infty}^{\infty} |x(t)|^2 \, \mathrm{d}t < \infty \tag{2.6.13}$$

and this is a sufficient condition for the existence of the Fourier transform of the signal.

Periodic signals, which are considered to extend to infinity in time, theoretically have infinite energy. But since the existence of a Fourier series requires that the signal be finite in power, P, the following condition must be met:

$$P = \lim_{T \to \infty} \frac{1}{2T} \int_{-T}^{T} |x(t)|^2 \, \mathrm{d}t < \infty. \tag{2.6.14}$$

Thus the modification of the Fourier transform for periodic signals results in a power spectrum, defined by

$$S(f) = \lim_{T \to \infty} \frac{1}{2T} \left| \int_{-T}^{T} x(t) \mathrm{e}^{-j2\pi f t} \, \mathrm{d}t \right|^2. \tag{2.6.15}$$

Convolution[4,5]

Consider Figure 2.13 in which a continuous signal $g(t)$ is applied to a system with an impulse response of $h(t)$. If the signal $g(t)$ is represented by a series of impulses $[g(t_n)]$, each separated by a time interval Δt, these then become the input to the system. Each of these impulses will produce a response from the system in the form of its impulse response scaled according to the amplitude of the impulse $g(t_n)$. The output of the system at any instant t_0 is the sum of the individual impulse responses at that time. This can be written as

$$y(t_0) = \sum_{n=-\infty}^{\infty} y_n(t_0) = \sum_{n=-\infty}^{\infty} g(t_n) h(t_0 - t_n). \tag{2.6.16}$$

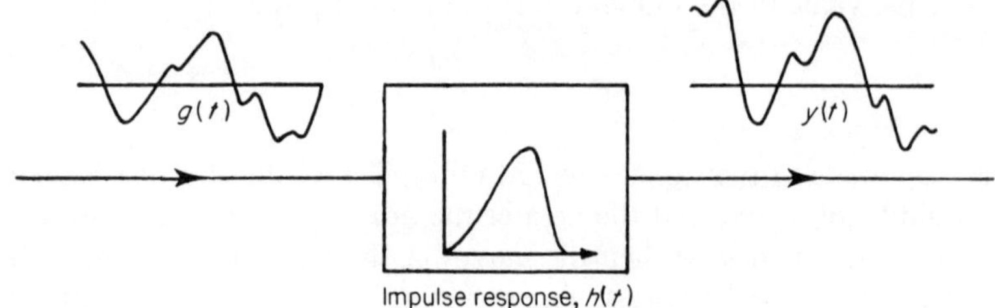

Figure 2.13. Convolution by a circuit of impulse response $h(t)$

More generally,

$$y(t)=\sum_{n=-\infty}^{\infty} g(t_n)h(t-t_n), \tag{2.6.17}$$

which in the limit as $t\to 0$ tends to

$$y(t)=\int_{-\infty}^{\infty} g(\tau)h(t-\tau)\mathrm{d}\tau. \tag{2.6.18}$$

This is the convolution integral and is conventionally written as

$$y(t)=g(t)*h(t), \tag{2.6.19}$$

using '$*$' to mean 'convolved with'.

The Fourier transform has the effect of transforming a convolution in the time domain into a multiplication in the frequency domain.

Let the forward transforms of $g(t)$, $h(t)$ and $y(t)$ be

$$G(f)=\int_{-\infty}^{\infty} g(t)\mathrm{e}^{-j2\pi ft}\mathrm{d}t, \tag{2.6.20}$$

$$H(f)=\int_{-\infty}^{\infty} h(t)\mathrm{e}^{-j2\pi ft}\mathrm{d}t \tag{2.6.21}$$

and

$$Y(f)=\int_{-\infty}^{\infty} y(t)\mathrm{e}^{-j2\pi ft}\mathrm{d}t; \tag{2.6.22}$$

as before

$$y(t)=g(t)*h(t)=\int_{-\infty}^{\infty} g(\tau)h(t-\tau)\mathrm{d}\tau. \tag{2.6.23}$$

Using equations (2.6.22) and (2.6.23),

$$Y(f)=\int_{-\infty}^{\infty}\left[\int_{-\infty}^{\infty} g(\tau)h(t-\tau)\mathrm{d}\tau\right]\mathrm{e}^{-j2\pi ft}\mathrm{d}t; \tag{2.6.24}$$

re-arranging and putting $u = (t - \tau)$,

$$Y(f) = \left[\int_{-\infty}^{\infty} g(\tau)e^{-j2\pi f\tau}d\tau\right]\left[\int_{-\infty}^{\infty} h(u)e^{-j2\pi fu}du\right]; \qquad (2.6.25)$$

hence

$$Y(f) = G(f)H(f). \qquad (2.6.26)$$

A similar argument can be used with the inverse transform to show that if

$$Y(f) = G(f) * H(f) \qquad (2.6.27)$$

then

$$y(t) = g(t)h(t). \qquad (2.6.28)$$

In general, convolution in one domain leads to multiplication in the other. Graphically, Figure 2.14 represents convolution by folding $h(\tau)$ about the $\tau = 0$

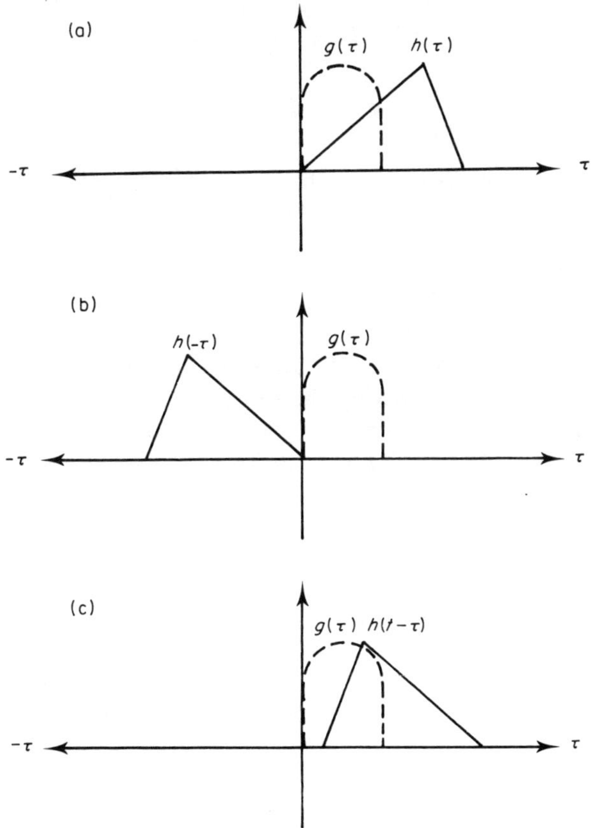

Figure 2.14. Graphical representation of the convolution of two signals in the time domain: (a) actual signal; (b) folding $h(\tau)$ about the $\tau = 0$ axis; (c) shifting $h(-\tau)$ along the τ axis.

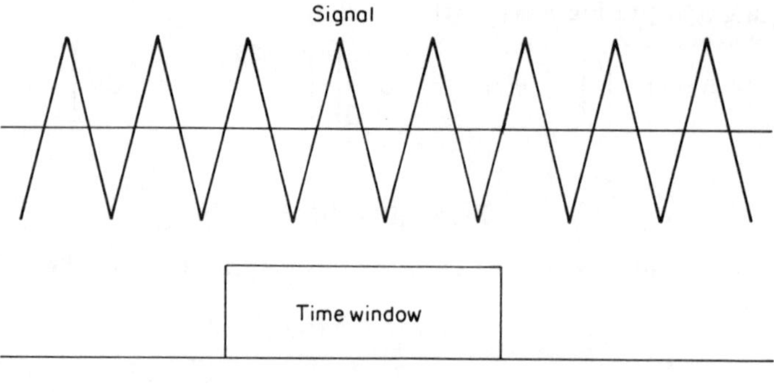

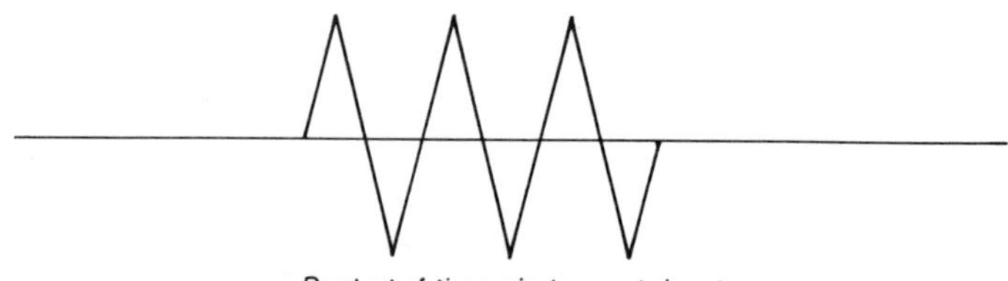

Figure 2.15. Influence of viewing a continuous function through a rectangular time window

axis (Figure 2.14(b)) and then moving the result in the positive direction along the τ axis (Figure 2.14(c)).

The value of $y(t)$ is then the area under the product of $g(\tau)$ and $h(t-\tau)$ for each τ.

In a practical system, restricting the sampling period to an interval of T seconds is equivalent to multiplying the signal in the time domain by a rectangular pulse of length T (Figure 2.15). This corresponds to the convolution in the frequency domain of their respective frequency spectra.

2.7 SAMPLED TIME FUNCTIONS[4,5]

With an increase in the digital processing of data, functions are often recorded by samples in the time domain. Thus the signal can be represented as in Figure 2.16, where $f_s = 1/t_1$ is the frequency of the sampling. In this case the Fourier transform of the signal is expressed as the summation of the discrete signal where each sample is multiplied by $e^{-j2\pi f n t_1}$; that is

$$X(f) = \sum_{n=-\infty}^{\infty} x(nt_1) e^{-j2\pi f n t_1}. \qquad (2.7.1)$$

The frequency domain spectrum, shown in Figure 2.17, is periodic and continuous.

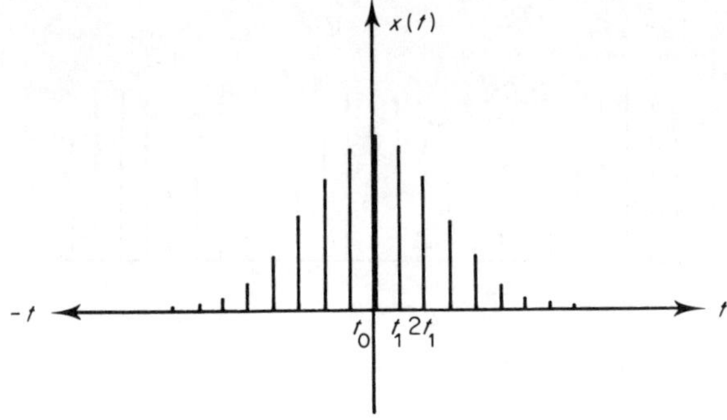

Figure 2.16. Sampled time domain function

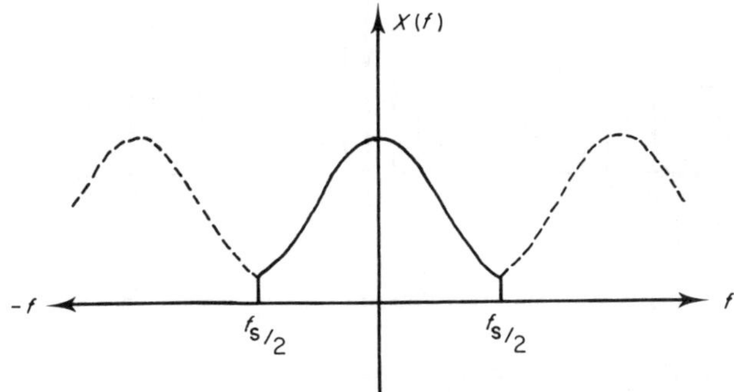

Figure 2.17. Frequency spectrum for discrete time domain function

The inverse Fourier transform is thus

$$x(t) = \frac{1}{f_s} \int_{-f_s/2}^{f_s/2} X(f) e^{j2\pi f n t_1} \, df. \tag{2.7.2}$$

2.8 DISCRETE FOURIER TRANSFORM[4,5]

In the case where the frequency domain spectrum is a sampled function, as well as the time domain function, we obtain a Fourier transform pair made up of discrete components

$$X(f_k) = \frac{1}{N} \sum_{n=0}^{N-1} x(t_n) e^{-j2\pi kn/N} \tag{2.8.1}$$

and

$$x(t_n) = \sum_{k=0}^{N-1} X(f_k) e^{j2\pi kn/N}. \tag{2.8.2}$$

Both the time domain function and the frequency domain spectrum are assumed periodic, as in Figure 2.18, with a total of N samples per period. It is in this discrete form that the Fourier transform is most suited to numerical evaluation by digital computation.

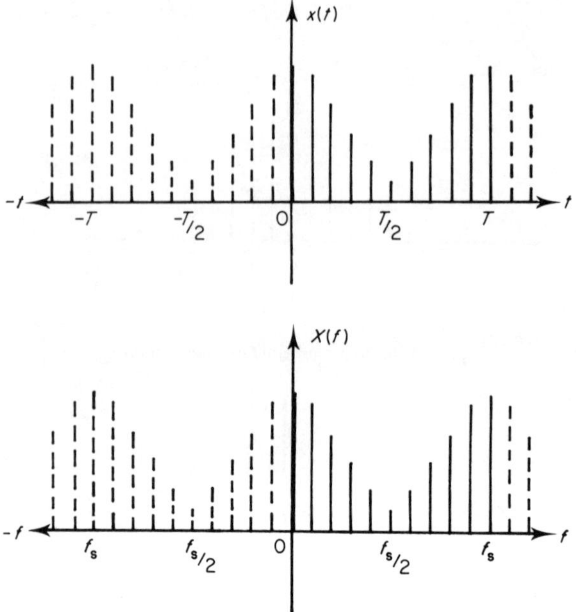

Figure 2.18. Discrete time and frequency domain functions

Consider equation (2.8.1) rewritten as

$$X(f_k) = \frac{1}{N} \sum_{n=0}^{N-1} x(t_n) W^{kn}, \tag{2.8.3}$$

where $W = e^{-j2\pi/N}$.

Over all the frequency components, equation (2.8.3) becomes a matrix equation,

$$\begin{bmatrix} X(f_0) \\ X(f_1) \\ \vdots \\ X(f_k) \\ \vdots \\ X(f_{N-1}) \end{bmatrix} = \frac{1}{N} \begin{bmatrix} 1 & 1 & \cdots & 1 & \cdots & 1 \\ 1 & W & \cdots & W^k & \cdots & W^{N-1} \\ \vdots & \vdots & & \vdots & & \vdots \\ 1 & W^k & \cdots & W^{k^2} & \cdots & W^{k(N-1)} \\ \vdots & \vdots & & \vdots & & \vdots \\ 1 & W^{N-1} & \cdots & W^{(N-1)k} & \cdots & W^{(N-1)^2} \end{bmatrix} \begin{bmatrix} x(t_0) \\ x(t_1) \\ \vdots \\ x(t_n) \\ \vdots \\ x(t_{N-1}) \end{bmatrix}, \tag{2.8.4}$$

or in a condensed form

$$[X(f_k)] = \frac{1}{N} [W^{kn}][x(t_n)]. \tag{2.8.5}$$

In these equations, $[X(f_k)]$ is a vector representing the N components of the function in the frequency domain, while $[x(t_n)]$ is a vector representing the N samples of the function in the time domain.

Calculation of the N frequency components from the N time samples therefore requires a total of N^2 complex multiplications to implement in the above form.

Each element in the matrix $[W^{kn}]$ represents a unit vector with a clockwise rotation of $2p\pi/N(p=0,1,2,\ldots,(N-1))$ introduced between successive components. Depending on the value of N, a number of these elements are the same.

For example, if $N=8$ then

$$W = e^{-2\pi/8} = \cos\frac{\pi}{4} - j\sin\frac{\pi}{4}.$$

As a consequence,

$$\begin{aligned} W^0 &= -W^4 = 1,\\ W^1 &= -W^5 = (1/\sqrt{2} - j1/\sqrt{2}),\\ W^2 &= -W^6 = -j,\\ W^3 &= -W^7 = -(1/\sqrt{2} + j1/\sqrt{2}). \end{aligned}$$

These can also be thought of as unit vectors rotated through $\pm 0°$, $\pm 45°$, $\pm 90°$ and $\pm 135°$ respectively.

Furthermore, W^8 is a complete rotation and hence equal to 1. The value of the elements of W^{kn} for $kn > 8$ can thus be obtained by subtracting full rotations, to leave only a fraction of a rotation, the values for which are shown above. For example, if $k=5$ and $n=6$, then $kn=30$ and $W^{30} = W^{3\times 8+6} = W^6 = j$.

Thus there are only four unique absolute values of W^{kn} and the matrix $[W^{kn}]$, for the case $N=8$, becomes

1	1	1	1	1	1	1	1
1	W	$-j$	W^3	-1	$-W$	j	$-W^3$
1	$-j$	-1	j	1	$-j$	-1	j
1	W^3	j	W	-1	$-W^3$	$-j$	$-W$
1	-1	1	-1	1	-1	1	-1
1	$-W$	$-j$	$-W^3$	-1	W	j	W^3
1	j	-1	$-j$	1	j	-1	$-j$
1	$-W^3$	j	$-W$	-1	W^3	$-j$	W

It can be observed that the d.c. component of the frequency spectrum, $X(f_0)$, obtained by the algebraic addition of all the time domain samples, divided by the number of samples, is the average value of all the samples.

Subsequent rows show that each time sample is weighted by a rotation dependent on the row number. Thus for $X(f_1)$ each successive time sample is rotated by $1/N$ of a revolution; for $X(f_2)$ each sample is rotated by $2/N$ revolutions, and so on.

2.9 FAST FOURIER TRANSFORM[4–8]

For large values of N, the computational time and cost of executing the N^2 complex multiplications of the discrete Fourier transform (DFT) can become prohibitive.

Instead, a calculation procedure known as the fast Fourier transform (FFT), which takes advantage of the similarity of many of the elements in the matrix $[W^{kn}]$, produces the same frequency components using only $(N/2)\log_2 N$ multiplications to execute the solution of equation (2.8.5). Thus, for the case $N = 1024 = 2^{10}$, there is a saving in computation time by a factor of over 200. This is achieved by factorizing the $[W^{kn}]$ matrix of equation (2.8.5) into $\log_2 N$ individual or factor matrices such that there are only two non-zero elements in each row of these matrices, one of which is always unity. Thus, when multiplying by any factor matrix only N operations are required.

The reduction in the number of multiplications required, to $(N/2)\log_2 N$, is obtained by recognizing that

$$W^{N/2} = -W^0, \qquad W^{(N+2)/2} = -W^1, \qquad \text{etc.}$$

To obtain the factor matrices, it is first necessary to re-order the rows of the full matrix. If rows are denoted by a binary representation, then the re-ordering is by bit reversal.

For the example where $N = 8$, row 5, represented as 100 in binary (row 1 is 000), now becomes row 2, or 001 in binary. Thus rows 2 and 5 are interchanged. Similarly, rows 4 and 7, represented as 011 and 110 respectively, are also interchanged. Rows 1, 3, 6 and 8 have binary representations which are symmetrical with respect to bit reversal and hence remain unchanged.

The corresponding matrix is now

1	1	1	1	1	1	1	1
1	-1	1	-1	1	-1	1	-1
1	$-j$	-1	j	1	$-j$	-1	j
1	j	-1	$-j$	1	j	-1	$-j$
1	W	$-j$	W^3	-1	$-W$	j	$-W^3$
1	$-W$	$-j$	$-W^3$	-1	W	j	W^3
1	W^3	j	W	-1	$-W^3$	$-j$	$-W$
1	$-W^3$	j	$-W$	-1	W^3	$-j$	W

.

This new matrix can be separated into $\log_2 8$ $(=3)$ factor matrices:

1	1						
1	-1						
		1	$-j$				
		1	j				
				1	W		
				1	$-W$		
						1	W^3
						1	$-W^3$

,

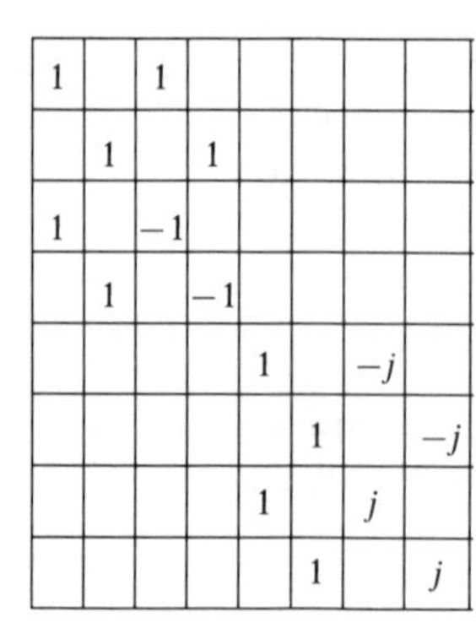

1		1					
	1		1				
1		-1					
	1		-1				
				1		$-j$	
					1		$-j$
				1		j	
					1		j

,

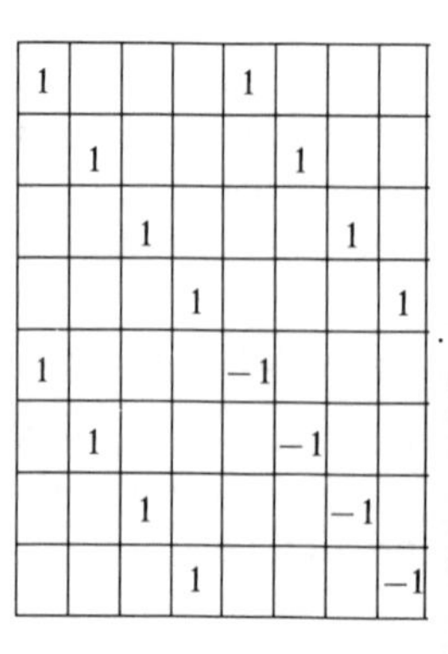

1				1			
	1				1		
		1				1	
			1				1
1				-1			
	1				-1		
		1				-1	
			1				-1

.

As previously stated, each factor matrix has only two non-zero elements per row, the first of which is unity.

The re-ordering of the $[W^{kn}]$ matrix results in a frequency spectrum which is also re-ordered. To obtain the natural order of frequencies it is necessary to reverse the previous bit-reversal.

In practice, a mathematical algorithm implicitly giving factor matrix operations is used for the solution of an FFT.[9]

Using $N = 2^m$ it is possible to represent n and k by m bit binary numbers such that

$$n = n_{m-1}2^{m-1} + n_{m-1}2^{m-2} + \cdots + 4n_2 + 2n_1 + n_0, \tag{2.9.1}$$

$$k = k_{m-1}2^{m-1} + k_{m-2}2^{m-2} + \cdots + 4k_2 + 2k_1 + k_0, \tag{2.9.2}$$

where $n_i = 0, 1$ and $k_i = 0, 1$. For $N = 8$,

$$n = 4n_2 + 2n_1 + n_0 \quad \text{and} \quad k = 4k_2 + 2k_1 + k_0,$$

where n_2, n_1, n_0 and k_2, k_1, k_0 are binary bits (n_2, k_2 most significant and n_0, k_0 least significant).

Equation (2.8.3) can now be re-written as

$$X(k_2, k_1, k_0) = \sum_{n_2=0}^{1} \sum_{n_1=0}^{1} \sum_{n_0=0}^{1} \frac{1}{N} x(n_2, n_1, n_0) W^{kn}. \tag{2.9.3}$$

Defining n and k in this way enables the computation of equation (2.8.3) to be performed in three independent stages, computing in turn

$$A_1(k_0, n_1, n_0) = \sum_{n_2=0}^{1} \frac{1}{N} x(n_2, n_1, n_0) W^{4k_0 n_2}, \tag{2.9.4}$$

$$A_2(k_0, k_1, n_0) = \sum_{n_1=0}^{1} A_1(k_0, n_1, n_0) W^{2(k_0+2k_1)n_1}, \tag{2.9.5}$$

$$A_3(k_0, k_1, k_2) = \sum_{n_0=0}^{1} A_2(k_0, k_1, n_0) W^{(k_0+2k_1+4k_2)n_0}, \tag{2.9.6}$$

The sequence of operations involved in the implementation of this algorithm is shown in Figure 2.19. Each operation is referred to as a 'butterfly' and has been implemented on special integrated circuits.

From Figure 2.19 and equation (2.9.6) it is seen that the A_3 coefficients contain the required $X(k)$ coefficients but in reverse binary order:

order of A_3 in binary form is $k_0 k_1 k_2$;

order of $X(k)$ in binary form is $k_2 k_1 k_0$.

Hence

Binary   *Reversed*

$$A_3(3) = A_3(011) = X(110) = X(6),$$

$$A_3(4) = A_3(100) = X(001) = X(1),$$

$$A_3(5) = A_3(101) = X(101) = X(5).$$

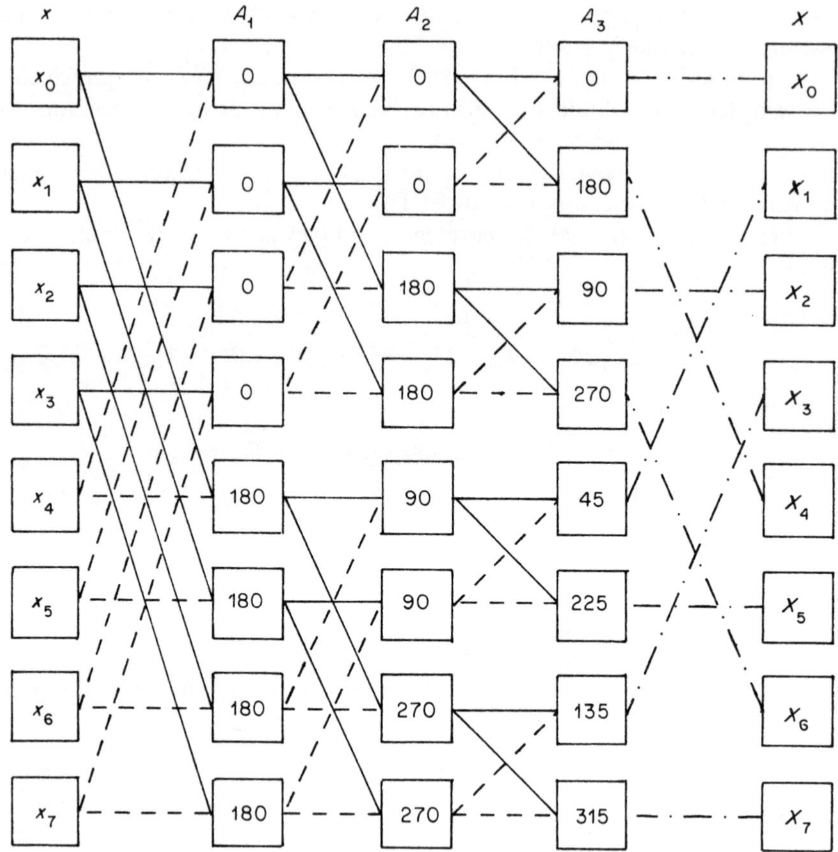

Figure 2.19. Flow chart for solution of a fast Fourier transform.——, Add only; ------, rotate through angle shown and add

The fast Fourier transform of a band-limited signal can be implemented off-line in a computer using a high level language program, where the signal is sampled and stored digitally in a recorder which can be accessed remotely from the computer. Alternatively, there is increasing use of spectrum analysers which perform the FFT algorithm using dedicated hardware such as the 'butterflies' referred to above.

The operation of such analysers is discussed in more detail in Chapter 6.

2.10 CHIRP Z TRANSFORM[10]

An alternative formulation of the discrete Fourier transform, the Chirp Z transform, leads to an arrangement which can be implemented cheaply using hybrid analogue/digital hardware.

The factor $2kn$ in the exponent of the forward discrete fourier transform of

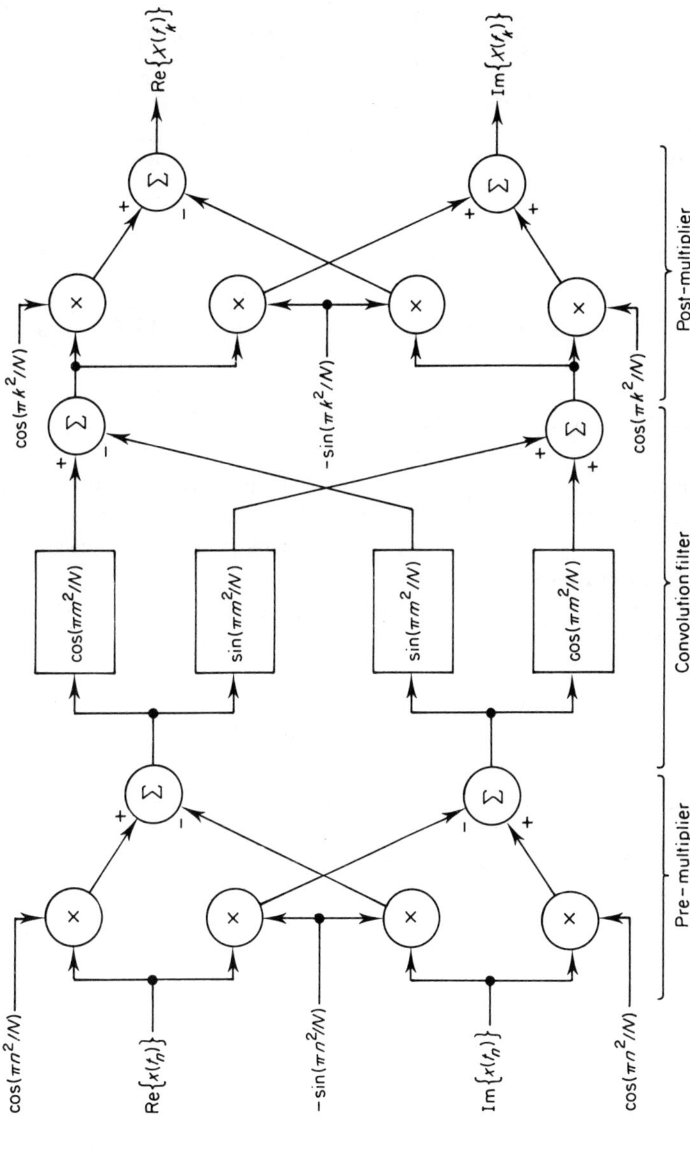

Figure 2.20. Chirp Z-transform solution of the discrete Fourier transform

equation (2.8.1) can be replaced by its seemingly more complicated equivalent:

$$2kn = k^2 + n^2 - (k-n)^2. \tag{2.10.1}$$

Thus equation (2.8.1) can be rewritten as

$$X(f_k) = \frac{1}{N} e^{-j\pi k^2/N} \sum_{n=0}^{N-1} x(t_n) e^{-j\pi n^2/N} e^{j\pi(k-n)^2/N}. \tag{2.10.2}$$

The solution of this equation requires three operations:

(i) Multiply each sample of the discrete time domain signal, $x(t_n)$, by

$$e^{-j\pi n^2/N} = \cos(\pi n^2/N) + j\sin(\pi n^2/N).$$

(ii) Perform a discrete convolution between the sequence of the combination obtained in (i) with $e^{j\pi n^2/N}$.

(iii) Multiply the resulting summation by the factor $e^{-j\pi k^2/N}$ for each term of $X(f_k)$.

This sequence of operations can be visualized in Figure 2.20 for a complex unit, with the real and imaginary parts separated.

The usable input frequency band for this transform method is from zero up to the sampling frequency f_s. No aliasing (see Section 2.11) occurs as there are effectively two times the sampling frequency values per second, since each input sample has two components.

In the case of real inputs, two of the pre-multipliers can be deleted, thus saving on hardware. However, the frequency band in this case is one half the sampling frequency as a real input is made up of both positive and negative complex frequency components whose imaginary portions cancel, and the negative frequency components appear as an alias in the $f_s/2 \rightarrow f_s$ band.

2.11 THE NYQUIST FREQUENCY AND ALIASING[4,5]

With regard to equation (2.8.4) for the discrete Fourier transform and the matrix $[W^{kn}]$ it can be observed that for the rows $N/2$ to N, the rotations applied to each time sample are the negative of those in rows $N/2$ to 1. Thus frequency components above $k = N/2$ can be considered as negative frequencies since the unit vector is being rotated through increments greater than π between successive components. In the example of $N = 8$, the elements of row 3 are successively rotated through $-\pi/2$. The elements of the row 7 are similarly rotated through $-3\pi/2$; or in negative frequency form through $\pi/2$. More generally, a rotation through $2\pi(N/2+p)/N$ radians for $p = 1,2,3,\ldots,(N/2-1)$, with N even, corresponds to a negative rotation of $-2\pi(N/2-p)/N$ radians. Hence $-X(k)$ corresponds to $X(N-k)$ for $k = 1$ to $N/2$, as shown by Figure 2.21.

This is an interpretation of the sampling theorem which states that the sampling frequency must be at least twice the highest frequency contained in the original signal for a correct transfer of information to the sampled system.

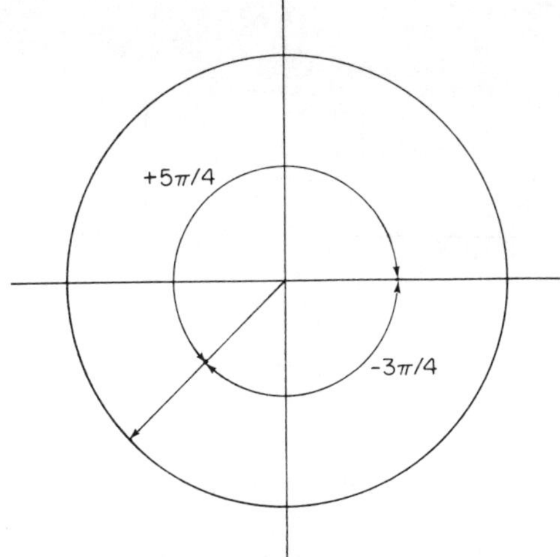

Figure 2.21. Correspondence of positive and negative angles

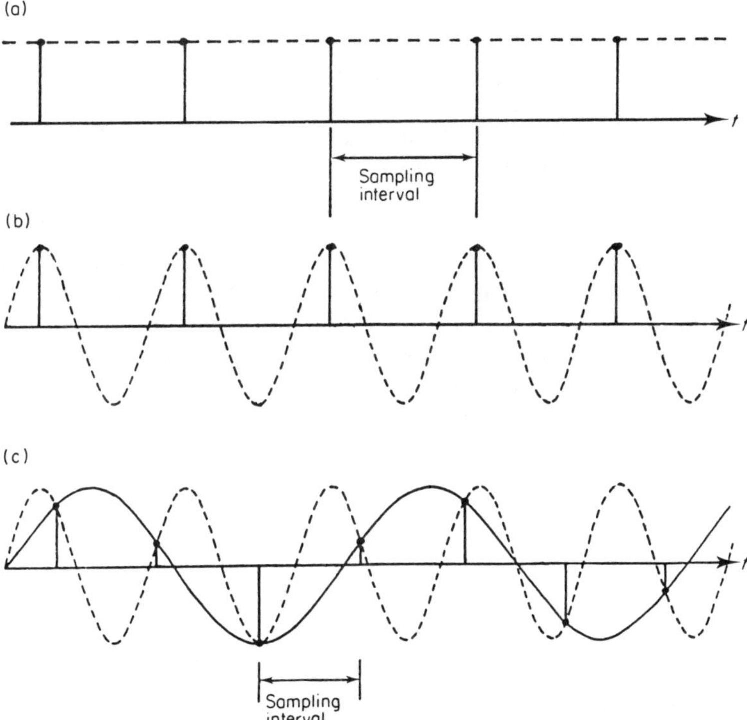

Figure 2.22. The effect of aliasing: (a) $x(t) = k$; (b) $x(t) = k \cos 2\pi nft$. For (a) and (b) both signals are interpreted as being d.c. In (c) the sampling can represent two different signals with frequencies above and below the Nyquist or sampling rate

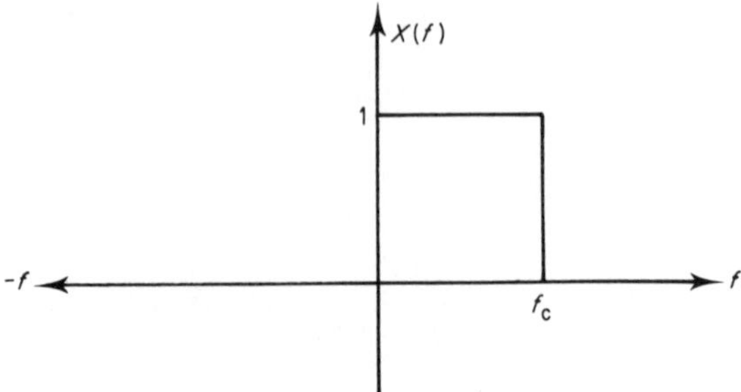

Figure 2.23. Frequency domain characteristic of an ideal low pass filter with cutoff frequency f_c.

The frequency component at half the sampling frequency is referred to as the Nyquist frequency.

The representation of frequencies above the Nyquist frequency as negative frequencies means that should the sampling rate be less than twice the highest frequency present in the sampled waveform then these higher frequency components can mimic components below the Nyquist frequency, introducing error into the analysis.

It is possible for high frequency components to complete many revolutions between samplings; however, since they are only sampled at discrete points in time, this information is lost.

This misinterpretation of frequencies above the Nyquist frequency, as being lower frequencies, is called 'aliasing' and is illustrated in Figure 2.22.

To prevent aliasing it is necessary to pass the analogue time domain signal through a band-limited low pass filter, the ideal characteristic of which is shown in Figure 2.23. The cutoff frequency, f_c, must be equal to the Nyquist frequency, although in practice this is limited to about 80% of the Nyquist frequency due to the non-ideal characteristics of a practical filter (Section 6.3).

Thus if sampling is undertaken on the filtered signal and the discrete Fourier transform applied, the frequency spectrum has no aliasing effect and is an accurate representation of the frequencies in the original signal that are below the Nyquist frequency. However, information on those frequencies above the Nyquist frequency is lost due to the filtering process.

2.12 WINDOW FUNCTIONS[11]

In any practical measurement of a time domain signal it is normal to limit the time duration over which the signal is observed. This process is known as 'windowing' and is particularly useful for the measurement of non-stationary signals which may be divided into short segments of a quasi-stationary nature with an implied infinite periodicity. Further more, in the digital analysis of waveforms, only a finite number of samples of the signal is recorded on which a spectral analysis is made. Thus even stationary signals are viewed from limited

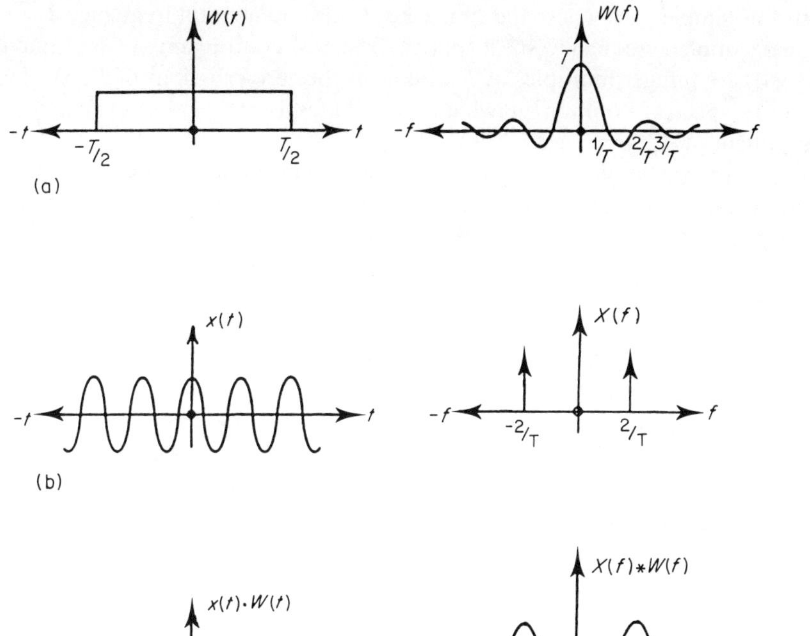

Figure 2.24. Infinite periodic function processed with a rectangular window function: (a) rectangular window function and frequency spectrum; (b) periodic function $x(t) = A\cos(4\pi t/T)$ and frequency spectrum; (c) infinite periodic function viewed through a rectangular time window.

time data and this can introduce errors in the frequency spectrum of the signal.

The effect of windowing can best be seen by defining a time domain function, which lies within finite time limits. Outside of these the function is zero. The simplest window function is the rectangular window of Figure 2.24. The frequency spectrum of this function, obtained in Section 2.6, is also included.

The application of a window function has the effect of multiplying each point of a time domain signal by the corresponding time point of the window function. Thus, within a rectangular window, the signal is just itself, but outside of this the signal is completely attenuated although a periodicity of the signal within the window is implied outside the defined window. This time domain multiplication has its equivalent in the frequency domain as the convolution of the spectra of the window function and the signal. This is illustrated for an infinite periodic function and a rectangular window function in Figure 2.24(b) and (c). It can be observed that there is significant power in the frequencies of the side lobes about the fundamental frequency, which is not present in the infinite fundamental frequency waveform.

In this simple case where the signal $x(t)$ is of fundamental frequency f_1 only, with a sampled frequency spectrum, the spectral components of the function $x(t)\cdot w(t)$ are integer multiples of f_1 and lie at the zero crossings of $X(f)$. Hence the only spectral component which contributes is that being evaluated, the component at f_1 in Figure 2.24(c).

However, it is highly likely that waveforms will be made up of many frequency components, not necessarily integer multiples of the fundamental window frequency f_1. Consequently, discontinuities will exist between the function at the start and finish of the window which will introduce uncertainty in the identification of the periodic components present, since Fourier analysis assumes periodicity of functions and continuity at the boundaries. The resulting error is known as spectral leakage and is the non-periodic noise contributing to each of the periodic spectral components present.

As an illustration of spectral leakage occurring when the duration of the rectangular time window is different from the fundamental of the actual waveform, consider a single frequency periodic waveform where the worst case sample gives the phase discontinuity as shown in Figure 2.25(b). Here the

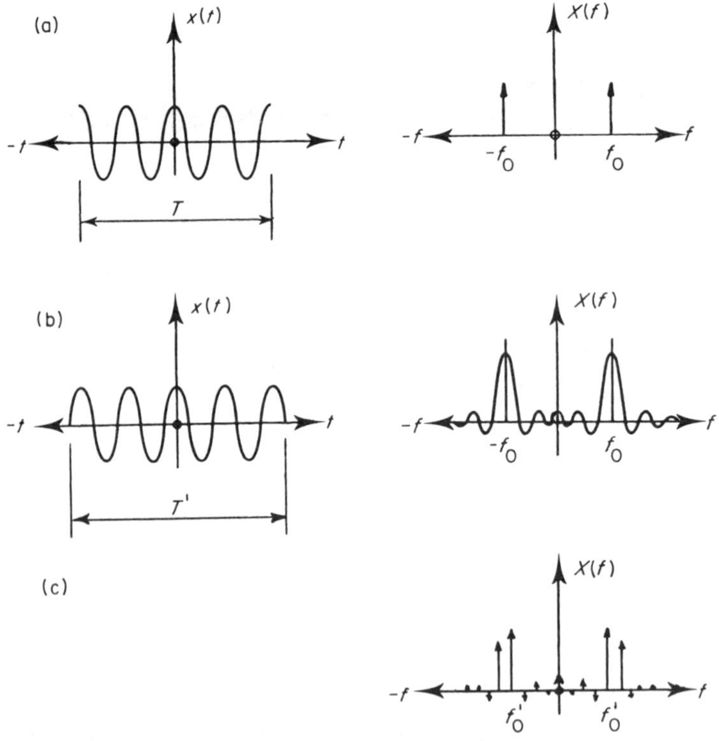

Figure 2.25. Infinite periodic signal viewed using different duration time windows with periodicity implied: (a) the time window is an exact multiple of the period of the waveform; (b) the time window is a $(2n + 1)/2$ multiple of the period of the waveform; (c) case (b) viewed as a discrete frequency spectrum

spectrum shows the main frequency component, the existence of high side lobes and a d.c. component.

The picket fence

The combination of the DFT and window function produces a response equivalent to filtering the time domain signal through a series of filters with centre frequencies at integer multiples of $1/T$, where T is the sampling period. The filter characteristic and the associated leakage is determined by the particular window function chosen. The resulting spectrum can therefore be considered as the true spectrum viewed through a picket fence with only frequencies at points corresponding to the gaps in the fence being visible.

When the signal being analysed is not one of these discrete, orthogonal frequencies then, because of the non-ideal nature of the DFT filter, it will be seen by more than one such filter, but at a reduced level in each. The effect can be reduced by adding a number of zeros, usually equivalent to the original record length, to the data to be analysed. This is called zero padding. This effective increase in the sampling period T introduces extra DFT filters at points between the original filters. The bandwidth of the individual filters still depends upon the original sample period and is therefore unchanged.

Spectral leakage reduction

The effect of spectral leakage can be reduced by changing the form of the window function. In particular, if the magnitude of the window function is reduced towards zero at the boundaries, any discontinuity in the original waveform is weighted to a very small value and thus the signal is effectively continuous at the boundaries. This implies a more periodic waveform which has a more discrete frequency spectrum.

A number of window functions are shown in Figure 2.26, along with their Fourier transforms, which give a measure of the attenuation of the side lobes that give rise to spectal leakage.

Choice of window function

The objective in choosing a window to minimize spectral leakage is to obtain a mainlobe width which is as narrow as possible, so that it only includes the spectral component of interest, with minimal sidelobe levels to reduce the contribution from interfering spectral components. These two specifications are inter-related for realizable windows and a compromise is made between mainlobe width compression and sidelobe level reduction.

The rectangular window function, defined by

$$W(t) = \begin{cases} 1 & \text{for } -T/2 < t < T/2, \\ 0 & \text{otherwise,} \end{cases} \tag{2.12.1}$$

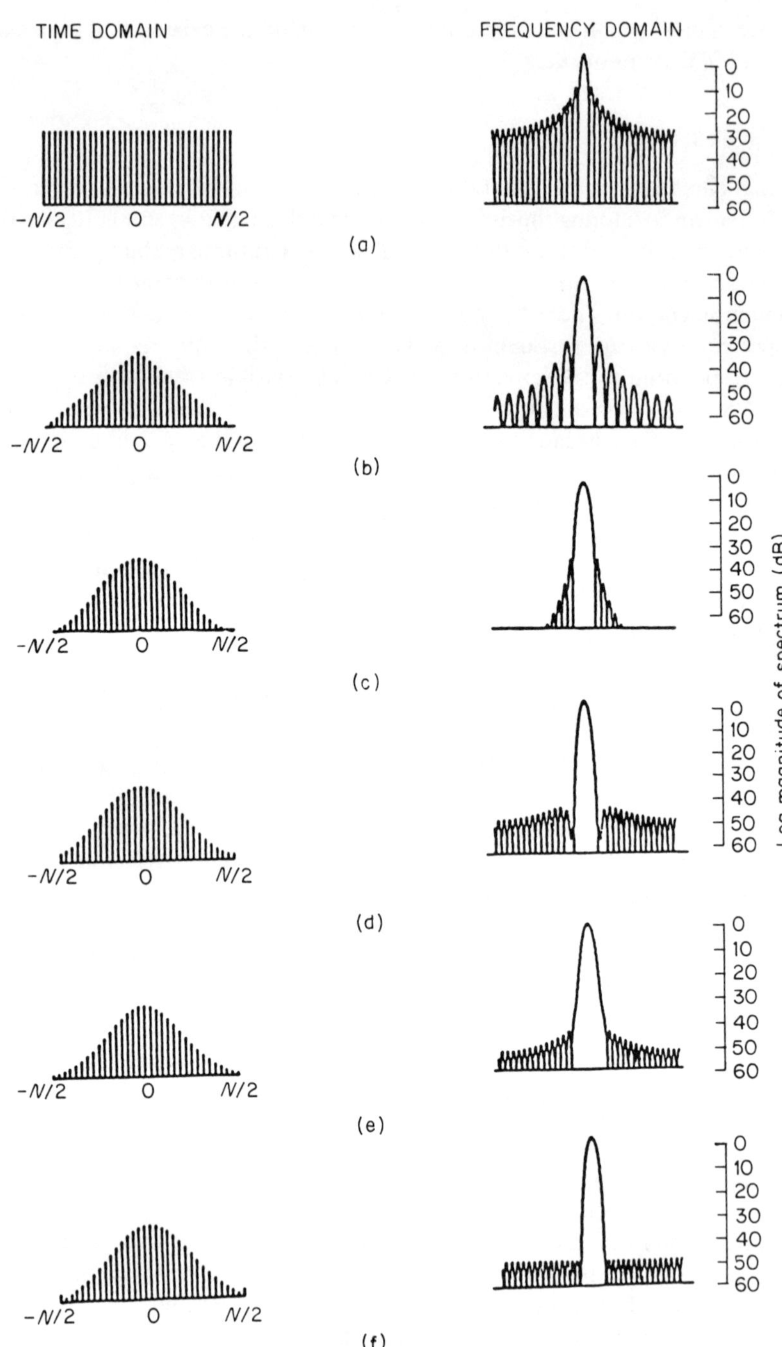

Figure 2.26. Fourier transform pairs of common window functions: (a) rectangular; (b) triangular; (c) cosine squared (Hanning); (d) Hamming; (e) Gaussian; (f) Dolph–Chebyshev

has a noise or effective bandwidth of $1/T$, where T is the window length; the sidelobe levels are large (-13 dB from the main lobe for the first side lobe), and their rate of decay with frequency is slow (being 20 dB per decade). This means that when evaluating the fundamental component of a signal, interfering spectral components near to it will be weighted heavily, contributing greater interference to the fundamental than for the other windows illustrated in Figure 2.26.

However, as mentioned previously, there is one situation where the rectangular window ideally results in zero spectral leakage and high spectral resolution. This situation occurs when the duration of the rectangular window is equal to an integer multiple of the period of a periodic signal. When the rectangular window spans exactly one period, the zeros in the spectrum of the window coincide with all the harmonics excepting one. This results in no spectral leakage under ideal conditions. Consequently, spectrum analysers often incorporate the rectangular window function facility for the analysis of periodic waveforms, to which the duration of the window can be matched. This is achieved through the use of a phase-locked loop (Section 6.5). This frequency matching gives the greatest resolution of the periodic frequency.

The triangular window, defined by

$$W(t)=\begin{cases} 1+2t/T & \text{for} \quad -T/2<t<0, \\ 1-2t/T & \text{for} \quad 0<t<T/2, \\ 0 & \text{otherwise,} \end{cases} \tag{2.12.2}$$

is a simple modification of the rectangular window, where the amplitude of the multiplying window is reduced linearly to zero from the window centre. The reduction of side lobe level is readily seen in Figure 2.26(b), but this is at the expense of mainlobe width and a consequent reduction in frequency resolution.

An international standard window function often incorporated into spectrum analysers is the cosine-squared or Hanning window, defined by

$$W(t)=\frac{1}{2}\left(1-\cos\frac{2\pi t}{T}\right) \quad \text{for} \quad -\frac{T}{2}<t<\frac{T}{2}, \tag{2.12.3}$$

and in which it is the power term that is cosine squared. This function is easily generated from sinusoidal signals and in FFT analysers a table of cosine values can be utilized for generating the window. The main lobe noise bandwidth is greater than that for the rectangular window, being $1.5/T$; however, the highest sidelobe is at -32 dB and the sidelobe fall-off rate is 60 dB per decade, thus reducing the effect of spectral leakage. This is illustrated in Figure 2.27 where the Hanning window is compared to the rectangular window for sidelobe level reduction.

By mounting the Hanning window on a small rectangular pedestal (but limiting the maximum of the function to unity) the Hamming window is obtained. This is described in its amplitude form as

$$W(t)=0.54-0.46\cos\frac{2\pi t}{T} \quad \text{for} \quad -\frac{T}{2}<t<\frac{T}{2}. \tag{2.12.4}$$

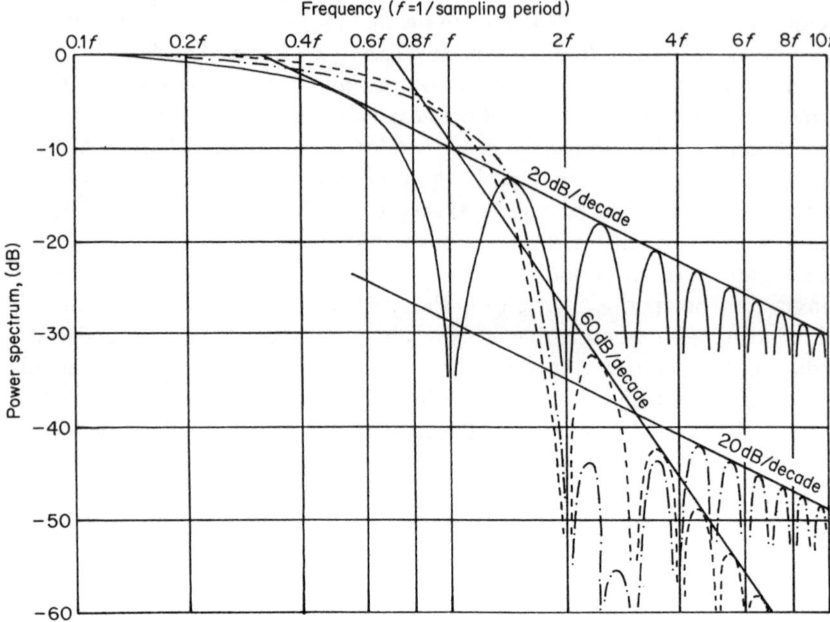

Figure 2.27. Power spectrum versus log(frequency) for selected window functions. ———, Rectangular window; ----, Hanning window; -·-·-, Hamming window

The second sidelobe of the rectangular function coincides with the first sidelobe of the Hanning function and since these are in opposite phase, they can be scaled to cancel each other. As a consequence the highest sidelobe level is -42 dB, Figure 2.27. The remaining sidelobes are dominated by the rectangular function and have a fall-off rate of 20 dB per decade. A slight improvement in mainlobe noise bandwidth (to $1.4/T$) is observed also.

The ideal window which has a single mainlobe and no sidelobes is in the form of a Gaussian function:

$$W(t) = \exp(-t^2/2\sigma^2). \tag{2.12.5}$$

The Gaussian function has the property of transforming, by the Fourier transform, to another Gaussian function. On a decibel scale, its shape is that of an inverted parabola, with a characteristic which becomes successively steeper. Theoretically the Gaussian function is defined between infinite time limits. For practical use, the function is truncated at three times the half-amplitude width, which is 7.06 times the standard deviation. As a consequence, sidelobes are established in the power spectrum but these are of the order of -44 dB down. The main lobe noise bandwidth is wider than the previous windows, being $1.9/T$.

The final window, presented in Figure 2.26(f), is the Dolph–Chebyshev function, the discrete form of which is defined as

$$W(t) = \frac{(-1)^r \cos[N\cos^{-1}[\beta\cos(\pi r/N)]]}{\cosh[n\cosh^{-1}(\beta)]} \quad \text{for} \quad 0 < r < N-1, \tag{2.12.6}$$

where r is an integer, N is the number of discrete samples of the window function,

$$\beta = \cosh\left[\frac{1}{N}\cosh^{-1}(10^{\alpha})\right],$$

and the inverse hyperbolic cosine is defined by

$$\cosh^{-1} x = \begin{cases} \pi/2 - \tan^{-1}[x/\sqrt{(1-x^2)}] & \text{for} \quad |x| < 1.0, \\ \ln[x + \sqrt{(x^2-1)}] & \text{for} \quad |x| > 1.0. \end{cases}$$

This function provides the narrowest possible mainlobe width for a given specified sidelobe level, which is constant on a decibel scale. The sidelobe levels are controlled by the parameter α in equation (2.12.6). With $\alpha = 4.0$ the sidelobes are at -80 dB (0.01%) with respect to the mainlobe.

Mainlobe width reduction

In obtaining low sidelobe levels to reduce spectral leakage, in all the window functions previously mentioned, there has been a sacrifice of mainlobe bandwidth. It is possible that in harmonic analysis, when evaluating for example the fundamental of a waveform, the resolution is such that the d.c. component and the second and third harmonics are included within the mainlobe. This causes considerable interference in the individual harmonic evaluations and restricts the identification of spectral leakage effects to higher order frequencies and noise, outside the mainlobe.

However, with the ability to change the sidelobe level with the Dolph–Chebyshev window, an algorithm presents itself to effectively reduce the mainlobe bandwidth.

Consider the complex fundamental Fourier component of the waveform, multiplied by the Dolph–Chebyshev window function, i.e. $[W(r)\cdot x(r)]_{r=1}^{N}$, obtained by the application of the discrete Fourier transform (using the FFT technique),

$$X_1 = W_0C_1 + W_1C_2 + W_2C_3 + W_{-2}C_{-1} + W_{-3}C_{-2} + W_{-4}C_{-3} + W_{-1}C_0 + \sum_{n=3}^{N/2} W_nC_{n+1} + \sum_{n=4}^{N/2-1} W_{-n+1}C_{-n}, \tag{2.12.7}$$

where W are the discrete window coefficients in the frequency domain and $C_n = C_{-n}$ are the complex periodic Fourier coefficients of harmonic order n.

By pre-processing the waveform, the d.c. component can be removed, thereby eliminating C_0 in equation (2.12.7). In addition, since the last two terms of this equation include all the higher order harmonics and noise, but are weighted with window coefficients of the order of 0.01%, they can also be neglected. Equation (2.12.7) can therefore be reduced to

$$X_1 = C_1(W_0 + W_{-2}) + C_2(W_1 + W_{-3}) + C_3(W_2 + W_{-4}). \tag{2.12.8}$$

Application of the discrete Fourier transform to the windowed discrete time domain waveform yields a value for X_1. The window coefficients in the frequency domain are known by the defining equation (2.12.4) and hence only the three harmonic terms C_1, C_2, C_3, are unknown.

The use of three windows, each with a different α parameter, and the application of the DFT three times to the same waveform, produces three simultaneous equations of the same form as equation (2.12.8). The solution of these leads directly to the values of C_1, C_2 and C_3, i.e. the fundamental component and second and third harmonics of the original waveform.

As an illustration of the effectiveness of this algorithm, consider the function defined by

$$x(t) = C_1 \cos(2\pi f_1 t + \phi_1) + \sum_{n=2}^{7} C_n \cos(n2\pi f_1 + \phi_n),$$

where $C_1 = 1.0$ and $C_n = 0.2$ for $n = 2$ to 7 for which 32 samples were available.

With window filtering the error introduced by the higher harmonics in the identification of C_1 was limited to 0.16%. However, when a non-periodic component of frequency $5.65f_1$ and magnitude 0.2 was introduced into $x(t)$, the error in identifying C_1 without window filtering was 2%. With the Dolph–Chebychev window and using α values of 3.2, 3.5 and 3.8, the error was reduced to 0.36%.

2.13 REFERENCES

1. Fourier, J. B. J. (1822). *Théorie analytique de la chaleur*, Paris.
2. Kreyszig, E. (1967). *Advanced Engineering Mathematics*, 2nd edn, John Wiley, New York.
3. Kuo, F. F. (1966). *Network Analysis and Synthesis*, John Wiley, New York.
4. Randall, R. B. (1977). *Application of B and K Equipment to Frequency Analysis*, Bruel and Kjaer.
5. Brigham, E. O. (1974). *The Fast Fourier Transform*, Prentice-Hall, Englewood Cliffs, N.J.
6. Cooley, J. W., and Tukey, J. W. (1965). 'An algorithm for machine calculation of complex Fourier series'. *Math. Comput.*, **19**, 297–301.
7. Cochran, W. T., Cooley, J. W., Favin, D. L., *et al.* (1967). 'What is the fast Fourier transform'. *Proc. IEEE*, **55**, 1664–1677.
8. Bergland, G. D. (1969). 'A guided tour of the fast Fourier transform'. *IEEE Spectrum*, July, 41–42.
9. Bergland, G. D. (1968). 'A fast Fourier transform algorithm for real-valued series', *Numer. Anal.*, **11**, 703–710.
10. Rabiner, L. R., Schafer, R. W., and Rader, C. M. (1969). 'The chirp Z-transform algorithm'. *IEEE Trans.*, **AU-17**, 86–92.
11. Harris, F. J. (1978). 'On the use of windows for harmonic analysis with the discrete Fourier transform'. *Proc. IEEE*, **66**, 51–83.
12. Turner, K. S., Heffernan, M. D., Arnold, C. P., and Arrillaga, J. (1981). 'Computation of a.c.–d.c. system disturbances. II. Derivation of power frequency variables from convertor transient response'. *IEEE Trans.*, **PAS-100**, 4349–4356.

3

Harmonic sources – the static convertor

3.1 INTRODUCTION

The derivation of the harmonic currents produced by static power convertors requires accurate information of the a.c. voltage waveforms at the convertor terminals, convertor configuration, type of control, a.c. system impedance and d.c. circuit parameters. However, the introduction of so many factors at the outset would obscure the basic principles involved. It is probably more appropriate to start by assessing the effect of the control philosophy and convertor configuration under idealized a.c. and d.c. system conditions and then introduce the other factors one by one.

According to the relative position of the firing instant from one valve to the next on the steady state, four basically different control principles are in common use:

(i) Constant phase-angle control produces consecutive valve firings equally spaced with reference to their respective commutating voltages.

(ii) Equidistant firing control produces consecutive firings at equal intervals of the supply frequency.

(iii) Modulated phase-angle control produces time-varying phase-modulated firings.

(iv) Integral cycle control selects an integer number of complete cycles or half cycles of the supply frequency.

Phase angle control is by far the most extensively used technique and most of this chapter is devoted to it. Moreover, the Fourier analysis described in Chapter 2 is directly applicable to the phase angle controlled and equidistant firing controlled waveforms. Modulated firing and integral cycle controls are less straightforward in this respect, and some consideration is given to their particular requirements in this chapter.

Forced commutated convertors, and in particular the invertor-fed a.c. drive, are also increasing in numbers and power rating. Their power source is normally the a.c. power system through a line-commutated rectifier and therefore the harmonic currents injected into the a.c. network are those discussed in the earlier part of the chapter. On the other hand the harmonic content of the inverted waveforms is rather different and will be given special consideration.

Constant phase angle is the type of control normally found in naturally commutated static convertors and a.c. voltage regulators.

A.C. voltage regulators, using back-to-back thyristor pairs in each phase, produce varying levels of harmonic content which, in the case of inductive loads, can include even-ordered harmonics and direct current. Although the use of thyristor-controlled a.c. voltage regulation is at present restricted to low power applications (such as light dimmers and small induction motors), with the growing interest in energy conservation, their use is likely to increase and may well be a source of harmonic problems in the future.

The main sources of harmonic current are at present the phase angle controlled rectifiers and invertors. These can be conveniently grouped into the following three broad areas of different harmonic behaviour: (i) large power convertors such as those used in the metal reduction industry and high voltage d.c. transmission; (ii) medium size convertors such as those used in the manufacturing industry for motor control and also in railway applications; (iii) low power rectification from single-phase supplies such as television sets and battery chargers.

The waveforms in group (i) are the closest to the ideal and will be used as a basis for the derivation of the characteristic harmonic content of the standard convertor configurations. This information is often used as a reference in the harmonic assessment of less ideal waveforms.

3.2 LARGE POWER CONVERTORS

Large power convertors (rated in megawatts) generally have much more inductance on the d.c. side than on the a.c. side. The direct current is thus reasonably constant and the convertor acts like a source of harmonic voltage on the d.c. side and of harmonic current on the a.c. side.

Moreover, with a perfectly symmetrical a.c. system the resulting currents are exactly equal in all the phases.

Harmonic components of the current waveform

The ideal p-phase one-way convertor, illustrated in Figure 3.1, has zero a.c. system impedance and infinite smoothing inductance. Under these conditions the phase currents consists of periodic positive rectangular pulses of width $w = 2\pi/p$, repeating at the supply frequency.

If in the analysis of the waveform of Figure 3.2, the origin is taken at the centre of the pulse, $F(\omega t)$ is shown to be an 'even' function (i.e. $f(x) = f(-x)$) and the Fourier series has only cosine terms. The relevant Fourier coefficients, with reference to a 1 per unit d.c. current, are

$$A_0 = \frac{1}{2\pi}\int_{-w/2}^{w/2} \mathrm{d}(\omega t) = \frac{w}{2\pi} = \frac{1}{p}, \tag{3.2.1}$$

$$A_n = \frac{1}{\pi}\int_{-w/2}^{w/2} \cos(n\omega t)\mathrm{d}(\omega t) = \frac{2}{\pi n}\sin\frac{nw}{2} = \frac{2}{\pi n}\sin\frac{\pi n}{p}. \tag{3.2.2}$$

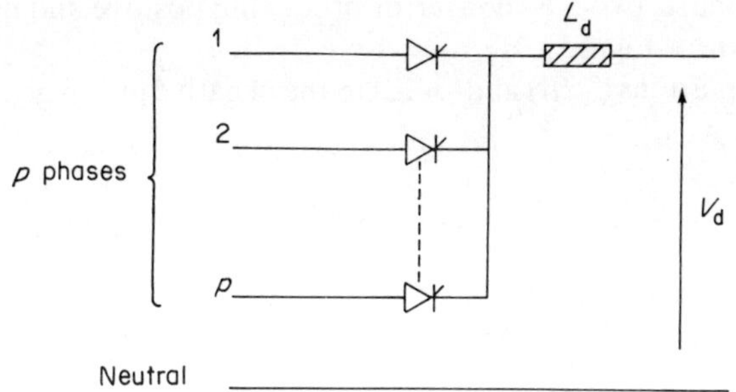

Figure 3.1. p-phase one-way convertor

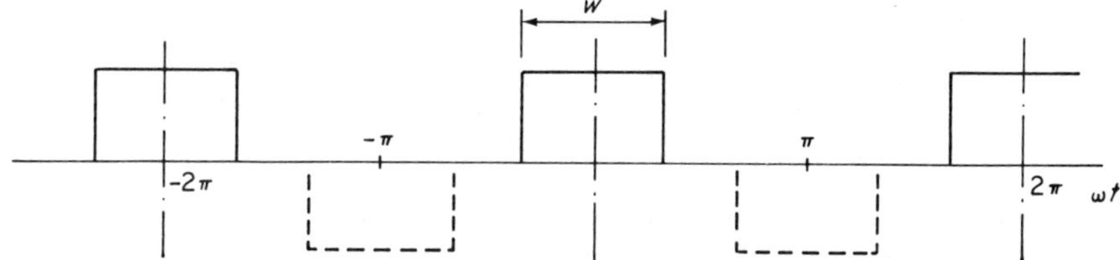

Figure 3.2. Trains of positive and negative pulses

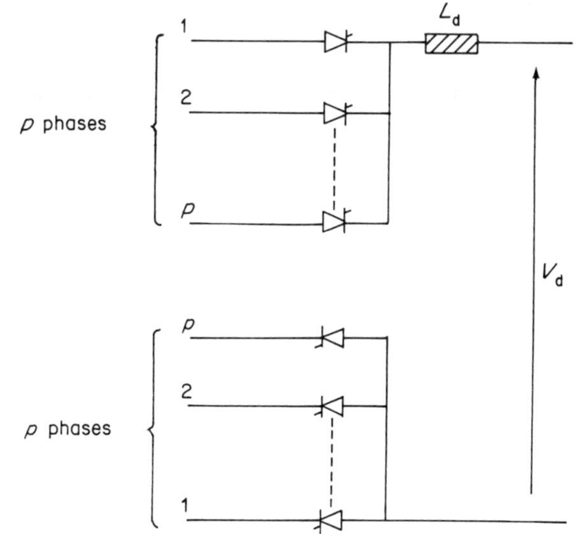

Figure 3.3. p-phase two-way convertor

The corresponding Fourier series for the positive current pulses is

$$F_p = \frac{2}{\pi}\left(\frac{w}{4} + \sin\frac{w}{2}\cos\omega t + \frac{1}{2}\sin\frac{2w}{2}\cos 2\omega t + \frac{1}{3}\sin\frac{3w}{2}\cos 3\omega t + \frac{1}{4}\sin\frac{4w}{2}\cos 4\omega t + \cdots\right). \tag{3.2.3}$$

An ideal p-phase, two-way convertor producing positive and negative current pulses is shown in Figure 3.3.

Applying equations (3.2.1) and (3.2.2) to the negative group gives the following Fourier series

$$F_n = \frac{2}{\pi}\left\{-\frac{w}{4} + \sin\frac{w}{2}\cos\omega t - \frac{1}{2}\sin\frac{2w}{2}\cos 2\omega t + \frac{1}{3}\sin\frac{3w}{2}\cos 3\omega t - \frac{1}{4}\sin\frac{4w}{2}\cos 4\omega t + \cdots\right\} \tag{3.2.4}$$

The phase current of the two-way configuration consists of alternate positive and negative pulses such that $F(\omega t + \pi) = -F(\omega t)$. Its Fourier series is obtained by combining equations (3.2.3) and (3.2.4)

$$F = F_p + F_n = \frac{4}{\pi}\left\{\sin\frac{w}{2}\cos\omega t + \frac{1}{3}\sin\frac{3w}{2}\cos 3\omega t + \frac{1}{5}\sin\frac{5w}{2}\cos 5\omega t + \cdots\right\} \tag{3.2.5}$$

In which the d.c. component and even ordered harmonics have been eliminated.

For the square wave of Figure 3.4(a) $w = \pi$ which on substituting into equation (3.2.5) gives as the equation for the waveform in the frequency domain

$$F(t) = \frac{4}{\pi}\{\cos(\omega t) - \tfrac{1}{3}\cos(3\omega t) + \tfrac{1}{5}\cos(5\omega t) - \tfrac{1}{7}\cos(7\omega t) + \cdots\} \tag{3.2.6}$$

in which harmonics of order $n = 1,5,9$, etc. are of positive sequence and those of order $n = 3,7,11$, etc. are of negative sequence.

The frequency domain representation of the square wave harmonic amplitudes is shown in Figure 3.4(b).

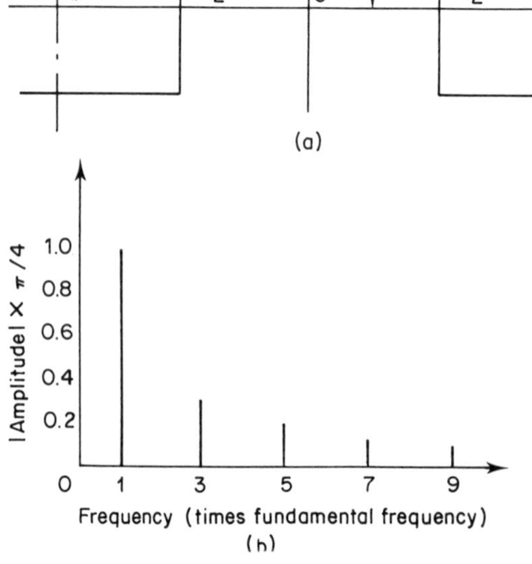

Figure 3.4. (a) Time domain representation of a square wave and (b) frequency domain representation of a square wave

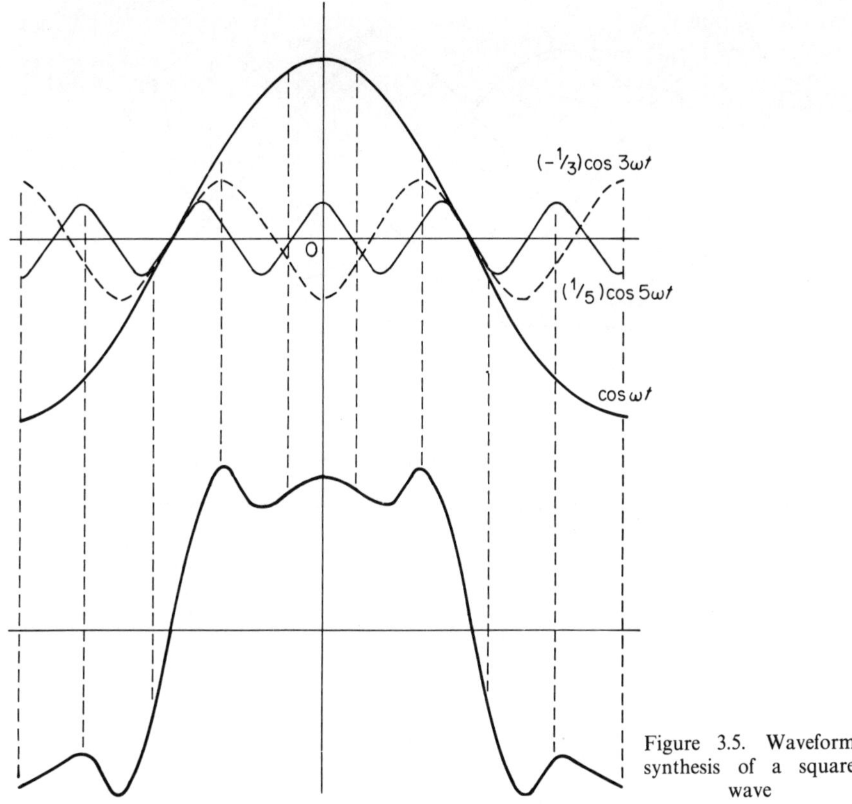

Figure 3.5. Waveform synthesis of a square wave

The time domain waveforms can also be synthesised from the combination of the time domain representation of the individual harmonics. Figure 3.5 shows this synthesis process for the square wave considered above. For clarity only the fundamental, third and fifth harmonic have been shown and the complex waveform produced is therefore not complete.

Six-pulse related harmonics

Six-pulse rectification (and inversion) is obtained from three-phase two-way configurations. Substituting $w = 2\pi/3$ in equation (3.2.5) and inserting the actual d.c. current I_d the frequency domain representation of the a.c. current in phase 'a' is

$$i_a = \frac{2\sqrt{3}}{\pi} I_d(\cos\omega t - \tfrac{1}{5}\cos 5\omega t + \tfrac{1}{7}\cos 7\omega t - \tfrac{1}{11}\cos 11\omega t$$

$$+ \tfrac{1}{13}\cos 13\omega t - \tfrac{1}{17}\cos 17\omega t + \tfrac{1}{19}\cos 19\omega t - \cdots). \tag{3.2.7}$$

The three phase currents are shown in Figure 3.6(b), (c) and (d) respectively.

Some useful observations can now be made from equation (3.2.7): (i) the absence of triple harmonics; (ii) the presence of harmonics of orders $6k \pm 1$ for integer values of k; (iii) those harmonics of orders $6k + 1$ are of positive sequence; (iv) those harmonics of orders $6k - 1$ are of negative sequence; (v) the r.m.s.

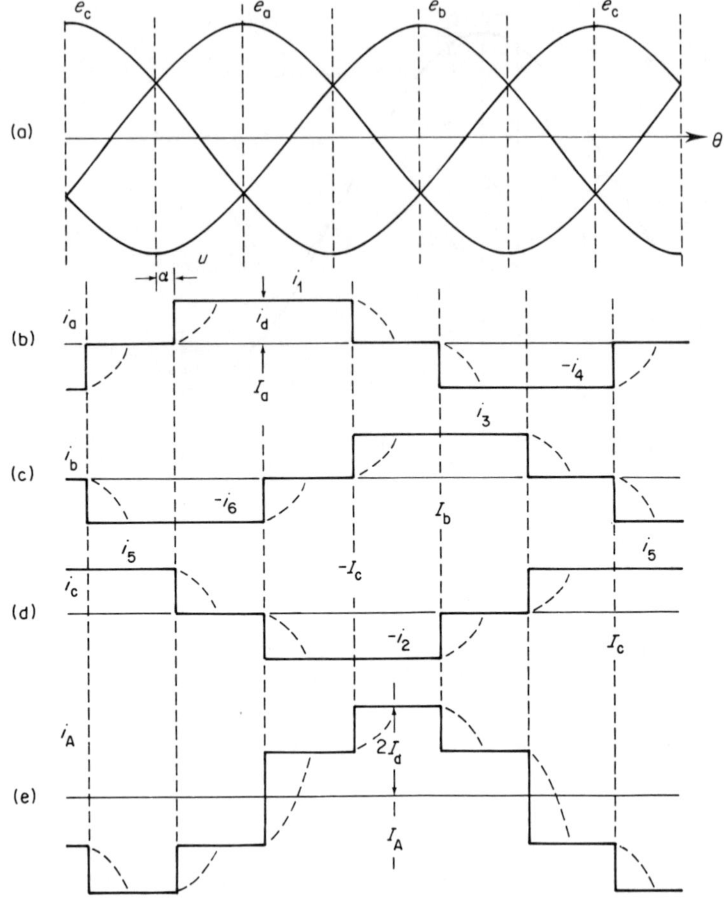

Figure 3.6. Six-pulse bridge waveforms: (a) phase to neutral voltages; (b)–(d) phase currents on the convertor side; (e) phase current on the system side with Δ-Y transformer

magnitude of the fundamental frequency is

$$I_1 = (1/\sqrt{2})(2\sqrt{3}/\pi)I_d = (\sqrt{6}/\pi)I_d; \tag{3.2.8}$$

(vi) the r.m.s. magnitude of the nth harmonic is

$$I_n = I_1/n. \tag{3.2.9}$$

Effect of transformer connection

If either the primary or secondary three-phase windings of the convertor transformer are connected in delta, the a.c. side current waveforms consist of the instantaneous differences between two rectangular secondary currents 120° apart as shown in Figure 3.6(e).

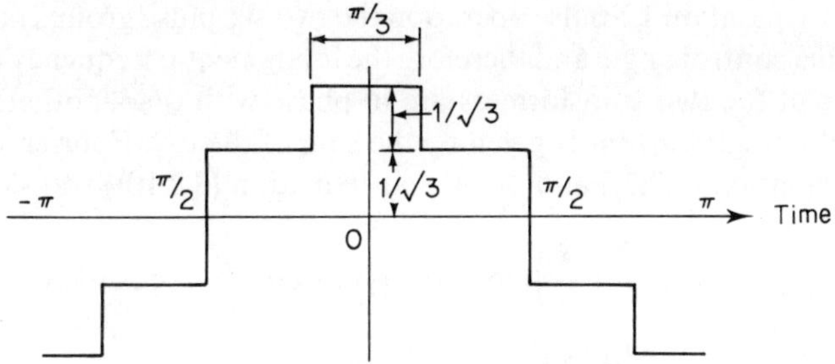

Figure 3.7. Time domain representation of a six-pulse waveform with delta–star

The Fourier series of the waveform shown in Figure 3.6(e) can be easily obtained from the general equation (3.2.5) by superimposing the results of two component pulses of widths π and $\pi/3$ respectively.

Moreover, to maintain the same primary and secondary voltages as for the star–star connection, a factor of $\sqrt{3}$ is introduced in the transformer ratio, and the current waveform is as shown in Figure 3.7.

The resulting Fourier series for the current in phase 'a' on the primary side is

$$i_a = \frac{2\sqrt{3}}{\pi} I_d (\cos \omega t + \tfrac{1}{5}\cos 5\omega t - \tfrac{1}{7}\cos 7\omega t - \tfrac{1}{11}\cos 11\omega t + \tfrac{1}{13}\cos 13\omega t$$
$$+ \tfrac{1}{17}\cos 17\omega t - \tfrac{1}{19}\cos 19\omega t - \cdots). \tag{3.2.10}$$

This series only differs from that of a star–star connected transformer by the sequence of rotation of harmonic orders $6k \pm 1$ for odd values of k, i.e. the fifth, seventh, 17th, 19th, etc.

Twelve-pulse related harmonics

Twelve-pulse configurations consist of two six-pulse groups fed from two sets of three-phase transformers in parallel, with their fundamental voltage equal and phase-shifted by 30°; a common 12-pulse configuration is shown in Figure 3.8.

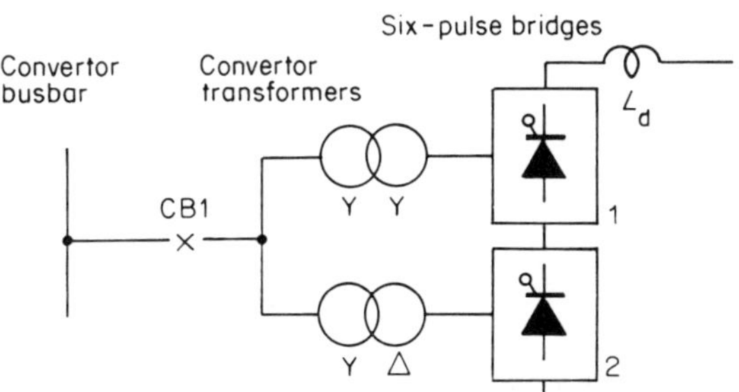

Figure 3.8. Twelve-pulse convertor configuration

Moreover, to maintain 12-pulse operation the two six-pulse groups must operate with the same control angle and therefore the fundamental frequency currents on the a.c. side of the two transformers are in phase with one another.

The resultant a.c. current is given by the sum of the two Fourier series of the star–star (equation (3.2.7)) and delta–star (equation (3.2.10)) transformers, i.e.

$$(i_a)_{12} = 2\left(\frac{2\sqrt{3}}{\pi}\right)\left(\cos \omega t - \tfrac{1}{11}\cos 11\omega t + \tfrac{1}{13}\cos 13\omega t - \tfrac{1}{23}\cos 23\omega t + \tfrac{1}{25}\cos 25\omega t - \cdots\right). \tag{3.2.11}$$

This series only contains harmonics of order $12k \pm 1$. The harmonic currents of orders $6k \pm 1$ (with k odd), i.e. $k = 5$, 7, 17, 19, etc., circulate between the two convertor transformers but do not penetrate the a.c. network. The time domain

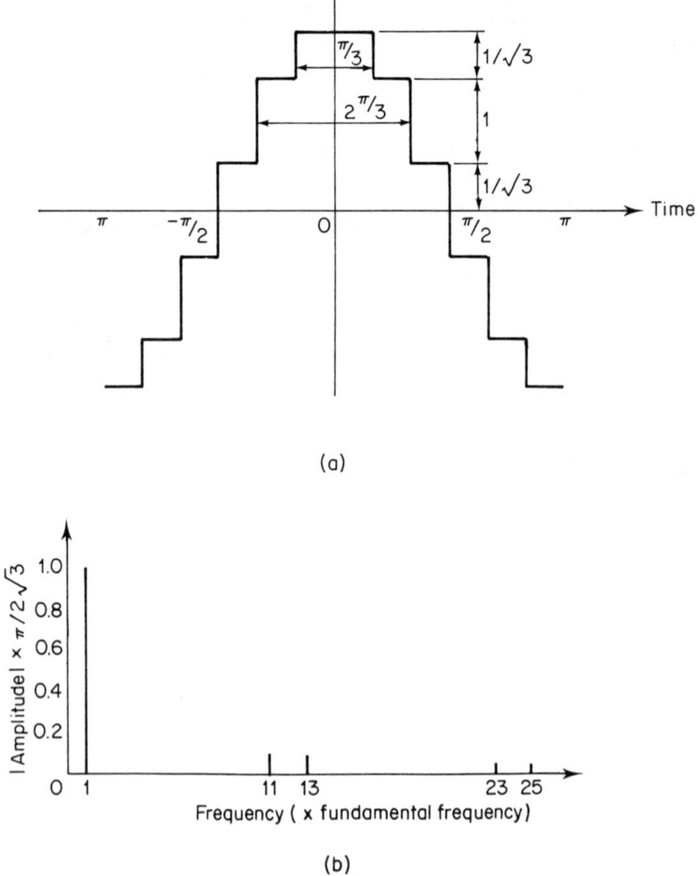

Figure 3.9. (a) Time domain representation of the 12-pulse phase current and (b) frequency domain representation of 12-pulse operation

representation of the 12-pulse waveform is shown in Figure 3.9(a) and the corresponding frequency domain representation in Figure 3.9(b).

Higher pulse configurations

In the last section, the use of two transformers with a 30° phase-shift has been shown to produce 12-pulse operation.

The addition of further appropriately shifted transformers in parallel provides the basis for increasing pulse configurations.

For instance 24-pulse operation is achieved by means of four transformers with 15° phase-shifts and 48-pulse operation requires eight transformers with 7.5° phase-shifts. Although theoretically possible, pulse numbers above 48 are rarely justified due to the practical levels of distortion found in the supply voltage waveforms, which can have as much influence on the voltage crossings as the theoretical phase-shifts.

Similarly to the case of the 12-pulse connection, the alternative phase-shifts involved in higher pulse configurations require the use of appropriate factors in the parallel transformer ratios to achieve common fundamental frequency voltages on their primary and secondary sides.

The theoretical harmonic currents are related to the pulse number (p) by the general expression $pk \pm 1$ and their magnitudes decrease in inverse proportion to the harmonic order. Generally harmonics above the 49th can be neglected as their amplitude is too small.

Effect of transformer and system impedance

In practice the existence of reactance in the commutation circuit causes conduction overlap of the in-coming and out-going phases.

As we have seen in previous sections, high-pulse configurations are combinations of three-pulse groups, i.e. the commutation overlaps are those of the three-pulse group as shown by the broken lines in Figure 3.6.

The current waveform has now lost the even symmetry with respect to the centre of the idealized rectangular pulse. Using as a reference the corresponding commutating voltage (i.e. the zero voltage crossing) and assuming a purely inductive commutation circuit, the following expression defines the commutating current.[(1)]

$$i_c = \frac{E}{\sqrt{2}X_c}(\cos\alpha - \cos\omega t), \tag{3.2.12}$$

where X_c is the reactance (per phase) of the commutation circuit, which is largely determined by the transformer leakage reactance.

At the end of the commutation $i_c = I_d$ and $\omega t = u$, and equation (3.2.12) becomes

$$I_d = \frac{E}{\sqrt{2}X_c}[\cos\alpha - \cos(\alpha + u)]. \tag{3.2.13}$$

Dividing (3.2.12) by (3.2.13),

$$i_c = I_d\left(\frac{\cos\alpha - \cos\omega t}{\cos\alpha - \cos(\alpha + u)}\right) \tag{3.2.14}$$

and this expression applies for $\alpha < \omega t < \alpha + u$.

The rest of the positive current pulse is defined by

$$i = I_d \quad \text{for} \quad \alpha + u < \omega t < \alpha + 2\pi/3 \tag{3.2.15}$$

and

$$i = I_d - I_d\left[\frac{\cos(\alpha + 2\pi/3) - \cos\omega t}{\cos(\alpha + 2\pi/3) - \cos(\alpha + 2\pi/3 + u)}\right] \quad \text{for} \quad \alpha + \frac{2\pi}{3} < \omega t < \alpha + \frac{2\pi}{3} + u. \tag{3.2.16}$$

The negative current pulse still possesses half-wave symmetry and therefore only odd-ordered harmonics are present. These can now be expressed in terms of the delay (firing) and overlap angles and their magnitudes, related to the fundamental components, are illustrated in Figures 3.10–3.13, for the fifth, seventh, 11th and 13th harmonics respectively.[1]

In summary, the existence of system impedance is seen to reduce the harmonic

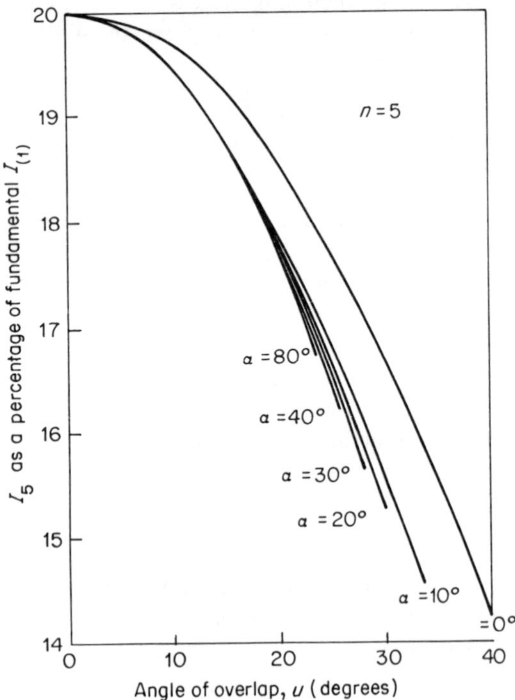

Figure 3.10. Variation of fifth harmonic current in relation to angle of delay and overlap

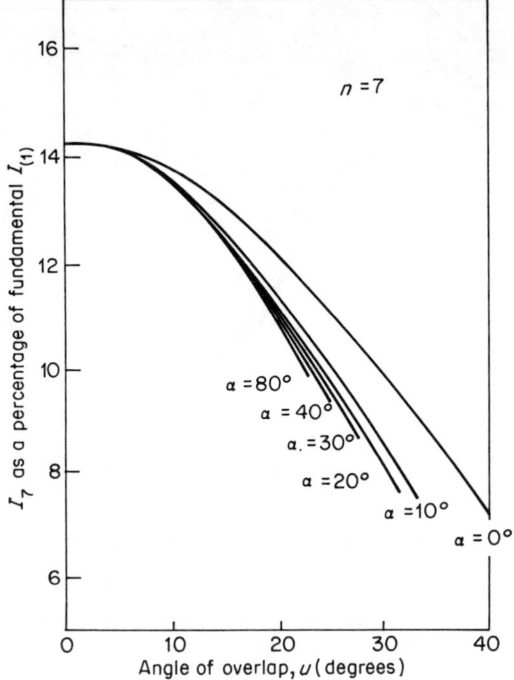

Figure 3.11. Variation of seventh harmonic current in relation to angle of delay and overlap

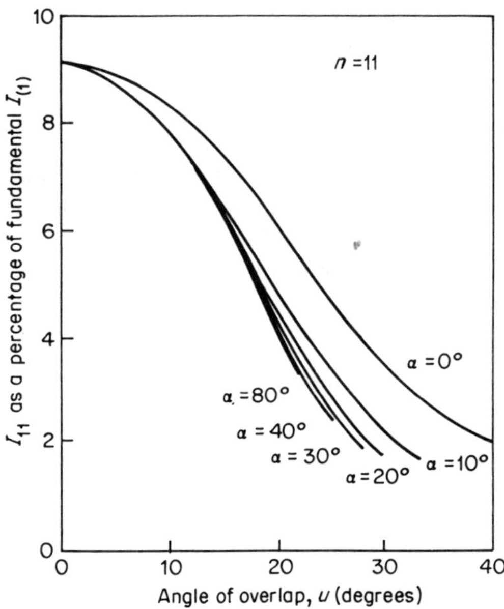

Figure 3.12. Variation of 11th harmonic current in relation to angle of delay and overlap

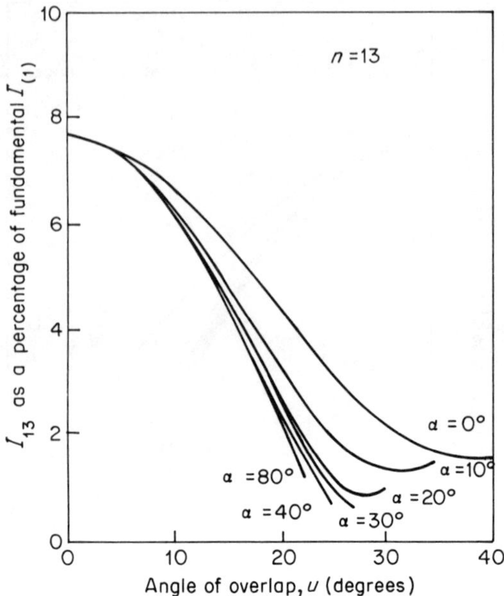

Figure 3.13. Variation of 13th harmonic current in relation to angle of delay and overlap

content of the current waveform, the effect being much more pronounced in the case of uncontrolled rectification. With large firing angles the current pulses are practically unaffected by a.c. system reactance.

Direct voltage harmonics

For the three-phase bridge configuration the orders of the harmonic voltages are $n = 6k$. The corresponding d.c. voltage waveforms are illustrated in Figure 3.14.

The repetition interval of the waveform shown in Figure 3.14(c) is $\pi/3$ and it contains the following three different functions with reference to voltage crossing C_1:

$$v_d = \sqrt{2}\,V_c \cos\left[\omega t + \frac{\pi}{6}\right] \quad \text{for} \quad 0 < \omega t < \alpha, \tag{3.2.17}$$

$$v_d = \sqrt{2}V_c \cos\left[\omega t + \frac{\pi}{6}\right] + \frac{1}{2}\sqrt{2}V_c \sin \omega t = \frac{\sqrt{6}}{2} V_c \cos \omega t \quad \text{for} \quad \alpha < \omega t < \alpha + u, \tag{3.2.18}$$

$$v_d = \sqrt{2}\,V_c \cos\left[\omega t - \frac{\pi}{6}\right] \quad \text{for} \quad \alpha + u < \omega t < \frac{\pi}{3}, \tag{3.2.19}$$

where V_c is the (commutating) phase to phase r.m.s. voltage.

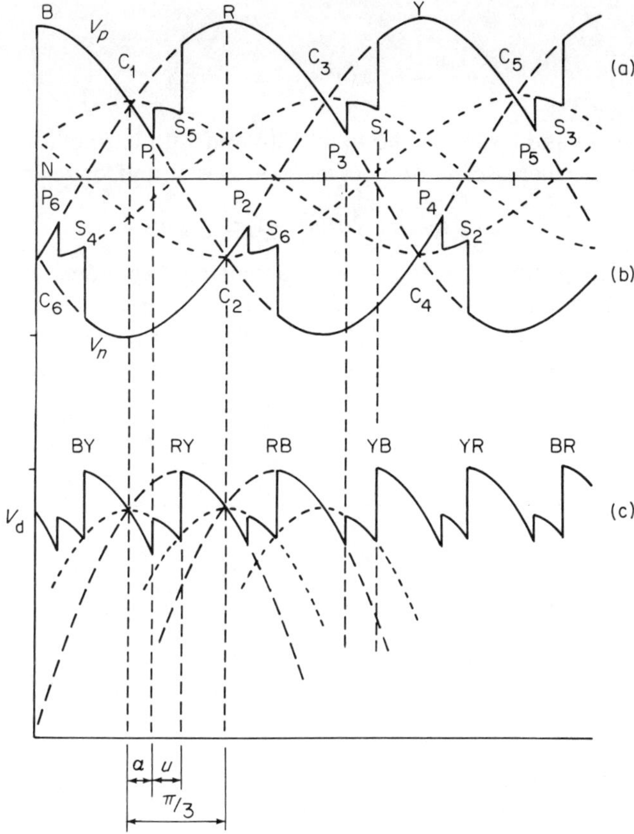

Figure 3.14. Six-pulse convertor d.c. voltage waveforms: (a) at the positive terminal; (b) at the negative terminal; (c) between output terminals

From equations (3.2.17), (3.2.18) and (3.2.19) the following expression is obtained for the r.m.s. magnitudes of the harmonic voltages of the d.c. voltage waveform:

$$V_n = \frac{V_{c0}}{\sqrt{2}(n^2-1)}\left\{(n-1)^2\cos^2\left[(n+1)\frac{u}{2}\right] + (n+1)^2\cos^2\left[(n-1)\frac{u}{2}\right] - 2(n-1)(n+1)\cos\left[(n+1)\frac{u}{2}\right]\cos\left[(n-1)\frac{u}{2}\right]\cos(2\alpha+u)\right\}^{1/2}. \quad (3.2.20)$$

Figure 3.15(a) and (b) illustrates the variation of the sixth and 12th harmonics[1] as a percentage of V_{c0}, the maximum average rectified voltage which for the six-pulse bridge convertor is $3(\sqrt{2})V_c/\pi$. These curves and equations show some interesting facts. Firstly, for $\alpha = 0$ and $u = 0$, equation (3.2.20) reduces to

$$V_{n0} = \sqrt{2}\,V_{c0}/(n^2-1) \quad (3.2.21)$$

or

$$\frac{V_{n0}}{V_{c0}} = \sqrt{2}/(n^2 - 1) \simeq \sqrt{2}/n^2, \tag{3.2.22}$$

giving 4.04, 0.99 and 0.44% for the sixth, 12th and 18th harmonics respectively.

Generally, as α increases, harmonics increase as well, and for $\alpha = \pi/2$ and $u = 0$,

$$V_n/V_{c0} = \sqrt{2}n/(n^2 - 1) \simeq \sqrt{2}/n, \tag{3.2.23}$$

which produces n times the harmonics content corresponding to $\alpha = 0$. This means that the higher harmonics increase faster with α. Equation (3.2.23) is of some importance as it represents the maximum proportion of harmonics in the system, particularly when it is considered that at $\alpha = 90°$ u is likely to be very small.

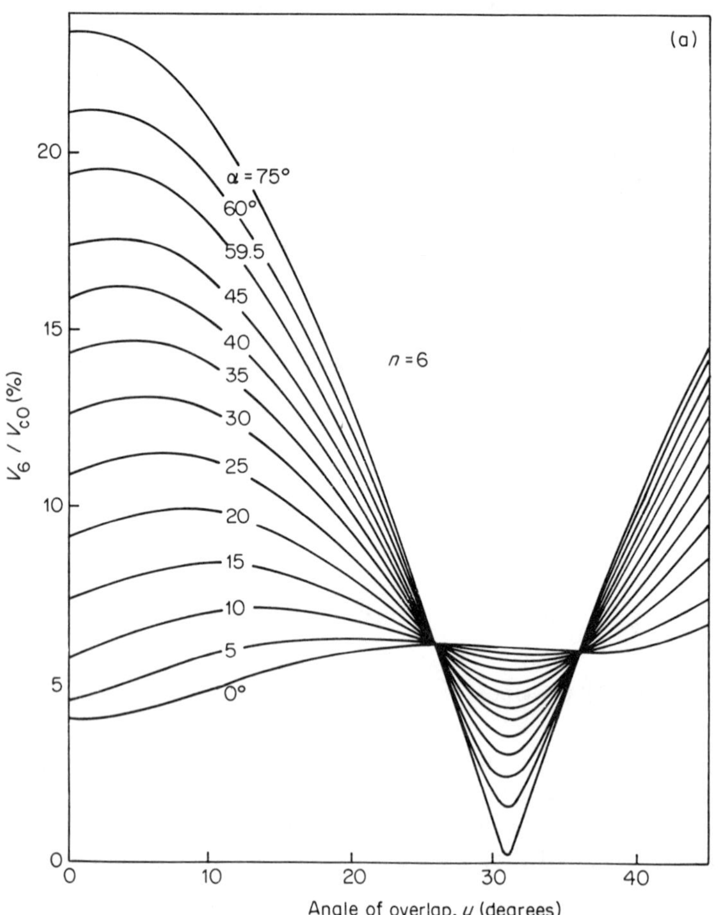

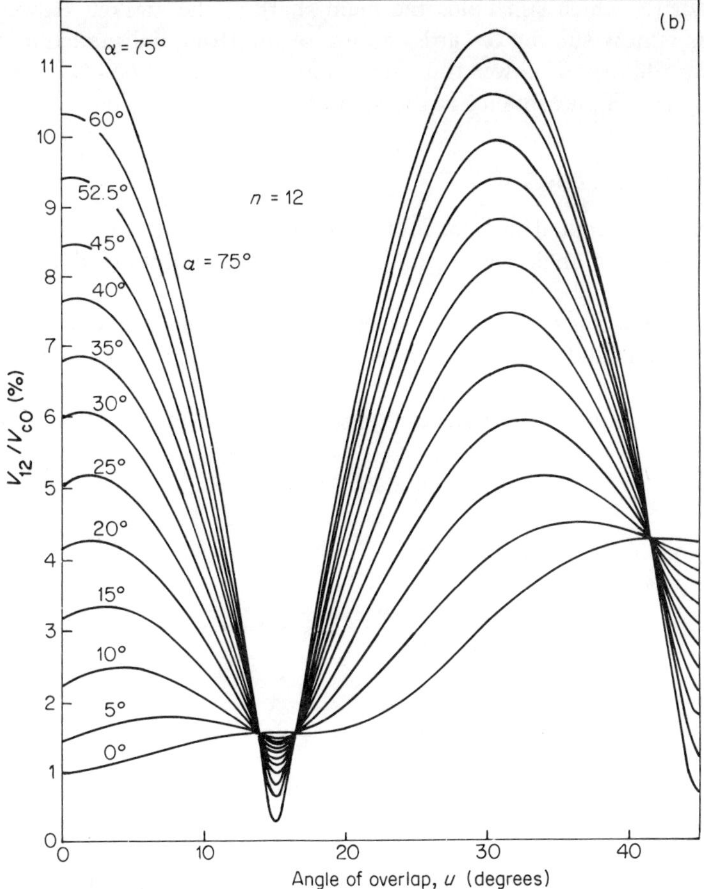

Figure 3.15. Variation of (a) sixth harmonic voltage and (b) 12th harmonic voltage in relation to angle of delay and overlap

If the convertor involves two bridges, one with a star–star or delta–delta transformer and the other with a delta–star or star–delta transformer, their respective voltages will be 30° out of phase and so the harmonics will accordingly be out of phase. Since 30° of mains frequency correspond to a half-cycle of the sixth harmonic, this harmonic will be in phase opposition in the two bridges. Similarly for the 12th harmonic, 30° corresponds to one cycle, giving harmonics in phase; for the 18th harmonic, 30° corresponds to one and a half cycles, giving harmonics in opposition and so on.

3.3 MEDIUM SIZE CONVERTORS

The number of medium size convertors (in tens and hundreds of kilowatts) is increasing fast throughout industry. Earlier applications were developed around the d.c. drive, which still holds the main share of the market. However, the emphasis is now shifting towards the use of invertors and induction motors. Moreover, the use of power transistors and gate turn off (GTO) thyristors is gradually gaining acceptance in the area of a.c. motor control.

Convertor-fed d.c. drives

Considering the limited inductance of the motor armature winding and the larger variation of firing angle, the constant d.c. current assumption of the large size convertors cannot be justified.

The d.c. load must now be represented as an equivalent circuit which in its simplest form includes resistance, inductance and back e.m.f. (as shown in Figure 3.16).

With sinusoidal supply voltage $V_m \sin \omega t$, the following equation applies:

$$V_m \sin \omega t = Ri + L\left(\frac{di}{dt}\right) + E, \tag{3.3.1}$$

and the load current has the form

$$i = K e^{-Rt/L} + \frac{V_m}{\sqrt{(R^2 + (\omega L)^2)}} \sin(\omega t - \phi) - \frac{E}{R}, \tag{3.3.2}$$

where

$$\phi = \tan^{-1}(\omega L/R)$$

and the constant K is derived from the particular initial conditions.

Under nominal loading the firing delay is kept low, but during motor start or light load conditions the delay increases substantially and the current may even be discontinuous. This extreme operating condition is illustrated in Fig. 3.17 for a six-pulse rectifier. Each phase consists of two positive and two negative current pulses, which are derived from the general expression (3.3.2) by using the appropriate voltage phase relationships with a common reference.

The current in phase R with reference to the instant when V_{RY} is maximum in

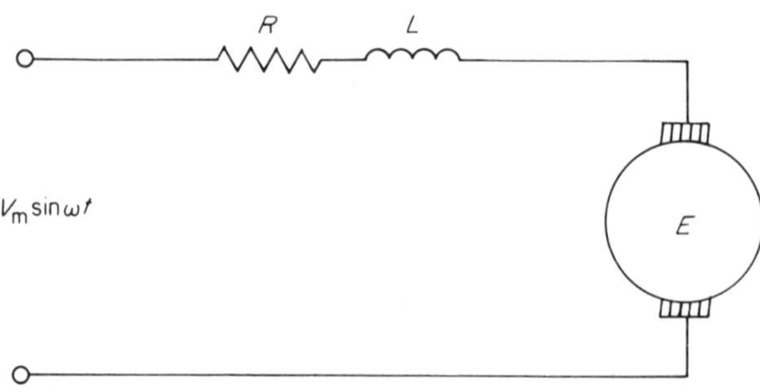

Figure 3.16. D.C. motor equivalent circuit

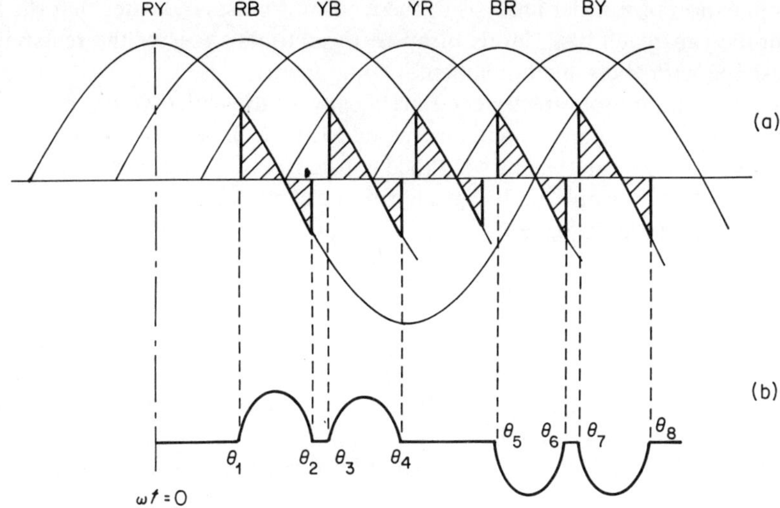

Figure 3.17. Discontinuous waveforms: (a) d.c. voltage; (b) a.c. current in phase R.

Figure 3.17 has the following components:

(i) Over the range $\theta_1 < \omega t < \theta_2$,

$$i = \frac{V_m}{R}\left\{\cos\phi\cos(\omega t - \phi) - \frac{E}{V_m} + \left[\frac{E}{V_m} - \cos\phi\cos(\theta_1 - \phi)\right]e^{(-R/\omega L)(\omega t - \theta_1)}\right\}. \tag{3.3.3}$$

(ii) When $\theta_3 < \omega t < \theta_4$, where $\theta_3 = (\theta_1 + \pi/3)$,

$$i = \frac{V_m}{R}\left\{\cos\phi\cos\left(\omega t - \frac{\pi}{3} - \phi\right) - \frac{E}{V_m} + \left[\frac{E}{V_m} - \cos\phi\cos(\theta_1 - \theta)\right]e^{(-R/\omega L)(\omega t - \pi/3 - \theta_1)}\right\}. \tag{3.3.4}$$

(iii) When $\theta_5 < \omega t < \theta_6$, where $\theta_5 = (\theta_1 + \pi)$,

$$i = -\frac{V_m}{R}\left\{\cos\phi\cos(\omega t - \pi - \phi) - \frac{E}{V_m} + \left[\frac{E}{V_m} - \cos\phi\cos(\theta_1 - \phi)\right]e^{(-R/\omega L)(\omega t - \pi - \theta_1)}\right\}. \tag{3.3.5}$$

(iv) When $\theta_7 < \omega t < \theta_8$, where $\theta_7 = (\theta_1 + 2\pi/3)$,

$$i = -\frac{V_m}{R}\left\{\cos\phi\cos\left(\omega t - \frac{2\pi}{3} - \phi\right) - \frac{E}{V_m} + \left[\frac{E}{V_m} - \cos\phi\cos(\theta_1 - \phi)\right]e^{(-R/\omega L)(\omega t - 2\pi/3 - \theta_1)}\right\}. \tag{3.3.6}$$

Application of Fourier analysis to these current pulses indicates that the fifth harmonic can reach peak levels of up to three times those of the rectangular waveshape with the same fundamental component.

When the d.c. motors are designed specifically for use with thyristor convertors their armature inductance is often increased to avoid current discontinuities and the above analysis can then be simplified considerably. An approximate method described by Dobinson[2] derives the harmonic components of the a.c. current in terms of the ripple ratio, i.e.

$$r = \frac{I_r}{I_d}, \tag{3.3.7}$$

where I_r is the alternating ripple of the direct current and I_d is the mean direct current, flowing in the motor armature circuit (Figure 3.18) at the relevant speed and load. The method ignores the effect of the commutation reactance, which at large delay angles is negligible.

With reference to Figure 3.18, a further approximation is made by assuming

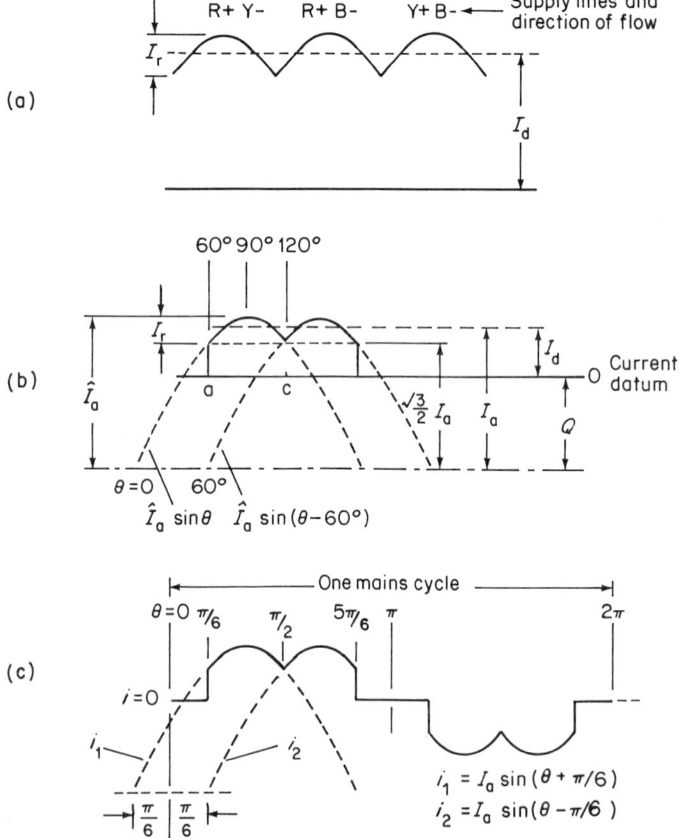

Figure 3.18. Convertor output and input current waveforms: (a) armature current; (b) thyristor current; (c) a.c. current (after Dobinson[2])

that the ripple is part of a sine wave displaced by a value θ relative to the zero direct-current level.

The information included in Figure 3.18(b) and (c) results in the following functions:

$$f(\theta)=0 \qquad \text{for} \quad 0<\theta<\pi/6, \tag{3.3.8}$$

$$f(\theta)=I_{\mathrm{d}}\left[7.46\,r\sin\left(\theta+\frac{\pi}{6}\right)-7.13\,r+1\right] \quad \text{for} \quad \pi/6<\theta<\pi/2, \tag{3.3.9}$$

$$f(\theta)=I_{\mathrm{d}}\left[7.46\,r\sin\left(\theta-\frac{\pi}{6}\right)-7.13\,r+1\right] \quad \text{for} \quad \pi/2<\theta<5\pi/6, \tag{3.3.10}$$

$$f(\theta)=0 \qquad \text{for} \quad 5\pi/6<\theta<\pi. \tag{3.3.11}$$

Application of Fourier analysis yields the following expression for the fundamental:

$$I_1=I_{\mathrm{d}}(1.102+0.014r), \tag{3.3.12}$$

also, the magnitudes of the characteristic harmonics (expressed as percentage of the fundamental) are

$$I_n=100\left(\frac{1}{2}+\frac{6.46r}{n-1}-\frac{7.13r}{n}\right)(-1)^k \quad \text{for} \quad n=kp-1 \tag{3.3.13}$$

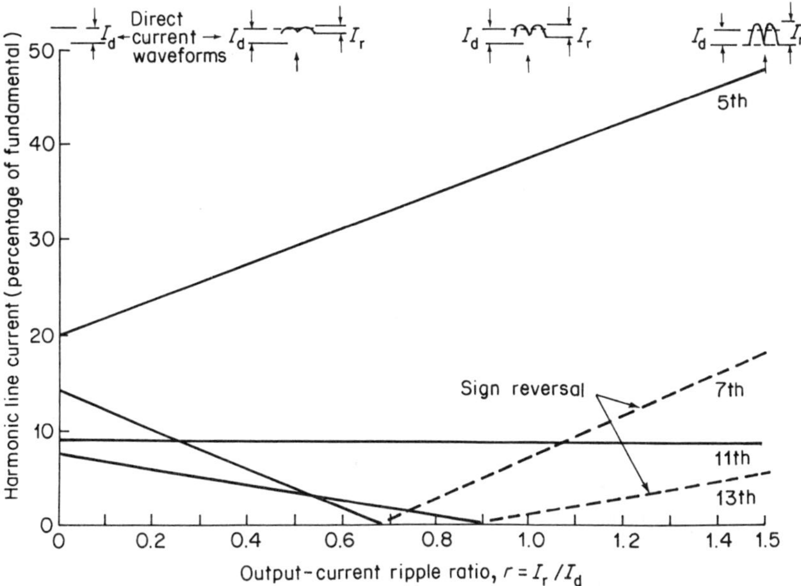

Figure 3.19. Harmonic content of supply current for six-pulse convertor with finite inductive load (after Dobinson[(2)])

and

$$I_n = 100\left(\frac{1}{n} + \frac{6.46r}{n+1} - \frac{7.13r}{n}\right)(-1)^k \quad \text{for} \quad n = kp + 1. \tag{3.3.14}$$

These are plotted in Figure 3.19 as a function of r covering the whole range from zero ripple (i.e. infinite inductance) to the limit case of continuous current (at $r = 1.5$).

Figure 3.19 shows that, although the fifth harmonic increases substantially with output current ripple, all the other harmonics reduce.

Half-controlled rectification

Because of the cheapness of its design the half-controlled version of the variable speed d.c. drive has been popular in some countries. When operating at full load

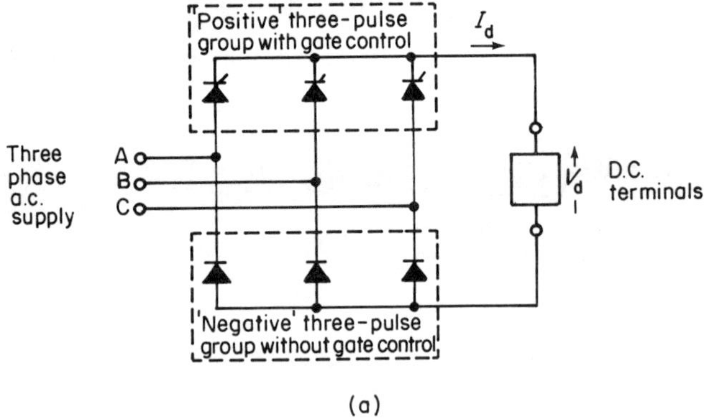

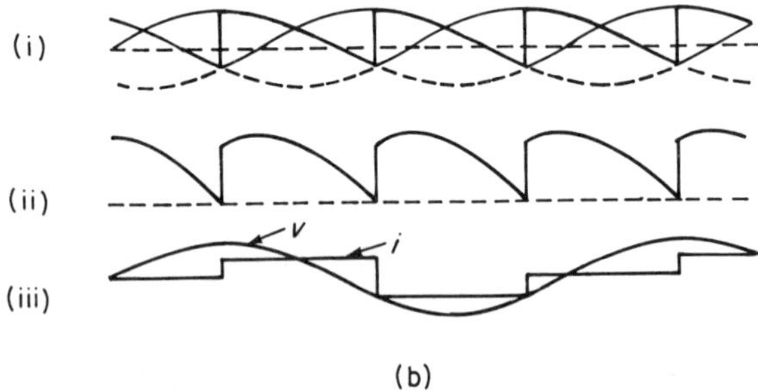

Figure 3.20. (a) Three-phase 'half-controlled' convertor and (b) theoretical waveforms for $\alpha = 60°$. In (b), (i) shows the voltage of a 'positive' group at the d.c. terminal with respect to supply neutral (——) and the voltage of a 'negative' group at the d.c. terminal with respect to supply neutral (-----); (ii) shows the d.c. terminal voltage of the bridge; (iii) shows the supply voltage and current of phase A

(i.e. with zero firing delay) these controllers produce virtually the same harmonic currents as the fully controlled convertor and operate very efficiently.

However, under operating conditions requiring firing delays, the half-wave symmetry of the current waveform is lost, as shown in Figure 3.20. At low loads these controllers not only have a very poor power factor but introduce severe waveform distortion particularly at even harmonics.

More often than not the controllers and motors initially installed are larger than required to cope with future expansions and operation is then at a fraction of the full load. Under these conditions the second harmonic component often reaches levels close to the fundamental current.

Individually controlled bridges

Individual bridge control is commonly used in railway traction configurations, such as that shown in Figure 3.21, which involves two groups of two bridge convertors connected in series to a parallel connection of two d.c. traction motors. At the start the back e.m.f. of the d.c. motors is zero, the supply d.c. voltage is low and the delay angle large. Therefore during the initial accelerating period, with maximum d.c. motor current, the bridge rectifier produces the worst harmonic currents and operates with the lowest power factor. To alleviate the situation at low speeds, one of the bridges is often bypassed and phase control exercised on the other. When the speed builds up, and the second bridge operates

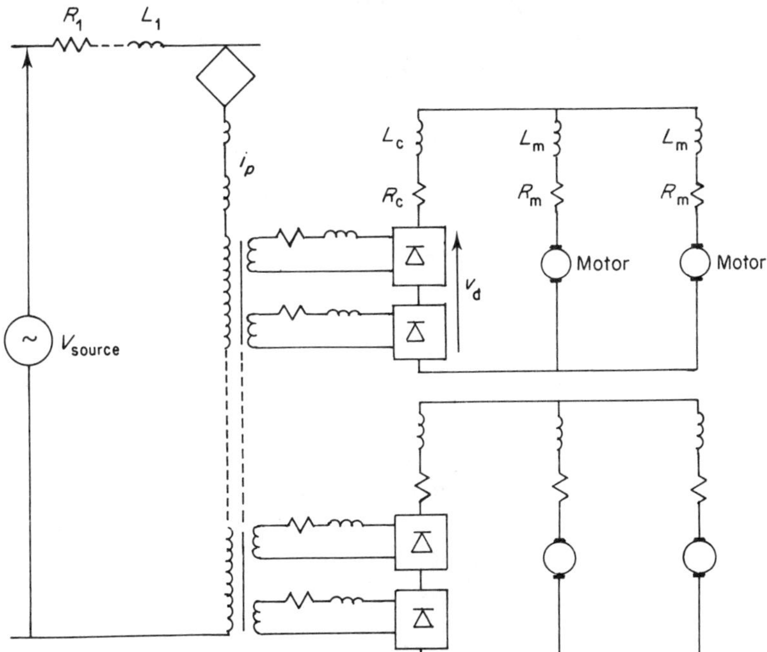

Figure 3.21. Typical locomotive power circuit

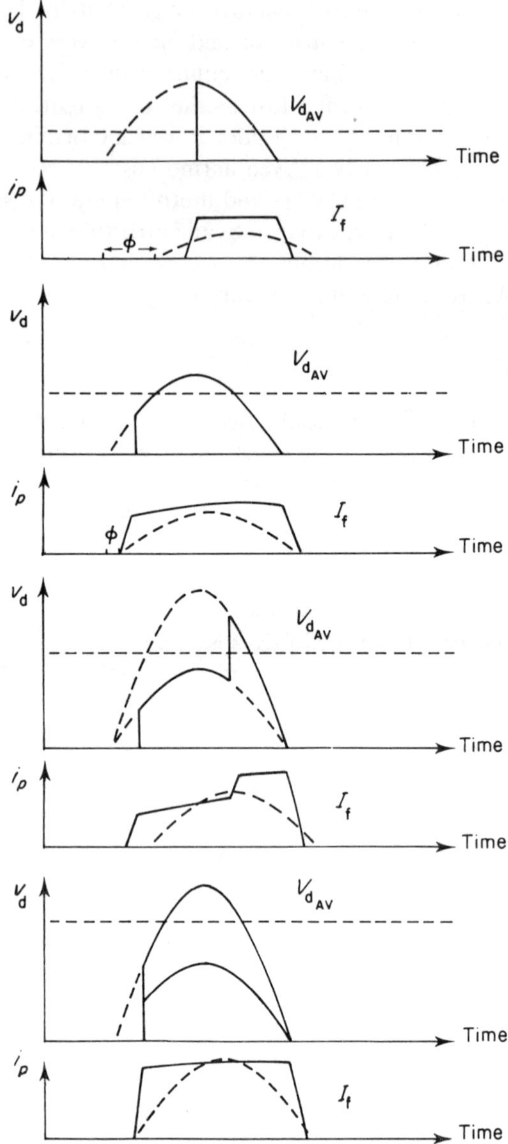

Figure 3.22. Voltage and current waveforms of a double-bridge individually controlled convertor

on minimum delay, phase control is exercised on the first bridge. The relevant waveforms are illustrated in Figure 3.22.

An indication[(3)] of the wide variation in harmonic current magnitudes corresponding to the waveform of Figure 3.22 is given in Figure 3.23.

Individual bridge control is rarely used in three-phase multibridge configur-

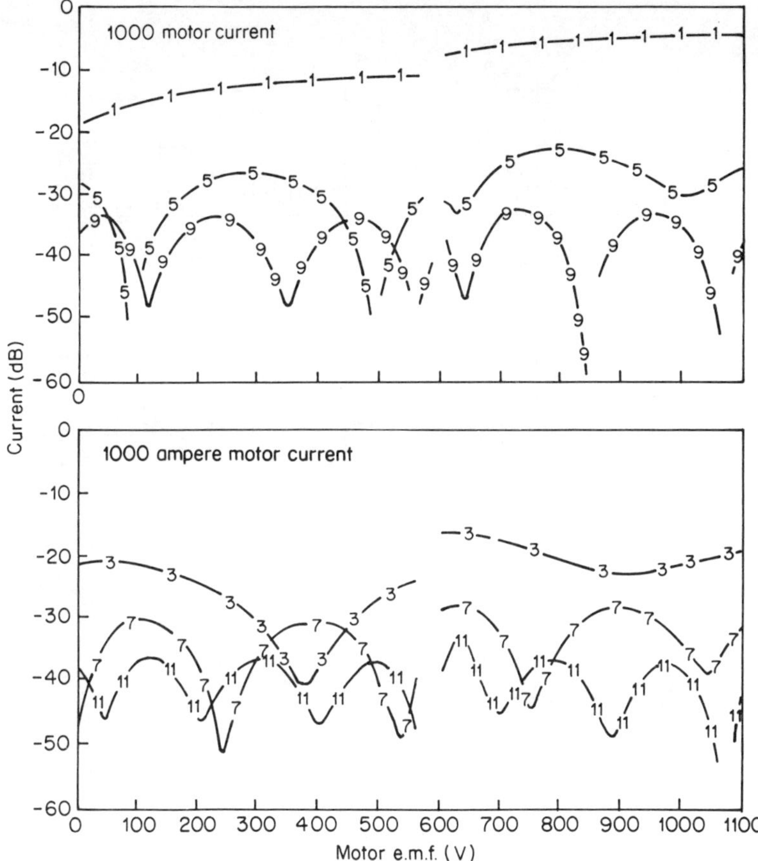

Figure 3.23. Variation in harmonic currents with locomotive operation

ations. It is instructive, however, to discuss the effect of firing diversity with reference to the nominally 12-pulse convertor configuration. Such effect is clearly illustrated[(4)] by the waveforms of Figure 3.24. The supply system alternatively sees 12-pulse and approximate six-pulse harmonics as the phase delay angle change. It is clear that fixed transformer phase-shifts are no guarantee of pulse multiplication in this case.

Invertor-fed a.c. drives

The basic three-phase invertor bridge commonly used in a.c. drive control is made up of six controlled semiconductors, each having a feedback diode connected in inverse parallel as shown in Figure 3.25; this figure does not include the auxiliary components required to force the commutations. The bridge is supplied from either a variable d.c. voltage, provided by a controlled rectifier, or a

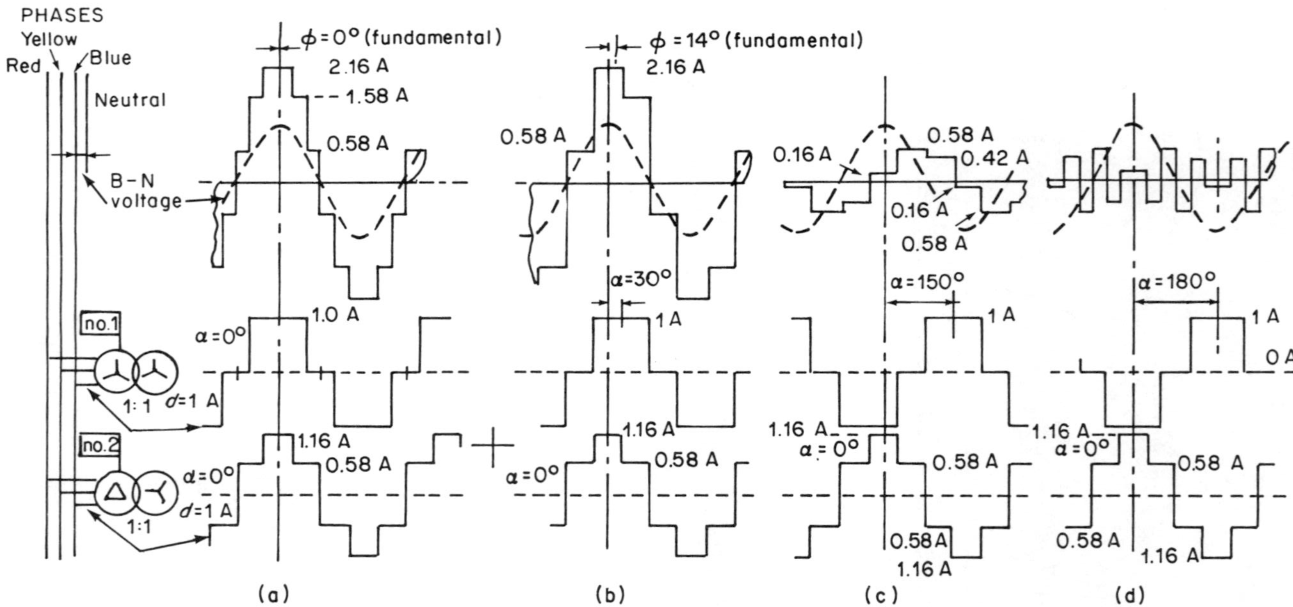

Figure 3.24. Input current for two similar thyristor equipments with independent firing control; ϕ is the displacement angle. (a) Identical phase-delay angle, true 12-pulse operation; (b) phase-delay angle difference of 30°, pseudo six-pulse operation; (c) phase-delay angle of 150°, pseudo six-pulse operation; (d) phase-delay angle of 180°, pseudo 12-pulse operation

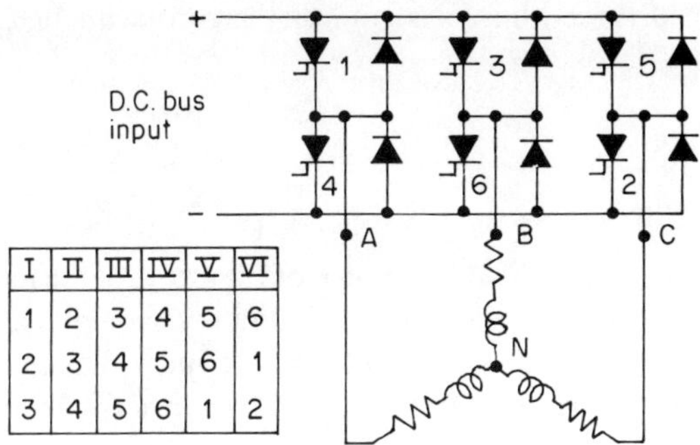

Figure 3.25. Basic three-phase invertor and balanced motor load showing an elementary switch-closing sequence

constant d.c. voltage. In each case the d.c. voltage level for a steady state operating condition can be considered constant.

MOTOR PHASE VOLTAGE

In the circuit of Figure 3.25 the invertor phase output voltage is always at one of two distinct voltage levels. The floating neutral voltage with respect to ground, expressed in terms of the invertor phase output voltage waveforms v_A, v_B, v_C, is

$$v_N = \tfrac{1}{3}(v_A + v_B + v_C), \tag{3.3.15}$$

so that a typical motor phase voltage is

$$v_{AN} = v_A - v_N = \tfrac{1}{3}(2v_A - v_B - v_C). \tag{3.3.16}$$

For a balanced linear, bilateral motor load impedance the motor phase voltage of a harmonic of order k can be expressed as

$$\begin{aligned} v_{AN(n)} &= \tfrac{1}{2}[2v_{A(n)} - v_{B(n)} - v_{C(n)}] \\ &= \tfrac{1}{3}[2v_{mn} \sin n\omega_1 t - v_{mn} \sin n(\omega_1 t + 2\pi/3) \\ &\quad - v_{mn} \sin n(\omega_1 t - 2\pi/3)] \\ &= \tfrac{2}{3} v_{mn} \sin n\omega_1 t\,[1 - \cos 2n\pi/3] \\ &= \tfrac{2}{3} v_{A(n)}[1 - \cos 2n\pi/3], \end{aligned} \tag{3.3.17}$$

and similarly for phases B and C:

$$v_{BN(n)} = \tfrac{2}{3} v_{B(n)}[1 - \cos 2n\pi/3]; \tag{3.3.18}$$

$$v_{CN(n)} = \tfrac{2}{3} v_{C(n)}[1 - \cos 2n\pi/3]. \tag{3.3.19}$$

Moreover, for all the positive and negative sequence harmonics

$$\cos 2n\pi/3 = -\frac{1}{2}$$

so that

$$v_{AN(n)} = v_{A(n)}, \qquad v_{BN(n)} = v_{B(n)}, \qquad v_{CN(n)} = v_{C(n)}, \tag{3.3.20}$$

and in the balanced three-phase system, for the zero sequence harmonics,

$$\cos 2n\pi/3 = 1$$

so that

$$v_{AN(n)} = v_{BN(n)} = v_{CN(n)} = 0. \qquad (3.3.21)$$

Thus, for the basic invertor-fed a.c. motor of Figure 3.25, the motor phase input voltage is identical to that of the corresponding invertor phase output voltage, with the exception that all invertor phase triplen harmonics have been eliminated. The effect of triplen harmonic elimination is shown in Figure 3.26 for

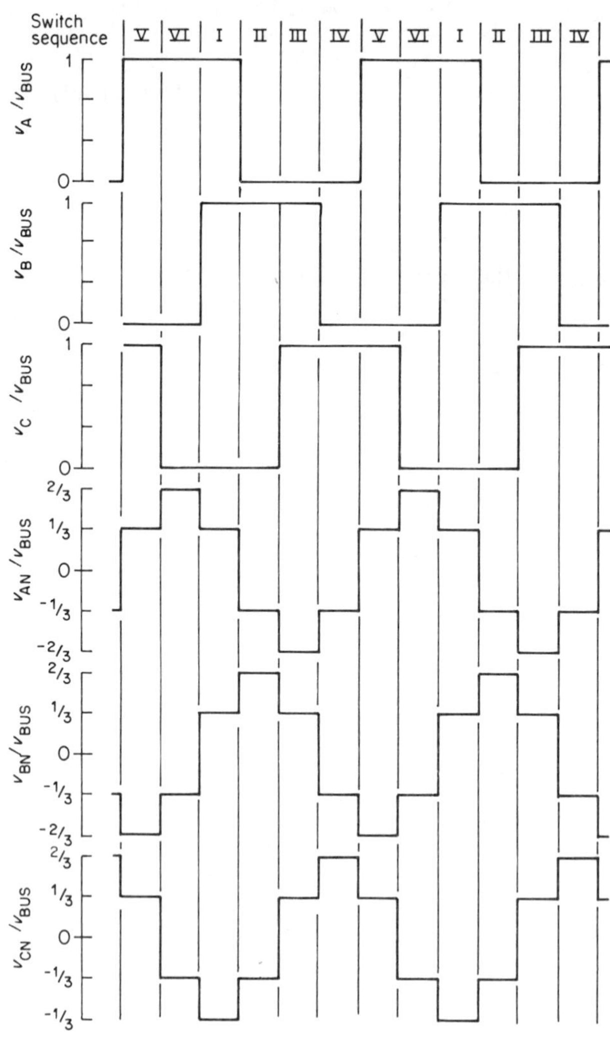

Figure 3.26. Basic six-step waveforms

the invertor output voltage waveforms resulting from the basic switch-closing sequence indicated in Figure 3.25. Since each motor phase input waveform results from square invertor phase voltage waveforms, the motor phase voltage in the frequency domain is

$$v_p = \frac{2}{\pi} v_{\mathrm{BUS}} \sum_{k=0}^{\infty} \left[\frac{1}{(6k+1)} \sin(6k+1)\omega_1 t + \frac{1}{(6k+5)} \sin(6k+5)\omega_1 t \right]. \quad (3.3.22)$$

MAGNETIZING CURRENT AND FLUX

The motor phase magnetizing inductance, L_{m}, acts as an integrating filter for the input voltage waveform, i.e.

$$i_p = \frac{1}{L_{\mathrm{m}}} \int_{t_0}^{t} v_p \, \mathrm{d}t. \quad (3.3.23)$$

Therefore for the nth harmonic phase voltage

$$v_{p(n)} = \frac{2}{\pi n} v_{\mathrm{BUS}} \sin n\omega_1 t, \quad (3.3.24)$$

$$i_{p(n)} = \frac{2 v_{\mathrm{BUS}}}{\pi n L_m} \int_{t_0}^{t} \sin n\omega_1 t \, \mathrm{d}t$$

$$= \frac{2 v_{\mathrm{BUS}}}{\pi n^2 \omega_1 L_m} \cos n\omega_1 t, \quad (3.3.25)$$

and the magnetizing current resulting from (3.3.22) is

$$i_p = \frac{2 v_{\mathrm{BUS}}}{\pi \omega_1 L_{\mathrm{m}}} \sum_{k=0}^{\infty} \left[\frac{1}{(6k+1)^2} \cos(6k+1)\omega_1 t + \frac{1}{(6k+5)^2} \cos(6k+5)\omega_1 t \right]. \quad (3.3.26)$$

For a motor phase magnetizing inductance of an equivalent N turns, the resultant motor phase air gap flux phasor is

$$\phi_p = \frac{2 v_{\mathrm{BUS}}}{\pi \omega_1 N} \sum_{k=0}^{\infty} \left[\frac{1}{(6k+1)^2} \cos(6k+1)\omega_1 t + \frac{1}{(6k+5)^2} \cos(6k+5)\omega_1 t \right]. \quad (3.3.27)$$

The relative amplitudes of these motor phase harmonic quantities are summarized in Table 3.1.

Table 3.1. Relative amplitudes of motor phase harmonics

Quantity	Harmonic number, n								
	1	5	7	11	13	17	19	23	25
v_p	1.000	0.200	0.143	0.091	0.077	0.059	0.053	0.043	0.040
i_p	1.000	0.040	0.020	0.008	0.006	0.004	0.003	0.002	0.002
ϕ_p	1.000	0.040	0.020	0.008	0.006	0.003	0.003	0.002	0.002

The peak amplitude of the fundamental frequency voltage in equation (3.3.22) is

$$v_{p(1)} = (2v_{BUS})/\pi$$

and from equation (3.3.27) the peak amplitude of the fundamental air-gap flux phasor is

$$\phi_{p(1)} = \frac{1}{\omega_1 N} \frac{2v_{BUS}}{\pi} = \frac{v_{p(1)}}{\omega_1 N}. \tag{3.3.28}$$

To maintain $\phi_{p(1)}$ constant when the fundamental frequency ω_1 varies, it is evident from (3.3.28) that $v_{p(1)}$ must be made a linear function of ω_1.

Some high-power invertor-fed a.c. motor speed controllers employ a separate d.c. chopper power supply in the d.c. bus to vary the voltage linearly with frequency. In this case the invertor output voltage waveforms are always square waves, as shown in Figure 3.26, and the air-gap harmonic flux vectors have the relative amplitudes indicated in Table 3.1.

An alternative to independent d.c. voltage control is the use of pulse width modulation, discussed in the next section.

PULSE WIDTH MODULATION (PWM)

PWM is becoming the most popular invertor technique to economize on power semiconductor switching stages. The operating principle consists of chopping the basic invertor square wave output voltage of Figure 3.26 in order to control the fundamental frequency voltages.

In its simplest form a sawtooth wave[5] is used to modulate the chops, as shown in Figure 3.27. The sawtooth has a frequency which is a multiple of three times the sine wave frequency, allowing symmetrical three-phase voltages to be generated from a three-phase sine wave set and one sawtooth waveform. This method controls line to line voltage from zero to full voltage by increasing the magnitude of the sawtooth, with little regard to the harmonics generated.

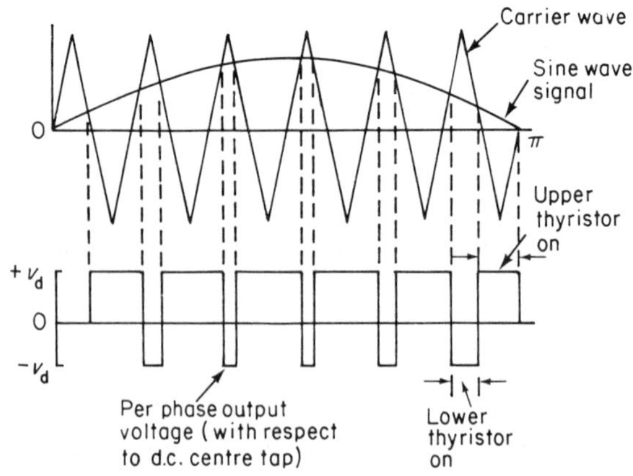

Figure 3.27. Principle of PWM.[5]

The most significant areas of voltage in the spectrum, apart from the fundamental, occur at the carrier frequency (sawtooth frequency) and its two sidebands, and to a significant extent at each multiple of these frequencies in the spectrum. Because the carrier frequency is six times the fundamental, therefore the triplen harmonics cancel in the system; however, the phase waveforms of Figure 3.27 do not have half-wave symmetry, hence even harmonics are present.

If the carrier frequency is a large multiple of the fundamental, the first large harmonics encountered are high in the harmonic spectrum. However, this effect is limited by the efficient switching capability of the invertor commutation circuits.

PWM WITH SELECTED HARMONIC ELIMINATION

More efficient PWM techniques have been developed to control the fundamental and harmonic voltages simultaneously.[6–9] For this purpose the chops can be created at predetermined angles of the square wave as shown in Figure 3.28. By

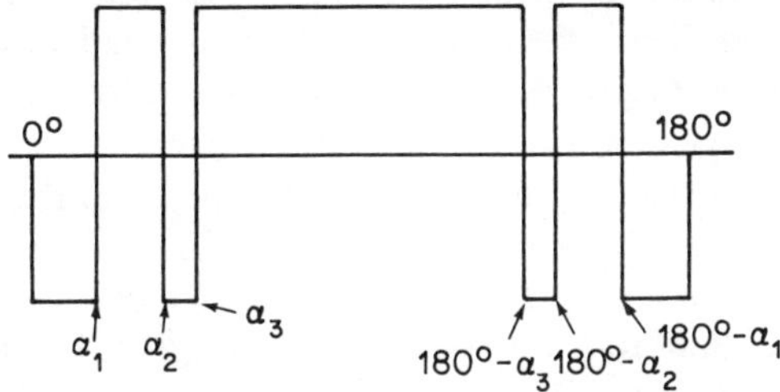

Figure 3.28. PWM wave on the basis of selected harmonic elimination look-up table

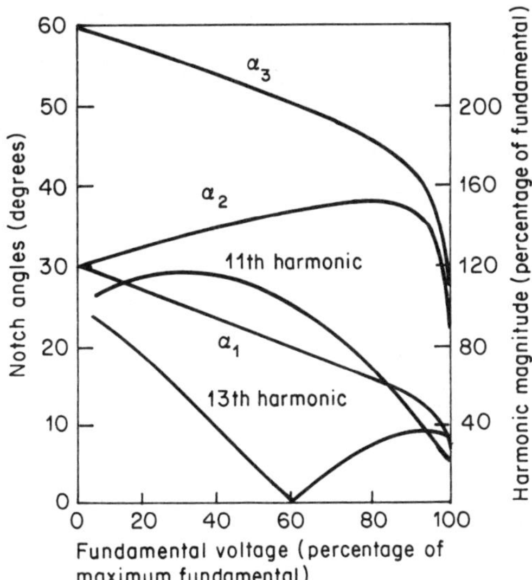

Figure 3.29. Notch angle curves with voltage control and selected harmonic elimination (fifth and seventh harmonics)

way of example, Figure 3.29 illustrates the case where the fifth and seventh harmonics are eliminated with the help of look-up table storing the angles required.

In the method of reference[9] the period is divided in six regions. If the second and fifth regions of each phase waveform are filled with a train of pulses, or chops, only these pulses appear in the line to line voltage.

In Figure 3.30, two chops per second and fifth period have been introduced into the red phase and the line to line waveform now contains only those chops. If the same is done to the three phases, the line to line voltage can be seen to be determined only by the markspace ratio of the chops. Hence the voltage the motors see is completely variable between the two limits 0 and 100%. The motor phase voltages, shown for a star-connected motor in Figure 3.30, are derived from the invertor line to line voltages.

Harmonic voltages again occur as multiples of the carrier frequency (i.e. the chop number (m) times six) with sidebands, which can be given by $L(6m \pm 1)$, where $L = 1, 3, 5, 7$ and m is the number of chops per half-wave in phase voltage.

Here also the carrier is a triplen harmonic and is cancelled out in a three-phase system. The phase waveforms have such symmetry that there are no even harmonics. The higher m is, the higher up the spectrum the harmonic voltages occur.

The number of invertor switchings per second, $F(2m + 1)$, limits the number of chops allowable as fundamental frequency increases, e.g. for eight chops there are 17 on/off switches per period.

In order to keep harmonic orders high in the spectrum, the number of chops is changed as fundamental frequency increases.

Further reduction of lower order harmonics can be achieved by the use of complex PWM control waveform strategies, at the cost of increasing the invertor

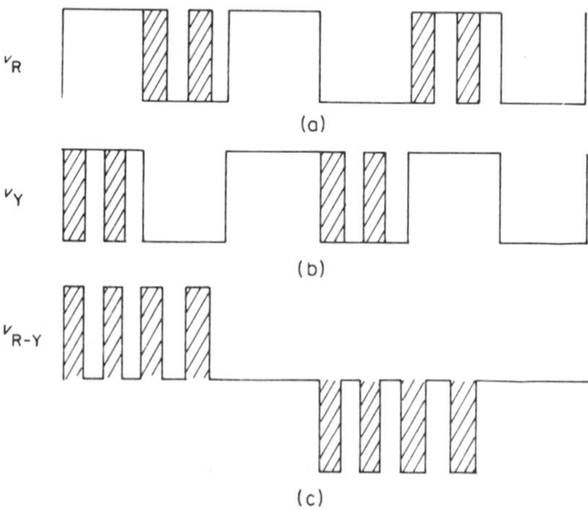

Figure 3.30. PWM voltage waveforms with two chops per half-way: (a) voltage in phase R; (b) voltage in phase Y; (c) motor voltage waveform

switching rate. For a given maximum invertor phase switching rate, the problem is to choose that PWM control strategy which will achieve the desired linear variation of fundamental voltage amplitude with frequency and reduce the effect of harmonic torques or minimize harmonic power losses within the motor.

Generally, at any fundamental switching frequency, each chop per half cycle of the invertor phase voltage waveform can eliminate one harmonic of the waveform or reduce a group of harmonic amplitudes.(6) Thus for m chops per half cycle one chop must be utilized to control the fundamental harmonic amplitude, so $m - 1$ degrees of freedom remain. The $m - 1$ degrees of freedom may be utilized to eliminate completely $m - 1$ specified low order harmonics or to minimize motor power losses caused by a specified range of harmonics within the motor.

At any fundamental frequency, elimination of the lower order harmonics from the phase waveforms will cause the portion of the r.m.s. which was provided by the eliminated harmonics to be spread over the remaining harmonic magnitudes. This occurs because the total harmonic r.m.s. voltage cannot change. The effect of this shifting motor performance needs to be determined, but the integrating filter characteristic of the motor should be more effective in reducing the current harmonics at higher orders.

3.4 LOW POWER CONVERTORS

As indicated in Section 3.1, under group (iii), two types of low power convertor loads need to be considered because of their contribution to harmonic distortion. One of them, the television set, has been a problem for some time. The second, the battery charger, is not a problem at the moment, but if the use of electric vehicles becomes generally accepted this load will be a source of considerable harmonic content.

Unlike the convertor loads previously discussed, where the power ratings are sufficiently large to be treated individually, the two loads now under consideration are only important when large numbers of individual units are simultaneous active.

The Monte Carlo technique has been used by Blommaert(10) to investigate the probability of exceeding a predetermined level of harmonics with reference to television sets. The same technique has been used by Orr(11) to investigate the harmonic problem caused by clusters of electric vehicle battery chargers.

Although the effect of these two problem areas is very similar, it is appropriate to consider them individually, given their rather different characteristics.

The contribution of television receivers

Television sets are generally supplied by a rectifier and a high smoothing capacitance. Some older generations of television receivers use half-wave rectification and thus produce considerable levels of d.c. and even-ordered harmonics. To illustrate the problem, Figure 3.31 displays the harmonic currents recorded a decade or so ago on the high voltage side of a distribution transformer

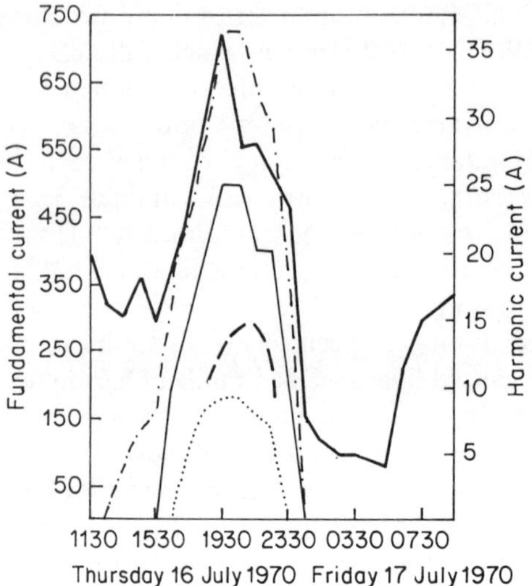

Figure 3.31. Red phase current measurements at a 1000 kVA distribution s/s (MV cable). ———, 50 Hz (fundamental); –·–·–, 100 Hz; ——, 150 Hz; ---, 200 Hz; ·····, 250 Hz

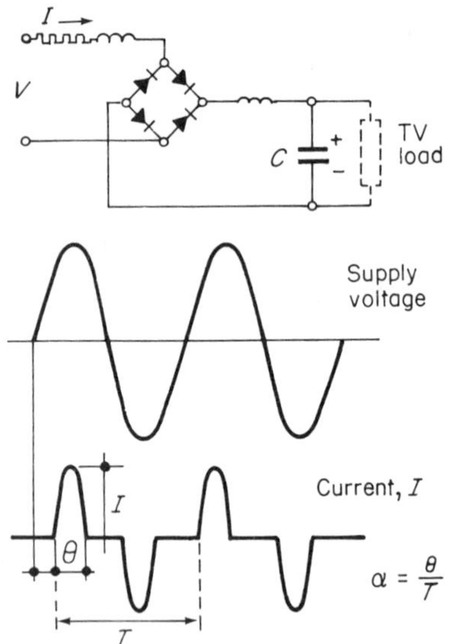

Figure 3.32. Four diodes bridge rectifier

at various times of the day.[12] The graphs show that all harmonics reached peak values at approximately 2100 hours, which was considered to be the peak viewing period, and when the programmes closed down all harmonics virtually disappeared.

Modern television sets use double-wave rectification, as shown in Figure 3.32,

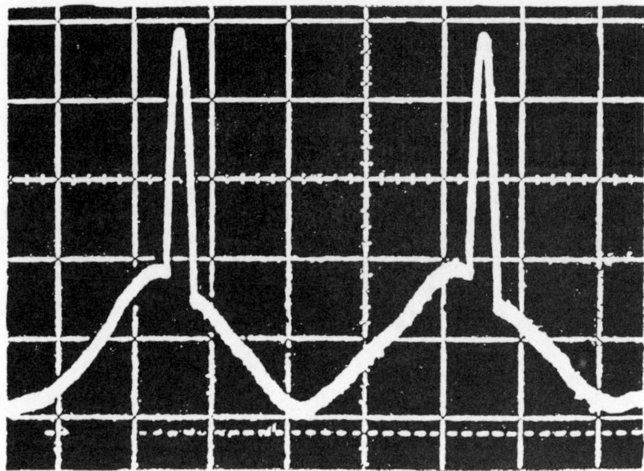

Figure 3.33. Current spike from colour television receivers

and more recently a thyristor has been added to this circuit; the thyristor is fired after the mains peak so that the circuitry following the thyristor is exposed to less than peak mains voltage. However, this results in higher peak currents, containing a correspondingly higher harmonic content. This effect is illustrated in Figure 3.33.

Colour receivers demand a peak current that is two to three times as large as that taken by a monochrome receiver. Applying Fourier analysis to a half-sinusoidal impulse produced by the rectifier of Figure 3.32 yields the following harmonic currents:

$$I_n = \frac{8\alpha I}{\pi} \sum_{n=1,3,5}^{\infty} \frac{\cos n\alpha\pi}{1 - n^2\alpha^2\pi^2} \cos n\omega t, \tag{3.4.1}$$

where I is the crest value of the current impulse and $\alpha = \theta/T$ its relative duration with respect to the period of the mains frequency.

The values of α are usually between 0.08 and 0.18 for television sets, regardless of their supply circuit.

Table 3.2 gives typical odd harmonic current components for various receivers.[10]

Table 3.2. Harmonic current (in amps)

Order of the harmonic	Type of receiver			
	Black and white, tube	Black and white, transistor	Colour, diodebridge	Colour, thyristor
3	0.53	0.32	0.73	0.82
5	0.31	0.25	0.59	0.66
7	0.13	0.15	0.43	0.34
9	0.055	0.08	0.27	0.14
11	0.045	0.04	0.15	0.090
15	0.03	0.03	0.045	0.040

Table 3.3. Harmonic current (in amps)

			Harmonic order				
			1	3	5	7	9
1 receiver		Current: I_1 (A)	0.80	0.67	0.48	0.29	0.09
10 receivers (per phase)	Neutral	Current: I_{10} (A)	8.00	5.80	3.50	1.70	0.70
		$I_{10}/10I_1$	1.00	0.86	0.73	0.58	0.77
	Phase	Current: I_{10} (A)	1.00	17.40	0.70	0.60	2.10
		$I_{10}/10I_1$	0.12	2.60	0.14	0.20	2.30
80 receivers (per phase)	Neutral	Current: I_{80} (A)	64.00	37.60	13.20	3.80	1.70
		$I_{80}/80I_1$	1.00	0.70	0.34	0.16	0.23
	Phase	Current: I_{80} (A)	9.60	116.00	3.0	0.90	4.60
		$I_{80}/80I_1$	0.15	2.10	0.08	0.04	0.63

The statistical distribution of phase displacement among the harmonics produced by various receivers in parallel has been given some consideration by Eléctricité de France[10] and a typical set of experimental results is shown in Table 3.3.

In nearly all cases, the crest of the harmonic coincides with the crest of the fundamental, so that the harmonics from different sources reinforce. Table 3.3 also illustrates that the strongest harmonic produced, i.e. the third harmonic, adds up in the neutral circuit, giving rise to large undesirable currents in a circuit which ideally should have no current in it.

The trend in colour television sets is towards chopper transistor regulators or invertors with elaborate overvoltage and overcurrent protection, and improved circuit efficiency with lower power drain from the low level circuitry. However, the trend to more televisions will offset the reduction in magnitude of current drawn by each set. It is unlikely that the harmonic distribution from each set is going to change significantly, although the higher harmonics may be reduced significantly if pressure is brought to bear, and the high third harmonic component is unlikely to be reduced, except at great expense to the consumer.

Harmonic contribution of battery chargers

The basic circuit used in battery charging is shown in Figure 3.34(a). The individual harmonics generated by such circuit depend on the initial battery voltage, and the overall harmonic content produced by clusters of battery chargers on the same busbars varies according to time and involves random probability.

With reference to Figure 3.34(b), conduction begins when $V > E$ and therefore

$$\theta = \sin^{-1}(E/V). \qquad (3.4.2)$$

The charging current can be obtained from equation (3.3.2), for the initial

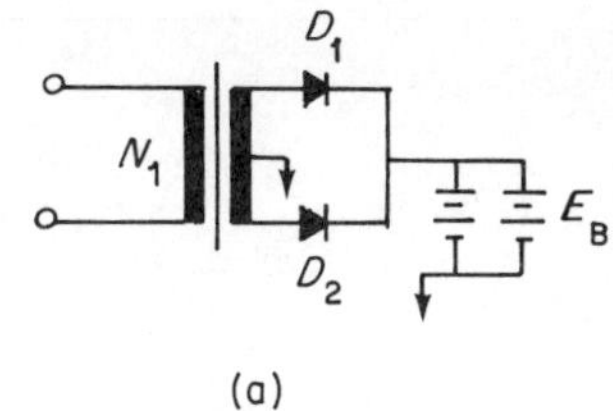

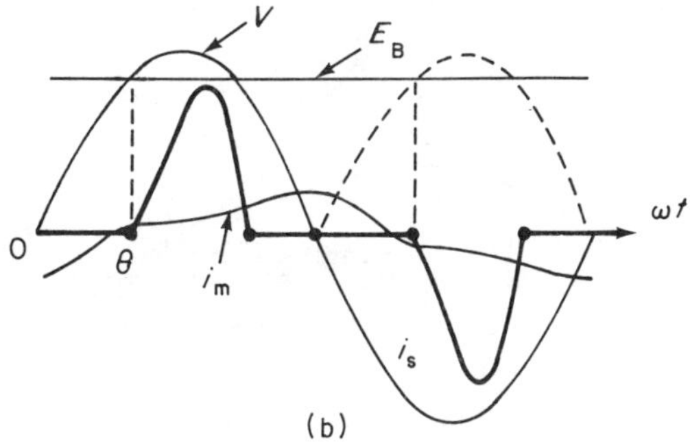

Figure 3.34. Electric vehicle battery charger: (a) connection diagram; (b) theoretical wave shape.[11]

condition $i = 0$ when $\omega t = \theta_1$; this yields the following expression:

$$I = \frac{V_m}{R}\left\{\sin(\omega t - \phi) - \sin(\theta - \phi) - \frac{\sin\theta}{\cos\phi}\exp[(\theta - \omega t)\cot\phi] - \frac{\sin\theta}{\cos\phi}\right\}$$

$$\text{for} \quad \theta < \omega t < \gamma. \tag{3.4.3}$$

Moreover, the primary current also carries the transformer magnetizing current, i_m.

The circuit of Figure 3.34(a) has been used to assess the harmonic content of clusters of battery chargers. The following parameters, experimentally determined, are used by Orr:[11]

$$\begin{aligned} \text{Transformer turns ratio} &= 2.437{:}1, \\ \omega L &= 0.189\,\Omega, \\ R &= 0.713\,\Omega. \\ i_m &= 0.08 \sin h\,(-5\cos\omega t) \quad \text{(in amps).} \end{aligned}$$

Figure 3.35 displays the magnitude of the lower current harmonic orders (per unit of the short-circuit current, V/Z, where $Z = \sqrt{(R^2 + \omega^2 L^2)}$) versus angle θ.

In common with television receivers, radios, stereos and other consumer articles employing direct current, the battery chargers produce high zero-sequence triplen harmonic content which overloads the neutral circuit. To make matters worse, fluorescent light appliances also produce triplen harmonic current with the same phase relationship. Moreover, the phase angle of the third

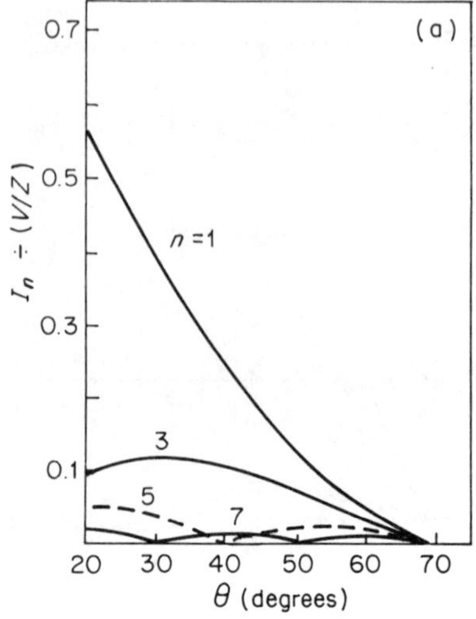

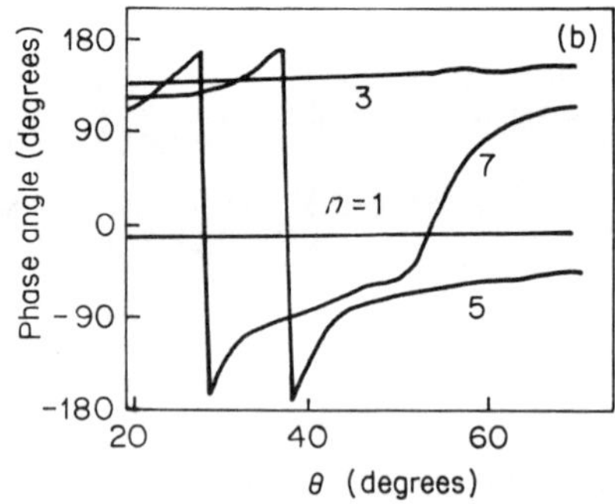

Figure 3.35. Variation of harmonic content of battery charger with angle θ:[11] (a) magnitude (per unit of short circuit current); (b) phase

harmonic doesn't vary enough to cause significant harmonic cancellation when a random group of chargers is in operation, so the third harmonics will add almost algebraically.

A Monte Carlo simulation procedure has also been used by Orr to predict amplitudes of harmonic currents injected into the distribution system by a cluster of batteries recharging on one bus. This procedure simulates (on a digital computer) a large number of actual recharge cycles using initial conditions on charge states and recharge start times conforming to statistical distributions. Using the known model of the battery and charger system, harmonic amplitudes versus time during each recharge cycle are determined and stored. After simulating many recharge cycles these results are used to determine expected levels of harmonic currents and their relative probabilities of occurrence.

The input data required for the simulation are: (i) battery parameters (e.m.f. versus state of change and ampere-hour capacity); (ii) charger parameters (transformer turns ratio, inductance, resistance, a.c. line voltage); (iii) statistical distributions of recharge start times and state of charge; (iv) number of vehicles charging on one bus.

To assess the effect of the harmonic currents injected it is necessary to know how frequently particular levels of harmonic amplitudes will be exceeded. This is normally expressed as a plot of the probability of given amplitudes being exceeded; the probability represents the proportion of a 24-hour period (in per cent) that the given amplitude is exceeded.

Figure 3.36 shows the results of a particular study(11) with clusters of one, five and 25 battery chargers connected to the same busbar for a typical state of charge distribution (related to expected daily driving distances) and with the standard deviation of start time, $\sigma_{ST} = 0.1$ h.

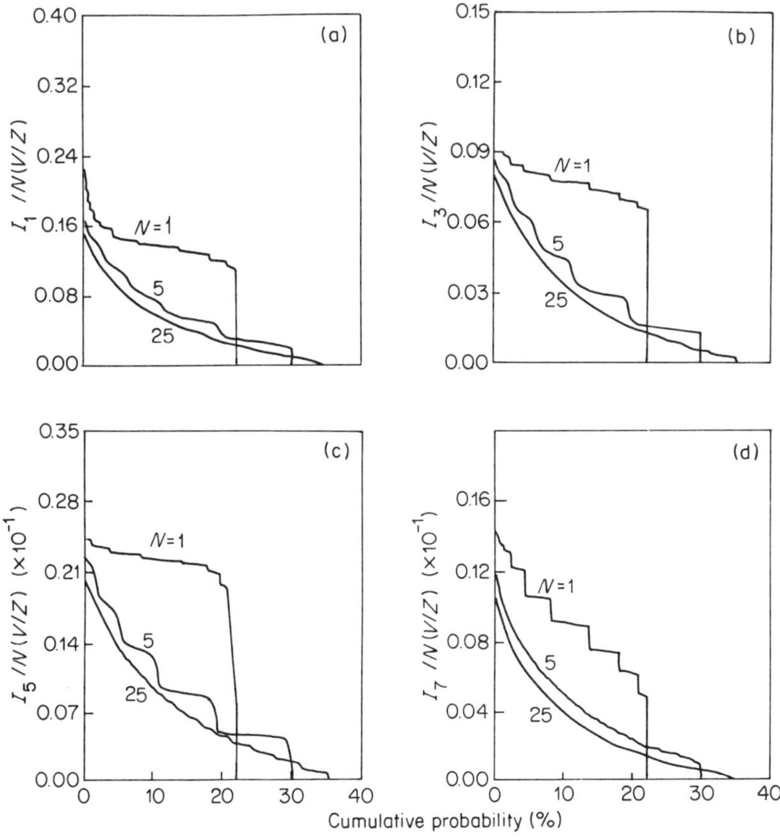

Figure 3.36. R.M.S. current harmonics probability distribution:(11) (a) fundamental; (b) third harmonic; (c) fifth harmonic; (d) seventh harmonic. $L\omega/R = 0.265$ and $V/Z = 162.6$A.(11)

The amplitudes are normalized to the short-circuit current and to one battery charger.

It must be emphasized that each particular situation will lead to different results, influenced by the distribution of vehicles in the network, the parameters of the individual electric vehicles, the statistical distributions of start times and state of charge, etc.

3.5 IMPERFECT SYSTEM CONDITIONS

The harmonic effects caused by imperfect system conditions encountered in practice cannot be derived from the idealized models described in Section 3.2.

In general each of the main three parts of the system is always in error to a lesser or greater extent:

(i) The a.c. system voltages are never perfectly balanced and undistorted nor the system impedances, in particular the convertor transformer, exactly equal in the three phases.
(ii) The d.c. current may be modulated from another convertor station in the case of a rectifier–invertor link.

Table 3.4. Harmonic measurements during back-to-back testing of the New Zealand high voltage d.c. convertors

Harmonic	400 A d.c. (one-third full load) current; phase-to-neutral voltages at Benmore on 220 kV		
	Red phase (%)	Yellow phase (%)	Blue phase (%)
1	100	100	100
2	0.5	0.7	1.0
3	2.9	0.3	1.0
4	0.6	0.3	0.4
5	0.25	0.15	0.25
6	0.25	0.30	0.35
7	0.15	0.15	0.1
8	0	0.05	0.1
9	0.05	0.05	0.15
10	0.05	0.05	0.05
11	0.1	0.15	0.1
12	0.15	0.05	0.15
13	0.05	0.05	0.05
14	0.05	0.05	0.05
15	0.15	0	0.2
16	0	0.1	0.15
17	0.3	0.3	0.3
18	0	0.05	0.1
19	0.3	0.3	0.7
20	—	—	—
21	—	—	—
22	0.2	0.2	0.5
23	0.4	0.2	0.3
25	0.2	0.2	0.15

(iii) The firing angles control systems often given rise to substantial errors in their implementation.

As a result the large static convertors often produce harmonic orders and magnitudes not predicted by the Fourier series of the idealized waveforms.

The uncertain nature of these 'uncharacteristic' harmonics makes it difficult to prevent them at the design stage. Filters are not normally provided for uncharacteristic harmonics and as a result their presence often causes more problems than the characteristic harmonics.

By way of example Table 3.4 shows the results of harmonic measurements during back-to-back testing of the New Zealand d.c. convertor station at Benmore. All the harmonic voltages are unbalanced, particularly the third and ninth. The table also illustrates the presence of all current harmonic orders, odd and even, with the uncharacteristic orders causing higher voltage distortion than the characteristic ones.

A realistic quantitative analysis of the uncharacteristic harmonic components can only be achieved by a complete three-phase computer model of the system behaviour with detailed representation of the convertor controls. A qualitative assessment of the main problem areas and the sensitivity of the system to small deviations from the ideal conditions are considered in this section.

Imperfect a.c. source

Deviations from the perfectly balanced sinusoidal supply can be caused by (i) presence of negative sequence fundamental frequency in the commutating voltage; (ii) harmonic voltage distortion of positive or negative sequence; (iii) asymmetries in the commutation reactances. In general an imperfect a.c. source produces asymmetrical firing references and d.c. current modulation. The first problem can be eliminated by using equidistant firing control but the second problem still remains. This effect, illustrated in Figure 3.37 for the case of an unrealistically high level of fundamental voltages asymmetry, produces considerable second harmonic content on the d.c. side and third harmonic on the a.c. side.

Under normal operating conditions the expected levels of asymmetry and distortion are small and their effects can be approximated with reasonable accuracy.

If a small positive or negative sequence signal V_n (per unit of the normal fundamental voltage) is added to the otherwise ideal three-phase supply of a 12-pulse convertor configuration, the order and maximum level (V_k) of uncharacteristic harmonic voltage at the rectified output come under one of the following categories:[14]

Case 1: If $n+k=12p_1+1$ and $n-k=12p_2+1$, where p_1 and p_2 are any integers, then

$$V_k=\begin{cases} V_n\left(\dfrac{n\sqrt{2}}{n^2-k^2}\right) & \text{if } n^2>k^2, \quad (3.5.1)\\[2ex] V_n\left(\dfrac{k\sqrt{2}}{k^2-n^2}\right) & \text{if } k^2>n^2. \quad (3.5.2)\end{cases}$$

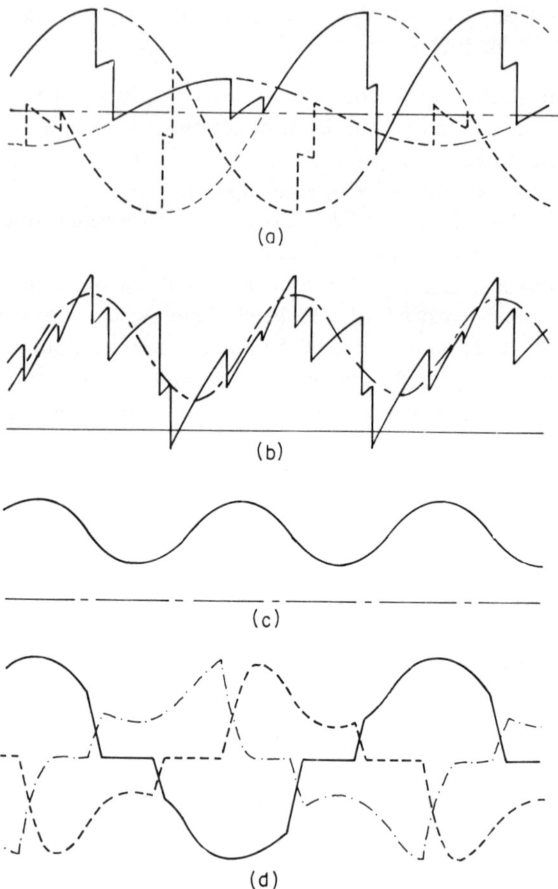

Figure 3.37. Sustained unbalanced voltages on a single convertor bridge: (a) three-phase voltages; (b) direct voltage; (c) direct current; (d) three-phase currents

Case 2: If $n + k = 12p_1 + 1$ but $n - k \neq 12p_2 + 1$, then

$$V_k = V_n \frac{1}{\sqrt{2}(n + k)}. \tag{3.5.3}$$

Case 3: If $n - k = 12p_1 + 1$ but $n + k \neq 12p_2 + 1$, then

$$V_k = V_n \frac{1}{\sqrt{2}(n - k)} \tag{3.5.4}$$

Case 4: If $n + k \neq 12p_1 + 1$ and $n - k \neq 12p_2 + 1$, then

$$V_k = 0 \tag{3.5.5}$$

A summary of all resulting harmonic voltages up to order 6 is given in Table 3.5 for interfering harmonics of orders -5 to $+5$. (There are also higher order

Table 3.5

Interfering a.c. voltage harmonic order, n	Harmonic voltage on d.c. side	
	Order, k	Amplitude, V_k/V_n
-1	2	0.707
$+2$	1	0.707
-2	3	0.707
$+3$	2	0.707
-3	4	0.707
-4	5	0.707
-5	6	0.707

harmonics in each case, not shown, of much lower amplitude, hence these are not practically important.)

The tabulated V_k is expressed per unit of the maximum average d.c. voltage and V_n per unit of the normal a.c. voltage.

The values given above are in practice applicable also to convertors of higher pulse number. Thus while characteristic harmonics can be reduced by using high pulse number, the uncharacteristic harmonics due to a.c. unbalance cannot.

It should be noted that the above is an approximation due to neglect of commutation reactance but is generally sufficiently valid at low harmonic orders (below the fifth).

The cause of the imperfection may also be some asymmetry in the commutation reactances, i.e.

$$X_a = X_0(1 + g_a),$$
$$X_b = X_0(1 + g_b),$$
$$X_c = X_0(1 + g_c),$$

where X_0 is the mean reactance and each value of g can vary between $\pm g_0$.

In this case the maximum level of uncharacteristic a.c. harmonic currents I_n of order n (in phase 'a') for the case of a six-pulse bridge occurs when

$$g_a = 0, \quad g_b = \pm g_0, \quad g_c = \mp g_0 \quad \text{for} \quad n = 3, 9\ 15, \text{etc.}$$

or

$$g_a = \pm g_0, \quad g_b = \mp g_0, \quad g_c = 0 \quad \text{for} \quad n = 5, 7, 11, 13, \text{etc.}$$

The maximum value of I_n neglecting changes of d.c. current and a.c. voltage is obtained from the expression

$$\begin{aligned} I_n = {} & \frac{I_1 g_0}{n(n^2 - 1) i_d X_0 \sqrt{3}} \\ & \times \{n^4[\cos(\alpha + u) - \cos\alpha]^2 + 2n^3 \sin\alpha \sin nu[\cos(\alpha + u) - \cos\alpha] \\ & + n^2[\sin^2\alpha + \sin^2(\alpha + u) + 2\cos nu(\cos^2\alpha - \cos u) \\ & + 2\cos\alpha(\cos(\alpha + u) - \cos\alpha)] \\ & + 2n\cos\alpha \sin nu(\sin\alpha + \sin(\alpha + u)) + 2\cos^2\alpha(1 - \cos n\mu)\}^{1/2} \end{aligned} \qquad (3.5.6)$$

Table 3.6

n	I_n(% of I_1)
3	0.70
5	0.33
7	0.29
9	0.50
11	0.22
13	0.19
15	0.31

for $n = 3, 9, 15$, etc., where I_1 is the fundamental r.m.s. current, i_d is the d.c. current per unit, X_0 is the mean commutation reactance per unit, α is the firing angle and u is the overlap (or commutation) angle. For $n = 5$, 7, 11, 13, etc., the above expression should be divided by 2.

Table 3.6 gives values for a typical case of $X_0 = 0.2$ per unit, $\alpha = 15°, g_0 = 0.075$.

Unequal commutation reactances also cause uncharacteristic voltages on the d.c. side. The highest magnitude of these occurs when $g_a = 0$, $g_b = +g_0$ and $g_c = -g_0$. Only even harmonics occur, given by

$$V_n(\text{max.}) = \frac{i_d X_0 g_0 V_{dio}}{2\sqrt{6}}, \tag{3.5.7}$$

where V_{dio} is the theoretical no-load d.c. voltage.

As an example for $i_d = 1$, $X_0 = 20\%$ and $g_0 = 0.075$, V_n(max.) is 0.31% of V_{dio} for $n = 2$, 4, 8, 10, 14, 16, etc., independent of harmonic order or of firing angle.

D.C. current modulation[14]

If we now assume a perfect three-phase supply and equidistant firing, the addition of a small current harmonic component I_k of order k on the d.c. side will generate a component I_n of different order but of the same sequence on the a.c. side, the maximum level of which is given in Table 3.7.

Table 3.7

Harmonic order, k, of modulating current on d.c. side	Harmonic current on a.c. side	
	Order n	Amplitude, I_n
1	0	0.707
	+2	0.707
2	−1	0.707
	+3	0.707
3	−2	0.707
4	−3	0.707
	+5	0.707

The amplitude I_n is in multiples of $I_1 I_k/I_d$, where I_1 is the r.m.s. fundamental current at the a.c. busbar, I_k is the r.m.s. interfering current on the d.c. side at order k and I_d is the d.c. current

Table 3.7 is independent of the prime cause of the d.c. current modulation. It is again approximate, valid at low frequencies only because of neglect of commutation reactance.

Control system imperfections

No general rules can be given in this case. By way of example, Ainsworth[14] describes the effect of modulating harmonic content of the control voltage applied to the oscillator of a control d.c. current system using the phase-locked oscillator principle,[15] assuming constant d.c. current and a.c. voltages, a V_c modulating harmonic signal of order n (per unit referred to the normal steady state control voltage), causes d.c. voltage components of orders $n_1 = \pm n \pm 12p$ in a 12-pulse convertor, where p is any integer.

The magnitude (per unit of maximum average rectified voltage) of the d.c. voltage modulation is

$$V = \frac{V_c \sin \alpha_0 \cos (n_1 u_0/2)}{n_1} \tag{3.5.8}$$

where α_0 and u_0 are the mean firing and overlap angles respectively.

The total a.c. current magnitude referred to one phase of one valve winding at harmonic order n_2 due to a similar excitation is

$$I_A(n_2) = \frac{I_d V_c \cdot 2\sqrt{3} \sin (n_2 u_0/2)}{n_2(\cos \alpha_0 - \cos (\alpha_0 + u_0))} \tag{3.5.9}$$

for $n_2 = \pm n \pm (1, 11, 13, \ldots)$ only.

Firing asymmetry

A.C. system imperfections or firing errors result in pulse-width deviations from the characteristic quasi-rectangular current waveform. Kimbark[16] describes the effect of late and early firings with reference to a six-pulse bridge convertor.

If the positive current pulses start early by an angle ε and the negative ones start late by the same angle, the non-conductive intervals are increased by 2ε. The even symmetry, which eliminates the even-ordered harmonics, is now lost and the use of equations shows the existence of even harmonics which for small overlap angles are given by the expression

$$\frac{I_n}{I_1} = \frac{2 \sin n\varepsilon}{2n \cos \varepsilon} \simeq \varepsilon, \tag{3.5.10}$$

e.g. for $\varepsilon = 1°$ the second and fourth harmonics are each approximately 1.74% of the fundamental current.

If the firings of the two valves connected to the same phase are late by ε then the positive and negative current pulses of that phase are ε degrees shorter than the normal. Moreover, the current pulses of one of the remaining phases (the leading phase) are increased by ε while those of the lagging phase remain unaltered. This produces triplen harmonic currents. On the assumption of zero overlap angle the ratio of the triplen harmonics ($h = 3q$) to the fundamental current are expressed by

$$\frac{I_n}{I_1} = \frac{\sin(q\pi \pm 1.5\,q\varepsilon)}{3q\sin(\pi/3 \pm \varepsilon/2)}. \tag{3.5.11}$$

For small values of ε the approximate levels of third harmonic are given by

$$\frac{I_3}{I_1} \simeq \frac{1.5\,q\varepsilon}{3q\sqrt{3}/2} = 0.577\,\varepsilon, \tag{3.5.12}$$

e.g. for $\varepsilon = 1^\circ$, $I_3 = 1\%$ of the fundamental.

3.6 MODULATED PHASE-CONTROL

The main application of modulated phase control is the cycloconvertor which provides static power conversion from one frequency to another. It consists of a dual-convertor configuration (shown by the simplified equivalent circuit of Figure 3.38) controlled through time-varying phase-modulated firing pulses so that it produces an alternating, instead of a direct, output voltage per phase as illustrated in Figure 3.39(a). The output current phase relationship and waveform (Figure 3.39(b)) depend on the load. The rectifier and invertor operating regions for each convertor are shown in Figure 3.39(c) and (d).

Similar to the phase-controlled covertor the input current will in general contain in-phase, quadrature and harmonic components.

Practical cycloconvertors operate with negligible internal circulating current between the two convertors and thus only the circulating current-free mode of operation is of importance with regard to harmonic assessment.

The output voltage waveform is formed by selected intervals of the three-phase input voltage supply and the input current in each phase by selected intervals of the output currents.

All practical convertor configurations used for large power ratings are combinations of the basic three-pulse group. Moreover, the harmonic content of these multipulse circuits can easily be derived from the basic harmonic series of the three-pulse phase-controlled convertor. The three-pulse waveform with arbitrary firing angle control should therefore provide the most general case for harmonic analysis.

The conventional Fourier analysis (described in Chapter 2) is not practical for the derivation of cycloconvertor harmonic components since the frequency spectrum of the output voltage and input current waveforms is related to both the main input and output frequencies; they produce 'beat frequencies' which are both the sums and differences of multiples of both these frequencies. Classical

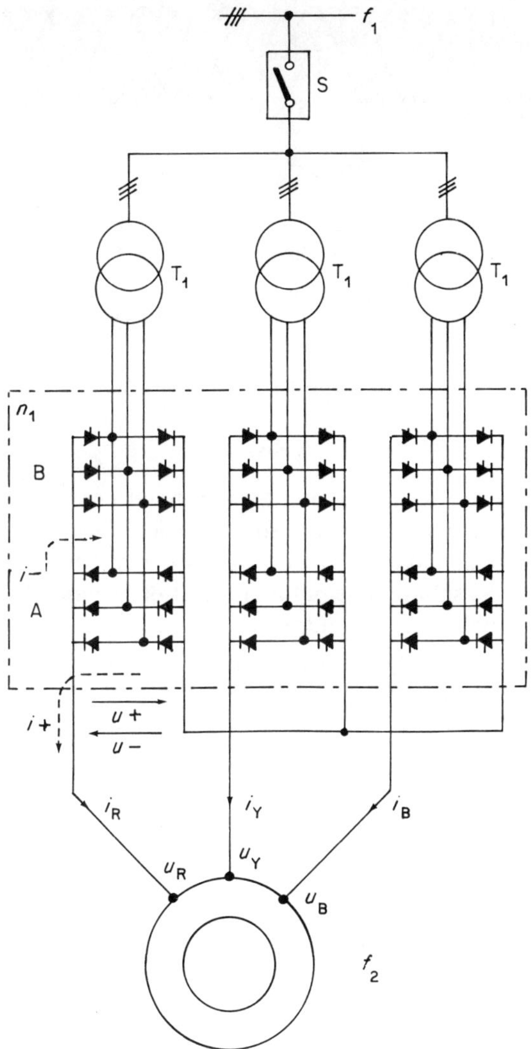

Figure 3.38. Basic circuit diagram of a six-pulse cycloconvertor-fed a.c. motor. A, B, three-phase bridges connected in anti-parallel; $i+$, $i-$, $u+$, $u-$, output current and voltage of the cycloconvertor; i_R, i_S, i_T, stator currents; u_R, u_Y, u_B, stator voltages; f_1, mains frequency; f_2, output frequency. From Pelly, *Thyristor Phase-Controlled Converters and Cycloconverters*. Copyright © 1971 Wiley–Interscience. Reprinted by permission of John Wiley & Sons Inc.

Fourier analysis resolves a periodical waveform into a fundamental component, the frequency of which is equal to the fundamental repetition frequency of the wave and a series of harmonic components which are multiples of the fundamental frequency.

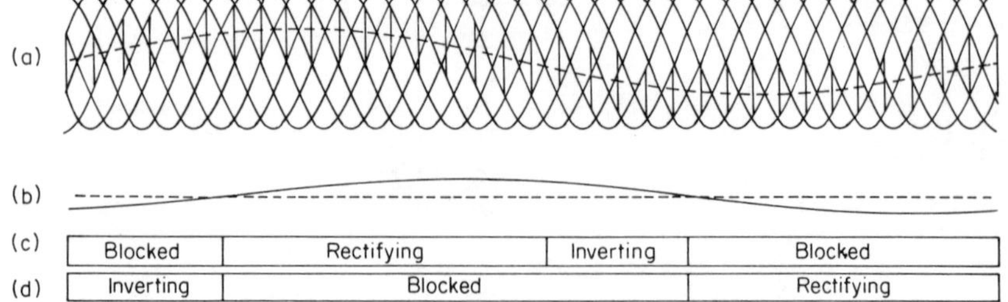

Figure 3.39. Theoretical waveforms for a six-pulse cycloconvertor on half-maximum output voltage ($r = 0.5$) with a displacement angle of 60° and one-sixth output frequency:[17] (a) output voltage; (b) output current; (c) operating mode of bridge A; (d) operating mode of bridge B. From Pelly, Thysistor Phase-controlled Converters and Cycloconverters. Copyright © 1971 Wiley–Interscience. Reprinted by permission of John Wiley & Sons, Inc.

The cycloconvertor waveforms, on the other hand, contain frequencies which are not integer multiples of the main output frequency. In fact there may not even be a clearly defined fundamental output frequency. Only when the output frequency is an exact submultiple of the product of the input frequency and the convertor pulse number is each output cycle identical with the next, i.e.

$$f_0 = \frac{3f_i}{k}, \tag{3.6.1}$$

where k is an integer.

A more general method is described in the next section which provides the output voltage and input current harmonic component in terms of each of the independent variables.

The switching function approach[17]

The general method is illustrated in Figure 3.40, where the effect of each thyristor switching is derived independently and the overall output waveform is then expressed as the addition of all the wave segments generated by the individual thyristors. The effect of each thyristor is expressed as the product of the appropriate input voltage (or output current) waveform and a 'switching function', of unity and zero amplitudes when the thyristor is ON and OFF respectively.

By expressing the switching function as a phase-modulated harmonic series, a general harmonic series can be derived for the output voltage (or input current) waveform in terms of the independent variables.

The quiescent firing (90°) is used as a reference for the modulated firings. The quiescent firing produces zero voltage in both the positive and negative convertors.

By way of illustration the quiescent voltage waveform of the positive convertor,

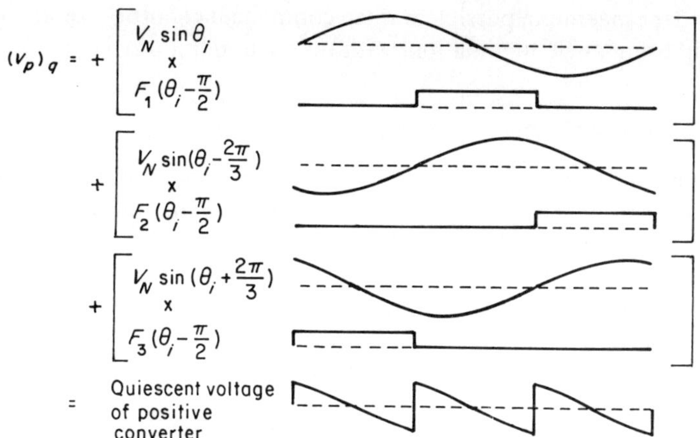

Figure 3.40. Derivation of voltage waveforms of the positive convertor for quiescent ($\alpha = 90°$) operation.[17] From Pelly, *Thyristor Phase-controlled Converters and Cycloconverters*. Copyright © 1971 Wiley–Interscience. Reprinted by permission of John Wiley & Sons Inc.

shown in Figure 3.40, is given by

$$(v_p)_q = V_N \sin\theta_i \cdot F_1\left(\theta_1 - \frac{\pi}{2}\right) + V_N \sin\left(\theta_1 - \frac{2\pi}{3}\right) \cdot F_2\left(\theta_i - \frac{\pi}{2}\right)$$

$$+ V_N \sin\left(\theta_i + \frac{2\pi}{3}\right) \cdot F_3\left(\theta_i - \frac{\pi}{2}\right). \tag{3.6.2}$$

The modulated firing control provides a 'to and fro' phase modulation $f(\theta_0)$ of the individual firings with respect to the quiescent firing.

In general the value of $f(\theta_0)$ will oscillate symmetrically to and from about zero, at a repetition frequency equal to the selected output frequency. The limits of control on either side of the quiescent point are then $\pm\pi/2$. Thus the general expressions for the switching function of the positive and negative convertors are

$$F\left(\theta_i - \frac{\pi}{2} + f(\theta_0)\right) \quad \text{and} \quad F\left(\theta_i + \frac{\pi}{2} - f(\theta_0)\right)$$

since the phase modulation of the firing angles of the positive and negative convertors is equal but of opposite sign.

Moreover, it has been shown[17] that the optimum output waveform, i.e. the minimum r.m.s. distortion, is achieved when the firing angle modulating function is derived by the 'cosine wave crossing' control. Under this type of control the phase of firing of each thyristor is shifted with respect to the quiescent position by

$$f(\theta_0) = \sin^{-1} r \sin\theta_0, \tag{3.6.3}$$

where r is the ratio of amplitude of wanted sinusoidal component of output voltage to the maximum possible wanted component of output voltage, obtained with 'full' firing angle modulation, with no commutation overlap.

Derivation of input current harmonics

For the derivation of the input current waveform it is more convenient to use two switching functions, i.e. the thyristor and the convertor (i.e. the conducting half of the dual convertor) switching functions.

It is also necessary to make the following approximations: (i) the output current is purely sinusoidal; (ii) the source impedance (including transformer leakage) is neglected. Considering first a single-phase output, illustrated in Figure 3.41, the current in each phase of the supply is given by

$$i_A = I_0 \sin(\theta_0 + \phi_0)\cdot F_1\left(\theta_i - \frac{\pi}{2} + f(\theta_0)\right)\cdot F_p(\theta_0)$$
$$+ I_0 \sin(\theta_0 + \phi_0)\cdot F_1\left(\theta_i + \frac{\pi}{2} - f(\theta_0)\right)\cdot F_N(\theta_0). \qquad (3.6.4)$$

From conventional Fourier analysis F_1, F_p and F_N can be expressed in terms of

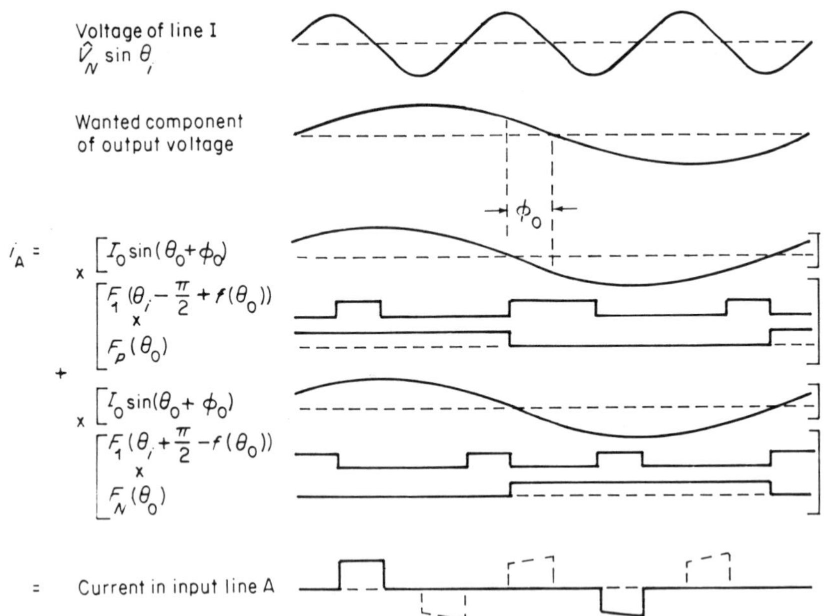

Figure 3.41. Derivation of the input line current of a cycloconvertor.[17] The input line current is shown in the bottom part of the figure as a continuous line for a single-phase load and as a broken line for a three-phase load. From Pelly, *Thyristor Phase-controlled Converters and Cycloconverters*. Copyright © 1971 Wiley–Interscience. Reprinted by permission of John Wiley & Sons. Inc.

the following series:

$$F_1\left(\theta_i \mp \frac{\pi}{2} \pm f(\theta_0)\right) = \frac{1}{3} + \frac{\sqrt{3}}{\pi}\left[\sin\left(\theta_i \pm \frac{\pi}{2} \mp f(\theta_0)\right) - \frac{1}{2}\cos 2\left(\theta_i \pm \frac{\pi}{2} \mp f(\theta_0)\right) - \frac{1}{4}\cos 4\left(\theta_i \pm \frac{\pi}{2} \mp f(\theta_0)\right)\right], \tag{3.6.5}$$

$$F_p(\theta_0) = \frac{1}{2} + \frac{2}{\pi}\left[\sin(\theta_0 + \phi_0) + \frac{1}{3}\sin 3(\theta_0 + \phi_0) + \frac{1}{5}\sin 5(\theta_0 + \phi_0) + \cdots\right], \tag{3.6.6}$$

$$F_N(\theta_0) = \frac{1}{2} - \frac{2}{\pi}[\sin(\theta_0 + \phi_0) + \tfrac{1}{3}\sin 3(\theta_0 + \phi_0) + \tfrac{1}{5}\sin 5(\theta_0 + \phi_0) + \cdots]. \tag{3.6.7}$$

Substituting in i_A and reducing,

$$\begin{aligned} i_A = I_0 \sin(\theta_0 + \phi_0)\Bigg\{&\frac{1}{3} + \frac{\sqrt{3}}{\pi}[\sin\theta_i \sin f(\theta_0) + \tfrac{1}{2}\cos 2\theta_i \cos 2f(\theta_0) \\ &- \tfrac{1}{4}\cos 4\theta_i \cos 4f(\theta_0) - \tfrac{1}{5}\sin 5\theta_i \sin 5f(\theta_0) + \cdots] \\ &+ \frac{4\sqrt{3}}{\pi^2}[-\cos\theta_i \cos f(\theta_0) - \tfrac{1}{2}\sin 2\theta_i \sin 2f(\theta_0) + \tfrac{1}{4}\sin 4\theta_i \sin 4f(\theta_0) \\ &+ \tfrac{1}{5}\cos 5\theta_i \cos 5f(\theta_0) \cdots][\sin(\theta_0 + \phi_0) + \tfrac{1}{3}\sin 3(\theta_0 + \phi_0) \\ &\tfrac{1}{5}\sin 5(\theta_0 + \phi_0) + \cdots]\Bigg\}. \end{aligned} \tag{3.6.8}$$

In the above expression $f(\theta_0) = \sin^{-1} r \sin\theta_0$ (see equation (3.6.3)) as explained above when the modulating function uses the cosine wave crossing control method.

In general, however, the output will also be three-phase and, assuming perfectly balanced input and output waveforms, each phase of the input will include the contribution of the three output currents, i.e. $i_A = i_{A1} + i_{A2} + i_{A3}$ and the corresponding waveform is illustrated by a broken line in Figure 3.41.

The above procedure can be extended to cases of three-phase output, under different transformer connections and different convertor configurations. This, however, is a very long and tedious task, thoroughly documented in Pelly's book,(17) and will not be detailed here. The chart of Figure 3.42 contains the main results of the harmonic analysis for the case of a balanced three-phase output; it gives the relationships which exist between the predominant harmonic frequencies present in the three-pulse input current waveform versus the output to input

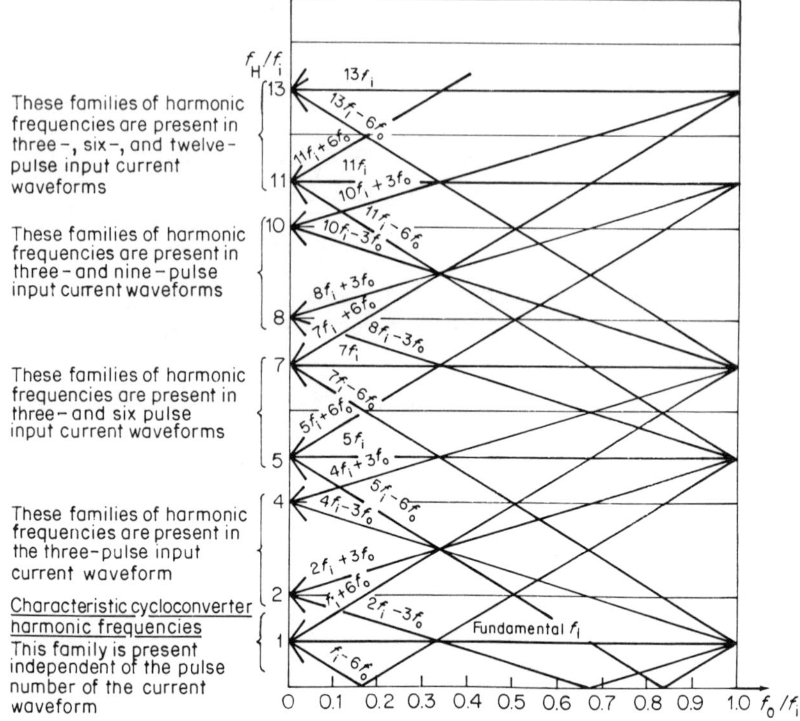

Figure 3.42. Relationships between the predominant harmonic frequencies of a cycloconvertor and the output to input frequency ratio.[17] From Pelly, *Thyristor Phase-controlled Converters and Cycloconverters.* Copyright © 1971 Wiley–Interscience. Reprinted by permission of John Wiley & Sons, Inc.

frequency ratio. It also indicates the groups of harmonics eliminated by the use of higher pulse numbers.

3.7 INTEGRAL CYCLE CONTROL

Instead of point on wave switching selection this type of control is based on the switching of entire voltage half-cycles. It is often called 'burst-firing' and has found application in long time constant loads (e.g. temperature control in electric ovens).

The fundamental supply frequency cannot be used as a basis for the Fourier analysis in this case, because the period of repetition, and thus the lowest frequency produced, is now a variable sub-harmonic frequency.

If the number of ON cycles is N and the number of cycles over which the pattern is repeated is M the period of repetition is M/f, where f is the supply frequency.

The lowest frequency, which now becomes the fundamental frequency, is f/M hertz.

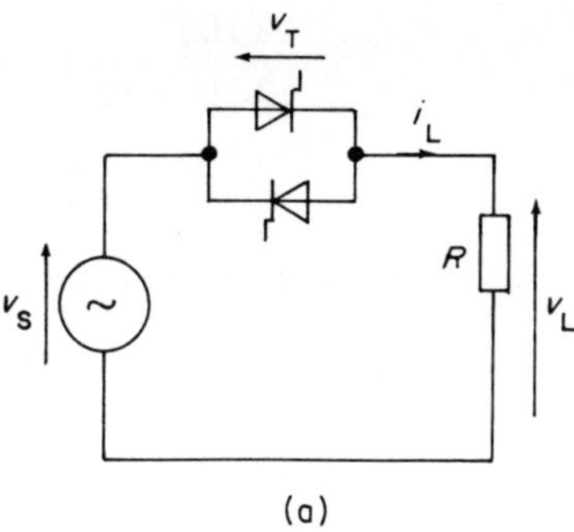

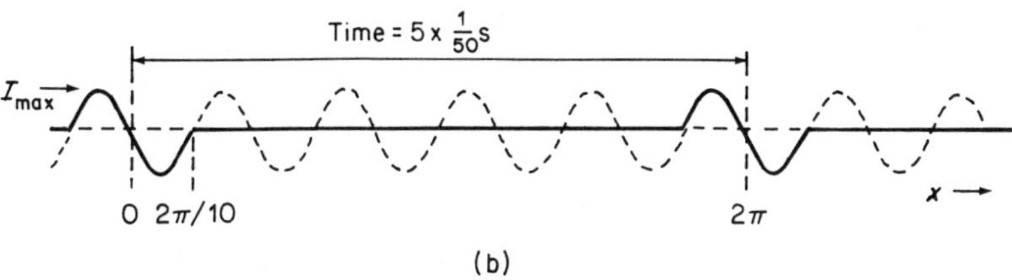

Figure 3.43. Integral cycle control: (a) basic circuit; (b) voltage waveform

With reference to this lowest frequency, the current being analysed can be expressed as

$$i = I_{max} \sin M\omega t. \tag{3.7.1}$$

Using the time reference indicated in Figure 3.43(b), the A_0 and A_n Fourier coefficients are zero, i.e. there are only sine terms, their general expression being

$$\begin{aligned} B_n &= \frac{2}{\pi} \int_0^{2\pi f/M} [-I_{max} \sin M\omega t \sin n\omega t] \, \mathrm{d}\omega t \\ &= -I_{max} \cdot \frac{2M}{\pi} \cdot \frac{\sin((N/M)n\pi)}{M^2 - n^2} \end{aligned} \tag{3.7.2}$$

Consider as an example the case where $M = 5$ and $N = 1$.

The lowest repetition frequency for a 50 Hz supply is therefore

$$f_1 = \frac{f}{M} = \frac{50}{5} = 10, \tag{3.7.3}$$

i.e. $n = 1$ corresponds with 10 Hz.

The per unit levels of the various frequencies present, obtained from equation (3.7.2), are

f_1 (10 Hz) = 0.087 $\quad$ f_8 (80 Hz) = 0.078

f_2 (20 Hz) = 0.14 $\quad$ f_9 (90 Hz) = 0.033

f_3 (30 Hz) = 0.189 f_{10} (100 Hz) = 0

f_4 (40 Hz) = 0.208 f_{11} (110 Hz) = 0.019

f_5 (50 Hz) = 0.2 f_{12} (120 Hz) = 0.025

f_6 (60 Hz) = 0.17 f_{13} (130 Hz) = 0.021

f_7 (70 Hz) = 0.126 f_{14} (140 Hz) = 0.011

When n is multiple of M the coefficients are zero, i.e. for 100 Hz, 150 Hz, etc. This clearly shows that integral cycle control produces no harmonic frequencies. It does, however, produce 'inter-harmonic' and subharmonic frequencies.

3.8 REFERENCES

1. Adamson, C., and Hingorani, N. G. (1960). *High Voltage Direct Current Power Transmission*, Garraway, London, Chap. 3.
2. Dobinson, L. G. (1975). 'Closer accord on harmonics'. *Electronics and Power*, May, 567–572.
3. Harker, B. J. (1983). 'Estimation of the harmonic currents entering the power system as a result of a.c. electrified railway traction'. Paper presented at the Annual General Meeting, IPENZ, New Zealand, February 1983.
4. Corbyn, D. B. (1972). 'This business of harmonics'. *Electronics and Power*, June, 219–223.
5. Schonung, A., and Stemmler, H. (1973). 'Reglage d'un motor triphase reversible a l'aide d'un convertisseur statique de frequence commande suivant le procede de la sous-oscillation'. *Revue Brown Boveri*, **51**, 557–576.
6. Patel, H. S., and Hoft, R. G. (1973). 'Generalised technique of harmonic elimination in voltage control in thyristor inverters. Part 1. Harmonic elimination'. *Trans. IEEE*, **IA-9** (3), 310–317.
7. Patel, H. S., and Hoft, R. G. (1974). 'Generalised technique of harmonic elimination and voltage control in thyristor invertors. Part 2. Voltage control techniques'. *IEEE Trans.*, **IA-10** (5), 666–673.
8. Buja, E. S., and Indri, G. B. (1977). 'Optimal pulse width modulation for feeding a.c. motors'. *IEEE Trans.*, **IA-13** (1), 38–44.
9. Byers, D. J., and Harman, R. T. C. (1976). 'Control of a.c. motors in a new concept of electric town car'. Paper presented to the Institute of Engineers, Australia, 1976.
10. Blommaert, J., de Vre, R., and Kniel, R. (1977). 'Analysis of harmonics in low voltage distribution networds caused by television receivers'. In *International Conference on Electricity Distribution*, Part 1, London, pp. 8–12.
11. Orr, J. A., Emanual, A. E., and Oberg, K. W. (1982). 'Current harmonics generated by a cluster of electric vehicle battery chargers'. *IEEE Trans.*, **PAS-101** (3), 691–700.
12. Lycett, J. D. (1975). 'Voltage disturbances and methods of maintaining the quality of supply'. Paper presented at the University Power Conference, Aston University, Birmingham.
13. Giesner, D. B., and Arrillaga, J. (1972). 'Behaviour of h.v.d.c. links under unbalanced a.c. fault conditions'. *Proc. IEE*, **119**, 209–215.
14. Ainsworth, J. D. (1981). 'Harmonic instabilities'. Paper presented at the conference on Harmonics in Power Systems, UMIST (Manchester), SSptember 1981.
15. Ainsworth, J. D. (1968). 'The phase-locked oscillator – a new control system for controlled static convertors', *Trans. IEEE*, **PAS-87**, 859–865.
16. Kimbark, E. W. (1971). *Direct Current Transmission*, Vol. I, John Wiley New York.
17. Pelly, B. R. (1971). *Thyristor Phase-controlled Converters and Cycloconverters*, Wiley–Interscience, New York.

4

Other harmonic sources

4.1 INTRODUCTION

Prior to the development of static convertor plant power system harmonic distortion was primarily associated with the design and operation of electric machines and transformers. Indeed the principal harmonic source present in the system in earlier days was the magnetizing current of the power transformers. Electric power generators provided the main secondary source since their practical economic design required that some departure from the ideal sinusoidal waveshape be accepted.

Modern transformers and rotating machines under normal steady state operating conditions do not of themselves cause significant distortion in the network. However, during transient disturbances and when operating outside their normal state range they can considerably increase their harmonic contribution.

Besides the static convertor there are two other non-linear loads that need to be considered because of their harmonic contribution; these are arc-furnaces and fluorescent lighting.

4.2 TRANSFORMER MAGNETIZATION NON-LINEARITIES

Normal excitation characteristics

At no-load the primary voltage of a transformer is practically balanced by the back e.m.f. because the effect of winding resistance and leakage reactance is negligible at low currents. At any instant, therefore, the impressed voltage v_1 for a sinusoidal supply is

$$v_1 = -e_1 = -E_m \sin \omega t = N_1 \frac{d\phi}{dt}. \tag{4.2.1}$$

From equation (4.2.1) the following expression is obtained for the main flux:

$$\phi = -\int \frac{e_1}{N_1} dt = \frac{E_m}{N_1 \omega} \cos \omega t = \phi_m \cos \omega t, \tag{4.2.2}$$

i.e. a sinusoidal primary voltage produces a sinusoidal flux at no-load. The primary current, however, will not be purely sinusoidal, because the flux is not linearly proportional to the magnetizing current, as explained in the next section.

Determination of the current waveshape

In an ideal core without hysteresis loss the flux ϕ and the magnetizing current i_m needed to produce it are related to each other by the magnetization curve of the steel used in the laminations as shown in Figure 4.1(a). In Figure 4.1(b), where ϕ represents the sinusoidal flux necessary to balance the primary voltage, the magnetizing current is plotted against time for each value of ϕ and the resulting waveform is far from sinusoidal.

When the hysteresis effect is included, as in the case of Figure 4.2, the non-sinusoidal magnetizing current wave is no longer symmetrical about its maximum value. In this case the current corresponding to any point on the flux

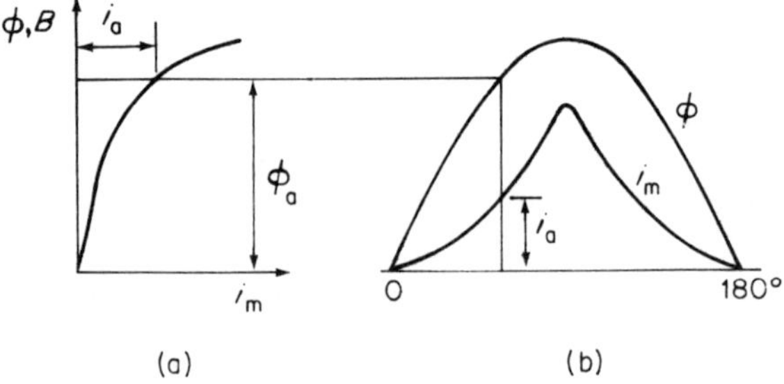

Figure 4.1. Transformer magnetization (without hysteresis): (a) magnetization curve; (b) flux and magnetization current waveforms

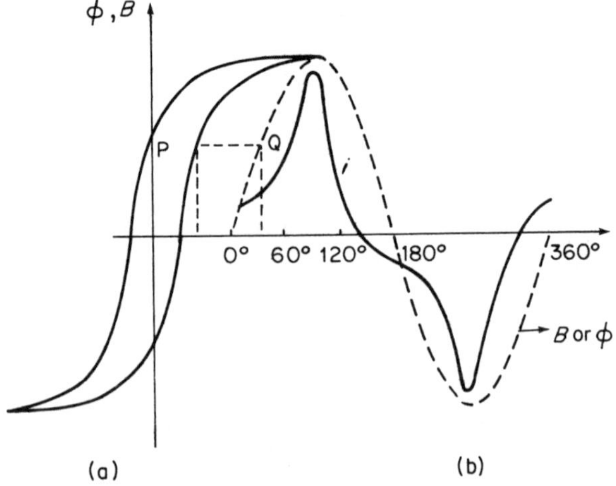

Figure 4.2. Transformer magnetization (including hysteresis): (a) magnetization curve; (b) flux and magnetization current waveforms

density wave of Figure 4.2(b) is determined from Figure 4.2(a), the ascending portion of the hysteresis being used for the ascending portion of the flux density wave, and the descending portion of the loop for the descending portion of the flux density wave.

The distortion illustrated in Figures 4.1 and 4.2 is mainly caused by triplen harmonics and particularly the third. Thus in order to maintain a reasonably sinusoidal voltage supply it is necessary to provide a path for the triplen harmonics and this is normally achieved by the use of delta-connected windings.

With three-limb transformers the triplen harmonic m.m.fs are all in phase and they act in each limb in the same direction. Hence the path of triplen harmonic flux must return through the air (or rather through the oil and transformer tank) and the higher reluctance of such path reduces the triplen harmonic flux to a very small value (about 10% of that appearing in independent core phases). Thus flux density and e.m.f. waveforms remain sinusoidal under all conditions in this case. The fifth and seventh harmonic components of the magnetizing current may also be large enough (5–10%) to produce visible distortion and cannot be ignored.

The magnetizing current harmonics often rise to their maximum levels in the early hours of the morning, i.e. when the system is lightly loaded and the voltage high.

Symmetrical overexcitation

For economic reasons transformers are normally designed to make good use of the magnetic properties of the core material. This means that a typical transformer using a good quality grain-oriented steel might be expected to run with a peak magnetic flux density in the steady state of perhaps 1.6–1.7 T. If a transformer running with this peak operating magnetic flux density is subjected

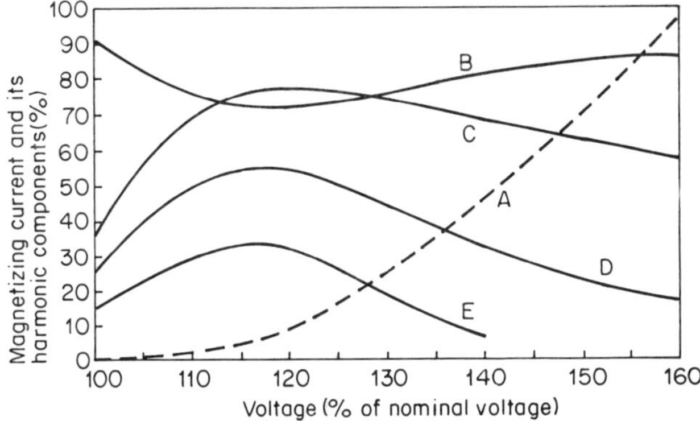

Figure 4.3. Harmonic components of transformer exciting current: curve A, magnetizing current (percentage of rated current); curve B, fundamental current (percentage of rated current); curve C, third harmonic current (percentage of fundamental current); curve D, fifth harmonic current (percentage of fundamental current); curve E, seventh harmonic current (percentage of fundamental current)

to a 30% rise in voltage, the core material may be subjected to a magnetic flux density of say 1.9–2.0 T which will produce considerable saturation.

The problem of overvoltage saturation is particularly onerous in the case of transformers connected to large rectifier plant following load rejection. It has been shown[1] that the voltage at the convertor terminals can reach a level of 1.43 per unit, thus driving the convertor transformer deep into saturation.

The symmetrical magnetizing current associated with a single transformer core saturation contains all the odd harmonics. If the fundamental component is ignored, and if it is assumed that all triplen harmonics are absorbed in delta windings, then the harmonics being generated are of orders 5, 7, 11, 13, 17, 19, etc., i.e. those of orders $6k \pm 1$, where k is an integer. In conventional, six-pulse rectifier schemes, it is usual to filter these harmonics from the a.c. busbars as they are exactly of the same order as the theoretical harmonics produced by a six-pulse convertor. If, however, a 12-pulse convertor configuration is used, the theoretical harmonics are of orders $12k \pm 1$, where k is an integer. In this case the fifth and seventh order harmonics produced by a saturated convertor transformer are not filtered and have to be absorbed by the a.c. system.

The level of the magnetizing current and its harmonic components versus the exciting voltage is typically as shown in Figure 4.3.

Inrush current harmonics

If a transformer is switched off it can be left with a residual flux density in the core of magnitude $+B_r$ or $-B_r$ (or under some circumstances zero). When the transformer is re-energized the flux density illustrated in Figure 4.4 can reach peak levels of $2B_{max}$ or $B_r + 2B_{max}$ (almost three times the working flux). For a normally designed transformer this can create peak flux densities of about 3.4 or 4.7 T respectively.[2] When this is compared to the saturation flux density levels of

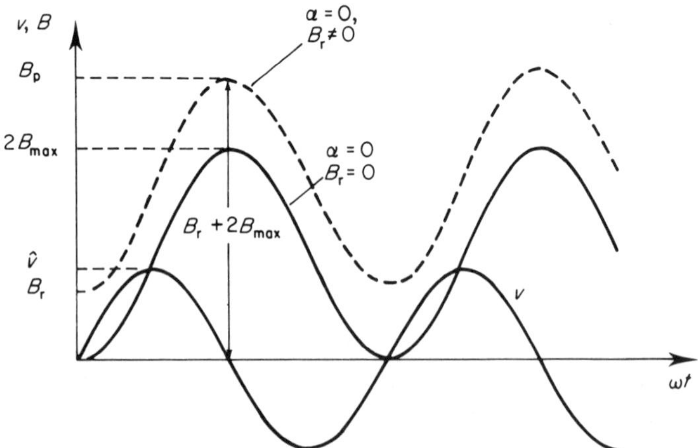

Figure 4.4. Flux density in a transformer at energization with both remnant flux B_r and zero remnant flux

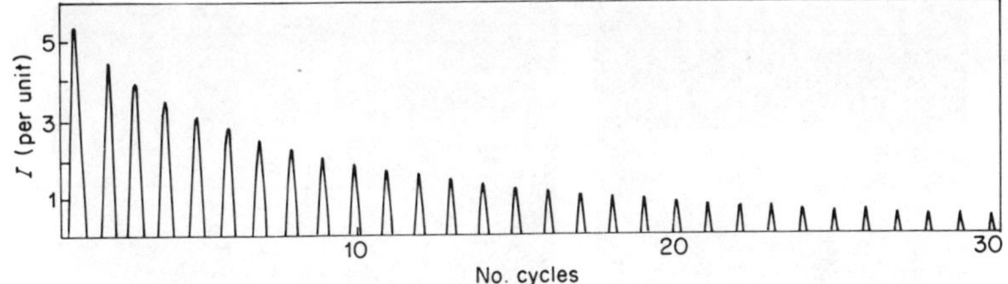

Figure 4.5. Inrush current of a 5 MVA transformer: $B_r = 1.3\,\text{T}$, $\alpha = 0$

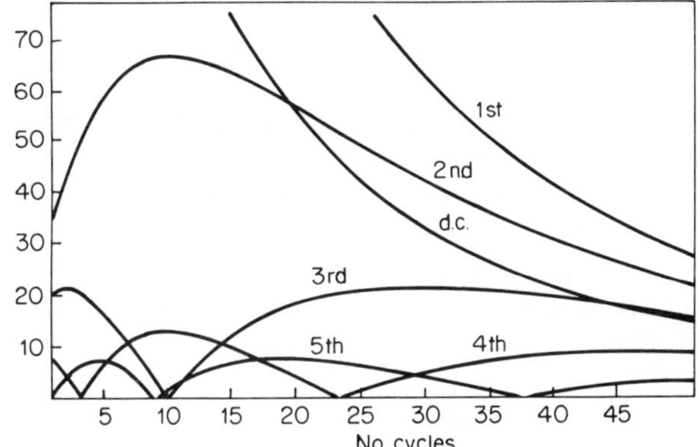

Figure 4.6. The variation of harmonic content (as a percentage of the rated current) with time

around 2.05 T to be expected from symmetrical overexcitation it can be seen that transformer core will be driven to extreme saturation levels and will thus produce excessive ampere-turns in the core. This effect gives rise to magnetizing currents of up to 5–10 per unit of the rating (as compared to the normal magnetizing current of a few per cent). Such an inrush current is shown in Figure 4.5.

The decrement of the inrush current with time is mainly a function of the primary winding resistance. For the larger transformers this inrush can go on for many seconds, because of their low resistance.

By way of illustration the Fourier series of the waveshape of Figure 4.5 yields the harmonic profile shown in Figure 4.6.

The harmonic content, shown as a percentage of the rated transformer current, varies with time and each harmonic has peaks and nulls.

D. C. Magnetization

It has been shown in previous sections that a transformer excited by sinusoidal voltage produces a symmetrical excitation current that contains only odd harmonics. If a linear or a non-linear load is connected to this transformer, the

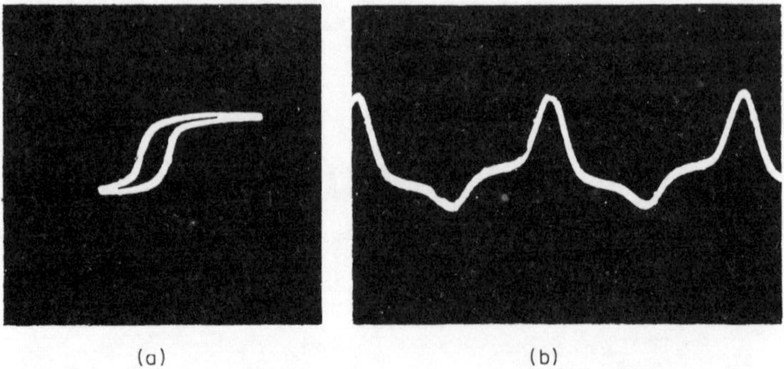

Figure 4.7. Excitation characteristic and current under asymmetrical magnetization: (a) excitation characteristic $\phi = f(i)$; (b) excitation characteristic $i = f(t)$

excitation current will again contain only odd harmonics, provided that the load does not produce a direct component of current.

Under magnetic imbalance the shape of the magnetizing characteristic and the excitation current are different from those under no-load conditions. If the flux is unbalanced, as shown in Figure 4.7(a), the core contains an average value of flux ϕ_{dc} and the a.c. flux component is offset by a value equal to ϕ_{dc}. The existence of an average flux implies that a direct component of excitation current is present in Figure 4.7(b).

Under such unbalanced conditions, the transformer excitation current contains both odd and even harmonic components. The asymmetry can be caused by any load connected to the secondary of the transformer, leading to a direct component of current, in addition to the sinusoidal terms. The direct current may be a feature of the design, as in a transformer feeding a halfwave rectifier, or may result from the unbalanced operation of some particular piece of equipment, such as a three-phase convertor with unbalanced firing.

It has been shown[3] that the magnitude of the harmonic components of the excitation current in the presence of direct current on the secondary side of the transformer increase almost linearly with the direct current content. The linearity is better for the lower order harmonics.

Moreover, the harmonics generated by the transformer under d.c. magnetization are largely independent of the a.c. excitation. Therefore there appears to be no advantage in designing a transformer to run 'underfluxed' in the presence of direct current. This independence is most noticeable at low levels of the direct current and for the lower harmonic orders.

4.3 ROTATING MACHINE HARMONICS

M.M.F. distribution of A.C. Windings

Figure 4.8 shows the m.m.f. and flux distribution in one phase of a fuil-pitched polyphase winding with one slot per pole per phase on the assumption of a constant air-gap and in the absence of iron saturation.

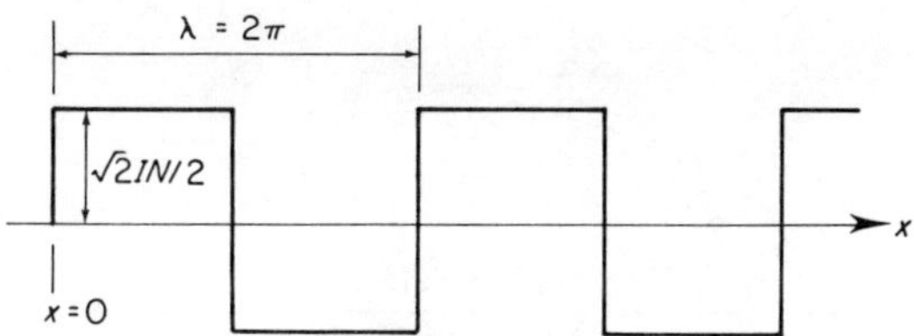

Figure 4.8. M.M.F. and flux distribution of full-pitch winding with one slot/pole

Under such idealized conditions the air-gap m.m.f. is uniform and has a maximum value $\sqrt{2}IN/2$, where I is the r.m.s. current per conductor and N is the number of conductors per slot.

The frequency domain representation of the rectangular m.m.f. distribution of Figure 4.8 is

$$F(x) = \frac{2\sqrt{2}IN}{\pi}\left\{\sin\frac{2\pi x}{\lambda} + \frac{1}{3}\sin 3\left(\frac{2\pi x}{\lambda}\right) + \frac{1}{5}\sin 5\left(\frac{2\pi x}{\lambda}\right) + \cdots\right\}. \quad (4.3.1)$$

Thus the rectangular m.m.f. distribution is reduced to a fundamental and harmonic component. The amplitude of the nth harmonic is $1/n$ times that of the fundamental. The pole pitch of the nth harmonic m.m.f. is $1/n$ times the fundamental pole pitch.

In general, for an alternating current of angular frequency $\omega = 2\pi f$, equation (4.3.1) becomes

$$F(x) = \frac{2\sqrt{2}IN}{\pi}\sin(\omega t)\sum_{n=1}^{\infty}\frac{1}{2}\sin n\left(\frac{2\pi x}{\lambda}\right) \quad \text{for} \quad n \text{ odd} \quad (4.3.2)$$

where λ is the wave length.

However, in practice the windings are distributed along the surface with g slots per pole per phase and the m.m.fs of the g coils are displaced from each other in space. Moreover, the displacement angle is different for the various harmonics since their pole pitches are different.

For an m-phase machine, the number of slots per pole is $Q = mg$ and the electrical angle between slots $\alpha = \pi/Q$.

The distribution factor is

$$k_d = \frac{\text{resultant m.m.f.}}{\text{sum of m.m.fs of individual coils}}.$$

From the geometry of Figure 4.9,

$$k_d = \frac{\sin(g\alpha/2)}{g\sin(\alpha/2)} \quad (4.3.3)$$

for the fundamental frequency and

$$k_{dn} = \frac{\sin(ng\alpha/2)}{g\sin(n\alpha/2)} \quad (4.3.4)$$

for the nth harmonic.

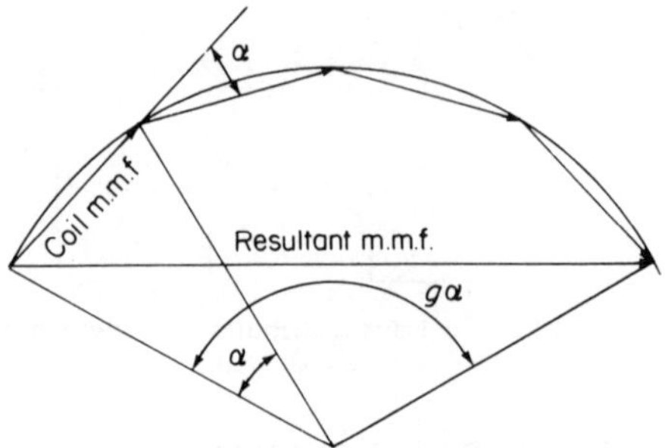

Figure 4.9. Determination of distribution factor

Hence the m.m.f. of one phase of a polyphase winding is

$$F(x)=\frac{2\sqrt{2}IN}{\pi}g\sin\omega t\sum_{n=1}^{\infty}\frac{k_{dn}}{n}\sin n\left(\frac{2\pi x}{\lambda}\right),\quad \text{for}\quad n\ \text{odd}. \tag{4.3.5}$$

Three-phase winding

The phase windings of a three-phase machine are displaced by $2\pi/3$ in space and the currents by $2\pi/3$ in time. The corresponding m.m.fs are

$$F_1(x)=\frac{2\sqrt{2}IN}{\pi}g\sin\omega t\left\{\sum_{n=1}^{\infty}\frac{k_{dn}}{n}\sin n\left(\frac{2\pi x}{\lambda}\right)\right\}, \tag{4.3.6}$$

$$F_2(x)=\frac{2\sqrt{2}IN}{\pi}g\sin\left(\omega t-\frac{2\pi}{3}\right)\left\{\sum_{n=1}^{\infty}\frac{k_{dn}}{n}\sin\left[n\left(\frac{2\pi x}{\lambda}-\frac{2\pi}{3}\right)\right]\right\}, \tag{4.3.7}$$

$$F_3(x)=\frac{2\sqrt{2}IN}{\pi}g\sin\left(\omega t-\frac{4\pi}{3}\right)\left\{\sum_{n=1}^{\infty}\frac{k_{dn}}{n}\sin\left[n\left(\frac{2\pi x}{\lambda}-\frac{4\pi}{3}\right)\right]\right\}. \tag{4.3.8}$$

The total m.m.f. is $F(x)=F_1(x)+F_2(x)+F_3(x)$, for which the nth harmonic term is

$$\begin{aligned}
&\frac{2\sqrt{2}IN}{\pi}g\frac{k_{dn}}{n}\left\{\sin n\left(\frac{2\pi x}{\lambda}\right)\sin\omega t+\sin\left[n\left(\frac{2\pi x}{\lambda}-\frac{2\pi}{3}\right)\right]\sin\left(\omega t-\frac{2\pi}{3}\right)\right.\\
&\qquad\left.+\sin\left[n\left(\frac{2\pi x}{\lambda}-\frac{4\pi}{3}\right)\right]\sin\left(\omega t-\frac{4\pi}{3}\right)\right\}\\
&=\frac{2\sqrt{2}IN}{\pi}g\frac{k_{dn}}{2n}\left\{\cos\left[\frac{2\pi nx}{\lambda}-\omega t\right]-\cos\left[\frac{2\pi nx}{\lambda}+\omega t\right]\right.\\
&\qquad+\cos\left[\frac{2\pi nx}{\lambda}-\omega t-(n-1)\frac{2\pi}{3}\right]-\cos\left[\frac{2\pi nx}{\lambda}+\omega t-(n+1)\frac{2\pi}{3}\right]\\
&\qquad\left.+\cos\left[\frac{2\pi nx}{\lambda}-\omega t-(n-1)\frac{4\pi}{3}\right]-\cos\left[\frac{2\pi nx}{\lambda}+\omega t-(n+1)\frac{2\pi}{3}\right]\right\}.
\end{aligned} \tag{4.3.9}$$

Putting in turn $n=1,3,5$, etc.,

$$F(x)=\frac{3\sqrt{2}IN}{\pi}g\left\{(k_{d1})\cos\left(\frac{2\pi x}{\lambda}-\omega t\right)+\left(\frac{k_{d5}}{5}\right)\cos\left(5\times\frac{2\pi x}{\lambda}+\omega t\right)\right.$$
$$\left.+\left(\frac{k_{d7}}{7}\right)\cos\left(7\times\frac{2\pi x}{\lambda}-\omega t\right)+\cdots\right\}. \tag{4.3.10}$$

It can be seen that the fundamental is a travelling wave moving in the positive direction, triplen harmonics (3, 9, 15, etc.) are absent, the fifth harmonic is a wave travelling in the negative direction, the seventh harmonic travels in the positive direction, etc.

Slot harmonics

If the machine has *mg* slots per pole (as shown in Figure 4.10), the variation of permeance in the air-gap can be approximated by

$$A_1+A_2\sin\left(2mg\frac{2\pi x}{\lambda}\right). \tag{4.3.11}$$

Since the fundamental m.m.f. varies as $B\sin(2\pi x/\lambda)$, the resultant flux density variation is

$$\left\{B\sin\frac{2\pi x}{\lambda}\right\}\left\{A_1+A_2\sin\left(2mg\frac{2\pi x}{\lambda}\right)\right\} \tag{4.3.12}$$

which has a fundamental frequency component, i.e.

$$A_1B\sin\frac{2\pi x}{\lambda}, \tag{4.3.13}$$

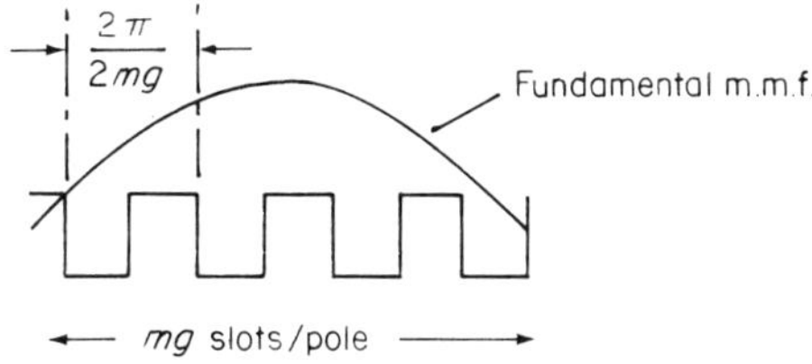

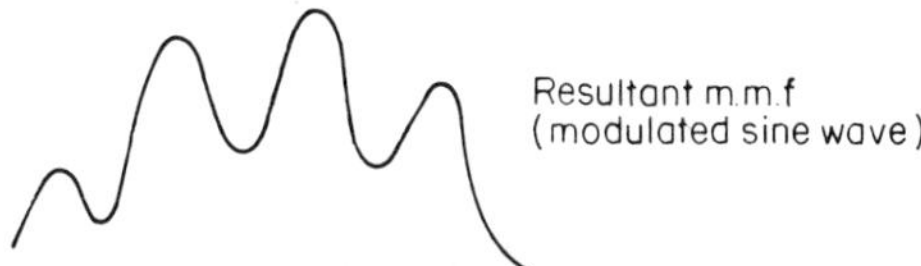

Figure 4.10. Slot harmonics

and frequency components expressed as

$$A_2 B \sin\frac{2\pi x}{\lambda}\sin\left(2mg\frac{2\pi x}{\lambda}\right)=\frac{A_2 B}{2}\left\{\cos\frac{2\pi x}{\lambda}(2mg-1)-\cos\frac{2\pi x}{\lambda}(2mg+1)\right\}. \tag{4.3.14}$$

Therefore slotting gives rise to harmonics of orders $2mg \pm 1$.

Voltage harmonics produced by synchronous machines

If the magnetic flux of the field system is distributed perfectly sinusoidally around the air-gap, then the e.m.f. generated in each full-pitched armature coil is $2\pi f \phi \sin \omega t$ volts per turn. Here ϕ is the total flux per pole and the frequency f is related to speed N (in revolutions per second) and pole pairs p by $f = Np$. However, the flux is never exactly distributed in this way, particularly in salient pole machines. A non-sinusoidal field distribution can be expressed as a harmonic series:

$$F(x) = F_1 \sin\frac{2\pi x}{\lambda} + F_3 \sin\frac{3\times 2\pi x}{\lambda} + F_5 \sin\frac{5\times 2\pi x}{\lambda} + \cdots \tag{4.3.15}$$

The machine can be considered to have $2p$ fundamental poles together with $6p$, $10p, \ldots, 2np$ harmonic poles, all individually sinusoidal and all generating e.m.fs. in an associated winding. The winding e.m.f. can be expressed as a harmonic series:

$$E(t) = E_1 \sin \omega t + E_3 \sin 3\omega t + E_5 \sin 5\omega t + \cdots \tag{4.3.16}$$

The magnitudes of the harmonic e.m.fs are determined by the harmonic fluxes, the effective electrical phase spread of the winding, the coil span, and the method of interphase connection.

For an integral slot winding with g slots per pole per phase and an electrical angle α between slots the distribution factor for the nth harmonic is

$$k_{dn} = \frac{\sin(ng\alpha/2)}{g\sin(n\alpha/2)}. \tag{4.3.17}$$

If the coils are chorded to cover $(\pi \pm \theta)$ electrical radians, the flux linked is reduced by $\cos(\theta/2)$ and the e.m.f. is reduced in proportion The effective chording angle for harmonics of order n is $n\theta$. Hence the general coil-span factor is

$$k_{sn} = \cos(n\theta/2). \tag{4.3.18}$$

By suitable choice of k_d and k_s many troublesome e.m.f. harmonics can be minimized or even eliminated. The triplen harmonics in a three-phase machine are usually eliminated by phase connection, and it is usual to select the coil-span to reduce fifth and seventh harmonics.

Standby generators with neutral require special consideration in this respect, illustrated by the following example.

A standby generator had to be designed to supply a 250kVA load consisting mainly of fluorescent light appliances. A neutral current of 40 A was considered sufficient for the generator design. However, in practice the machine generated 250 A of third harmonic zero sequence and had to be rewound with a two-thirds pitch (i.e. $k_{s3} = \cos(3 \times 60/2) = 0$).

Slotting (the slots are on the stator) produces variation of permeance which may be represented as $A_2 \sin[2mg(2\pi x/\lambda)]$. The fundamental rotor m.m.f. can be represented as a travelling wave $F_1 \cos[(2\pi x/\lambda) - \omega t]$. The slot ripple component of flux density is of the form

$$F_1 A_2 \sin\left(2mg\frac{2\pi x}{\lambda}\right)\cos\left(\frac{2\pi x}{\lambda} - \omega t\right). \tag{4.3.19}$$

This can be resolved into two counter-rotating components, i.e.

$$\frac{F_1 A_2}{2}\left\{\sin\left[(2mg+1)\frac{2\pi x}{\lambda} - \omega t\right] + \sin\left[(2mg-1)\frac{2\pi x}{\lambda} + \omega t\right]\right\} \tag{4.3.20}$$

which are slow moving multi-pole harmonics. Their wavelengths are $\lambda/(2mg \pm 1)$ and the corresponding velocities are $f\lambda/(2mg \pm 1)$. As the number of waves passing any point on the stator per second is (speed ÷ wavelength), obviously each component induces an e.m.f. of fundamental frequency in the armature.

Relative to the rotor, however, these two waves have different velocities. The rotor velocity being $f\lambda$, the waves travel at velocities $f\lambda - [f\lambda/(2mg+1)]$ and $f\lambda + [f\lambda/(2mg-1)]$ with respect to the rotor. In any closed rotor circuit they will each generate currents of frequency $2mgf$ (by considering the ratio of speed to wavelength) and these superimpose a time-varying m.m.f. at frequency $2mgf$ on the rotor fundamental m.m.f. This can be resolved into two counter-rotating components relative to the rotor, each travelling at high velocity $2mgf\lambda$, and therefore at $2mgf\lambda \pm f\lambda$ relative to the stator. The resultant stator e.m.fs have frequencies $(2mg \pm 1)f$.

Slot harmonics can be minimized by skewing the stator core, displacing the centre line of damper bars in successive pole faces, offsetting the pole shoes in successive pairs of poles, shaping the pole shoes, and the use of composite steel–bronze wedges for the slots of turboalternators.

It can be shown that the distribution factor for slot harmonics is the same as for the fundamental e.m.f.; they are not reduced by spreading the winding. Fractional instead of integral slotting should be used.

Voltage harmonics produced by induction motors

The speed of the synchronous rotating field of the stator of an induction motor is the fundamental frequency times the wavelength, i.e. $f_1\lambda$. For a slip s, the rotor speed is thus $f_1\lambda(1-s)$ and the frequency of the rotor currents sf_1.

Time harmonics are produced by induction motors as a result of the harmonic content of the m.m.f. distribution and are speed dependent.

A harmonic of order n in the rotor m.m.f. (i) has a wavelength λ/n; (ii) travels at a

Table 4.1. Typical harmonic currents produced by a wound-rotor induction motor

Frequency (Hz)	Current as % of fundamental	Cause
20	3.0	Pole unbalance
40	2.4	Rotor-phase unbalance
50	100.0	Fundamental mutual
80	2.3	Pole unbalance
220	2.9	Fifth and seventh harmonic mutuals
320	3.0	
490	0.3	11th and 13th harmonic mutuals
590	0.4	

speed $\pm(sf)\lambda/n$ with respect to the rotor; (iii) travels at a speed $f\lambda(1-s) \pm(sf)\lambda/n$ with respect to the stator.

This harmonic induces an e.m.f. in the stator at a frequency equal to the ratio speed ÷ wavelength, i.e.

$$f^1 = \frac{f\lambda(1-s) \pm (sf)(\lambda/n)}{\lambda/n} = f\{n - s(n \pm 1)\}, \tag{4.3.21}$$

the positive sign being taken when the harmonic rotor m.m.f. travels in the opposite direction to the fundamental.

Harmonics can also occur as a result of electrical asymmetry. Consider an electrically unbalanced rotor winding, the stator winding being balanced such that the supply voltage produces a pure rotating field travelling at speed $f\lambda$. Slip frequency e.m.f. is induced in the rotor but, since the rotor winding is unbalanced, both positive and negative phase sequence currents will flow, giving fields in forward and reverse directions. These travel at speed $\pm sf\lambda$ with respect to the rotor, and therefore at $f\lambda(1-s) \pm sf\lambda$ with respect to the stator. The frequencies of stator e.m.fs induced by these fields are f and $(1-2s)f$, the latter being considered here as a harmonic frequency. Interaction of harmonic and mains frequency currents results in beats at this low frequency $2sf$ being registered on connected meters.

An idea of the relative magnitudes of the harmonic current produced by a wound-rotor induction motor and their cause is given in Table 4.1, taken from reference 4, for a six-pole motor running at a speed of 0.9 per unit.

4.4 DISTORTION CAUSED BY ARC-FURNACES

A combination of arc ignition delays and the highly non-linear arc voltage–current characteristics introduces harmonics of the fundamental frequency.

In addition, the stochastic voltage changes due to sudden alterations of arc length produce a spread of frequencies, predominantly in the range 0.1–30 Hz[(5)] about each of the harmonics present. This effect is more evident during the melting phase, caused by continuous motion of the melting scrap and the interaction of electromagnetic forces between the arcs.

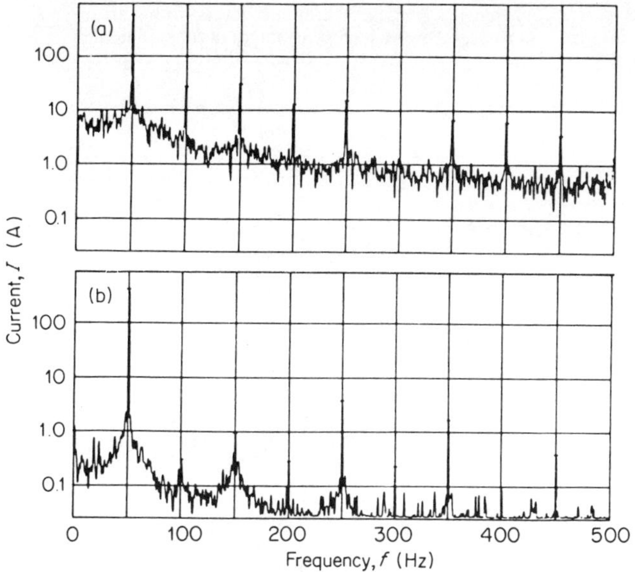

Figure 4.11. Frequency spectra for (a) melting and (b) refining periods[1]

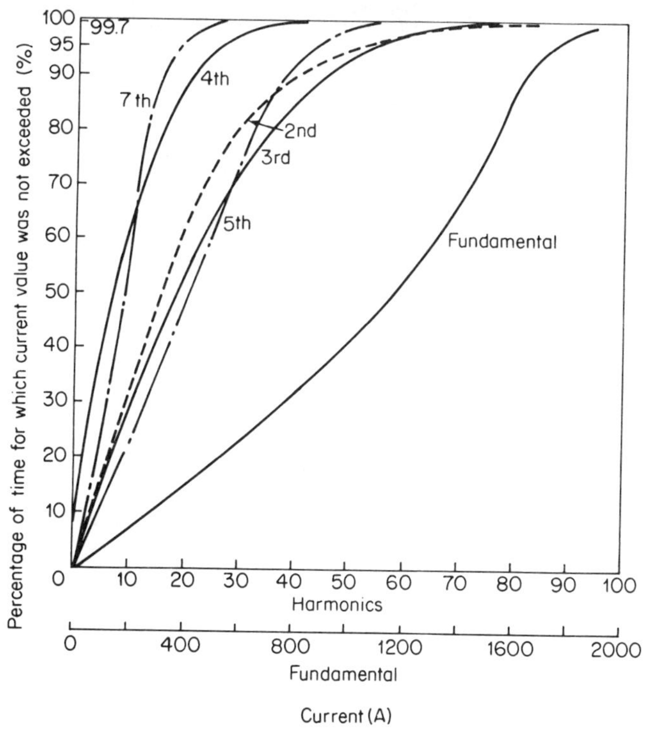

Figure 4.12. Probabilistic harmonic magnitude plots[2]

Table 4.2. Average harmonic levels from arc-furnaces expressed as percentages of fundamental

Order	Level		
	Ref. 7	Ref. 5	Ref. 6
2	3.2	4.1	4.5
3	4.0	4.5	4.7
4	1.1	1.8	2.8
5	3.2	2.1	4.5
6	0.6	Not given	1.7
7	1.3	1.0	1.6
8	0.4	1.0	1.1
9	0.5	0.6	1.0
10	>0.5	>0.5	>1.0

Note: Coates and Brewer[(6)] also found levels of 1% at the 22nd harmonic order.

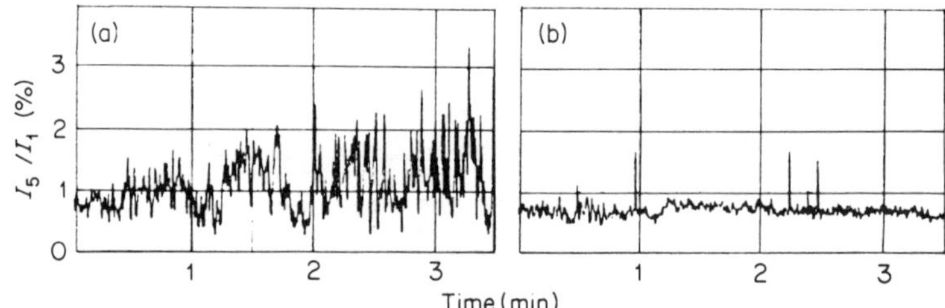

Figure 4.13. Fifth harmonic as a percentage of the fundamental with time:[(5)] (a) melting; (b) refining

During refining the arc is better behaved but there is still some modulation of arc length by waves on the surface of the molten metal.

Typical time-averaged frequency spectra of the melting and refining periods are shown in Figure 4.11.

However, the levels of harmonic currents vary markedly with time and are better displayed in the form of probabilistic plots, such as that shown in Figure 4.12. Three sets of averaged harmonic current levels obtained by different investigators are listed in Table 4.2 as percentages of the fundamental.

Finally the time variation of the harmonic currents is exemplified by the records shown in Figure 4.13 for the fifth harmonic, an important point being that the harmonic current not only varies with time but also in respect to the fundamental component.

4.5 FLUORESCENT LIGHTING HARMONICS

Luminous discharge lighting and in particular fluorescent tube appliances are highly non-linear and give rise to considerable odd-ordered harmonic currents.

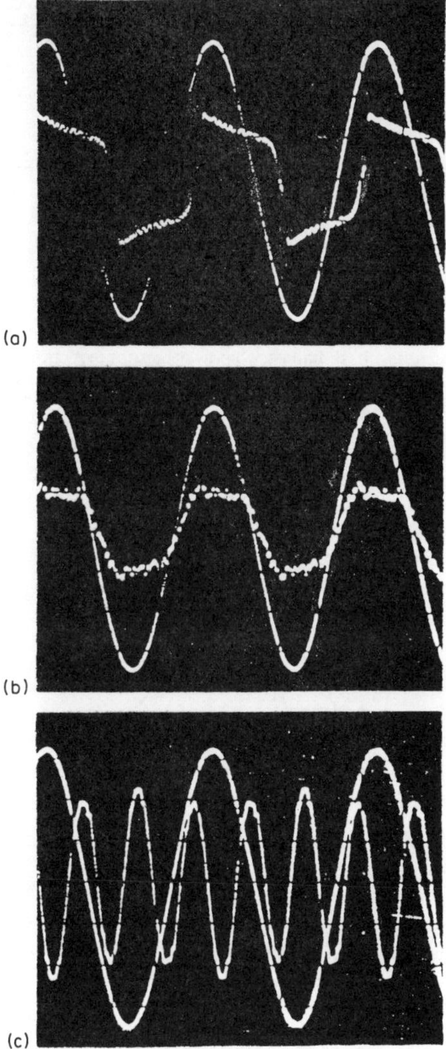

Figure 4.14. (a) Tube voltage; (b) phase current with capacitor, one lamp, 240 mA/division; (c) neutral current, three banks of three lamps in star, 240 mA/division

In a three-phase, four-wire load the triplens are basically additive in the neutral and the third is the most dominant.

With reference to the basic fluorescent circuit of Figure 4.15, a set of voltage and current oscillograms is displayed in Figure 4.14. These waveforms are shown with reference to the sinusoidal phase voltage supply. The voltage across the tube itself (Figure 4.14(a)) illustrates clearly the non-linearity. The waveform in Figure

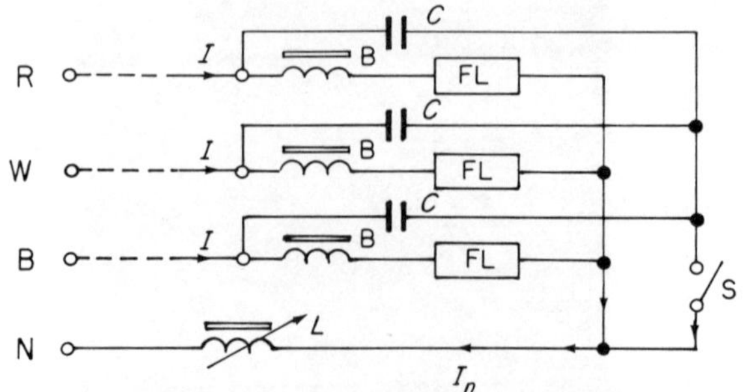

Figure 4.15. Three-phase fluorescent lighting test circuit. FL, fluorescent lamp; B, ballast; *C*, power factor correction capacitor; *L*, variable inductor; S, switch to isolate capacitor star point.

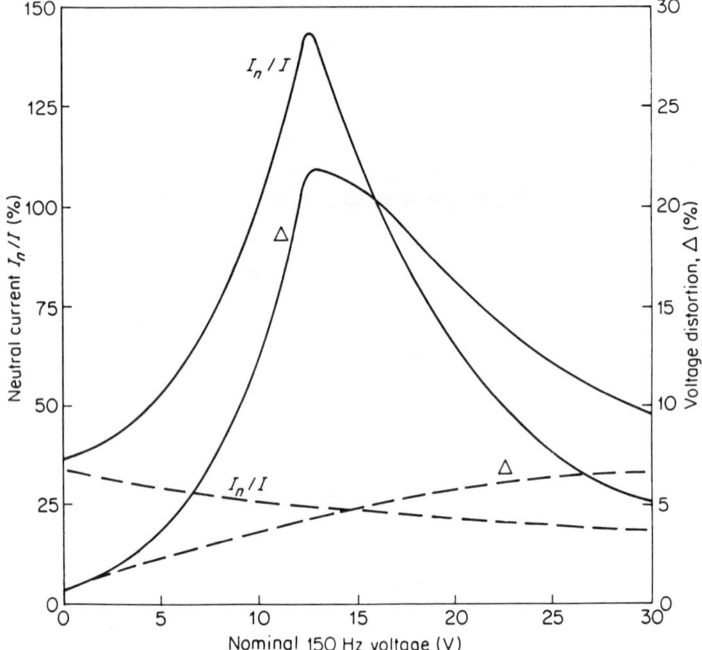

Figure 4.16. Characteristics of the fluorescent lighting test circuit. The nominal 150 Hz voltage is the calculated product of the 150 Hz lamp current per phase and the circuit zero-sequence impedance (150 Hz). ——, Switch S closed (star point to neutral); -----, switch S open (star point floating)

4.14(b) shows the phase current and the waveform in Figure 4.14(c) the neutral current for a case of three banks of three lamps connected in star. The latter consists almost exclusively of third harmonic.

Lighting circuits often involve long distances and have very little load diversity. With individual power factor correction capacitors the complex LC circuit can approach a condition of resonance at third harmonic. This effect has been illustrated by laboratory results on a balanced three-phase fluorescent lamp installation.[8] The results of greatest interest refer to the effects of increasing the neutral reactance and isolating the capacitor star point (see Figure 4.15). In the graph of Figure 4.16 the abscissa used is the nominal third harmonic voltage, i.e. the product of the lamp third harmonic current per phase and the corresponding circuit third harmonic zero-sequence impedance. It is seen that with the capacitor star point connected of neutral, the third harmonic neutral current can by far exceed the nominal value calculated by the conventional method based on three times the nominal lamp current. With the star point disconnected the neutral current is less than the nominal value.

The results demonstrated in the laboratory test were in fact a confirmation of actual field test results taken on a 900 kVA fluorescent installation. In this case the full load condition operated well above the resonant point; at about half load some neutral current values exceeded the corresponding phase current values, and the voltage distortion at some distribution boards exceeded 20%.

Whenever possible the design procedure recommended to avoid resonance is to try and avoid individual lamp compensation by providing capacitor banks adjacent to distribution boards connected either in star with floating neutral or in delta.

4.6 REFERENCES

1. Bowles, J. P. (1980). 'Alternative techniques and optimisation of voltage and reactive power control at h.v.d.c. convertor stations'. Paper presented at the IEEE conference on Overvoltages and Compensation on Integrated A.C.–D.C. Systems, Winnipeg.
2. Yacamini, R. (1971). 'Harmonics caused by transformer saturation'. Paper present at an international conference of Harmonics in Power System, UMIST, Manchester.
3. Yacamini, R., and De Oliveira, J. C. (1978). 'Harmonics produced by direct current in convertor transformers'. *Proc. IEE*, **125**, 873–878.
4. Wallace, A. K., Ward, E. S., and Wright, A. (1974). 'Sources of harmonic currents in slip-ring induction motors'. *Proc. IEE*, **121**, 1495–1500.
5. Jahn, H. H., and Käuferle, J. (1974). 'Measuring and evaluating current fluctuations of arc-furnaces'. *IEE Conf. Publ.*, **110**, 105–109.
6. Coates, R., and Brewer, G.L. (1974). 'The measurement and analysis of waveform distortion caused by a large multi-furnace arc-furnace installation'. *IEE Conf. Publ.* **110**, 135–143.
7. Lemoine, M. (1978). 'Résonances en présence des harmoniques creés par les convertisseurs de puissance et les fours à arcs associés à dés dispositits de compensation', *Revue Gen. Electricité*, **87**, 945–962.
8. Muntz, V. A., and Jones, R. M. (undated). 'Control of third harmonic current in three-phase neutrals, e.g. large fluorescent lighting installations'. Monograph MT, No. 3, State Electricity Commission of Victoria, Australia.

5

Harmonic effects

A Within the power system

5.1 INTRODUCTION

Once the harmonic sources and their magnitudes are clearly defined, they must be interpreted in terms of their effect on system and equipment operation. Individual elements of the power system must be examined for their sensitivity to harmonics as a basis for recommendations on the allowable levels.

The main effects of voltage and current harmonics within the power system are (i) amplification of harmonic levels resulting from series and parallel resonances, (ii) reduction of efficiency in power generation, transmission and utilization, (iii) ageing of the insulation of electrical plant components and thus shortening of their useful life, (iv) plant maloperation.

The extent to which these problems affect individual power plant components and systems is discussed in this part of the chapter.

5.2 RESONANCES

The presence of capacitors such as those used for power factor correction can result in local system resonances which lead in turn to excessive currents with subsequent damage to such capacitors.[1]

Parallel resonance

Parallel resonance results in a high impedance being presented to the harmonic source at the resonant frequency. Since the majority of harmonic sources can be considered as current sources, this results in increased harmonic voltages and high harmonic currents in each leg of the parallel impedance.

Parallel resonances can occur in a number of ways, the simplest perhaps being that where a capacitor is connected to the same busbar as the harmonic source. A parallel resonance can then occur between the source and the capacitor.

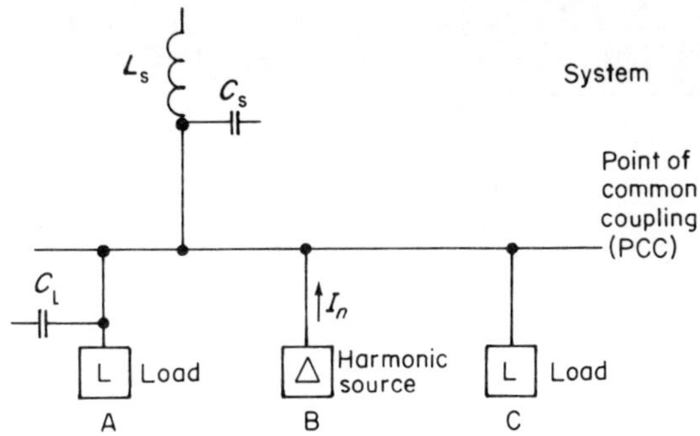

Figure 5.1. Parallel resonance at a point of common coupling

Assuming the source to be entirely inductive, the resonant frequency is

$$f_p = f\sqrt{\left(\frac{S_s}{S_c}\right)}, \tag{5.2.1}$$

where f is the fundamental frequency (Hz), f_p is the parallel resonant frequency (Hz), S_s is the source short circuit rating (VAr), S_c is the capacitor rating (VAr).

Further opportunities for parallel resonance occur when the system is considered. In Figure 5.1 the harmonic current from consumer B encounters a high harmonic impedance at the busbar. This may be due to a resonance between the system inductance (L_s) and either the system capacitance (C_s) or the load capacitance (C_l).

To determine which resonance condition exists it is necessary to measure the harmonic currents in each consumer load and the supply, together with the harmonic voltage at the busbar. In general, if the current flowing into the power system from the busbar is small while the harmonic voltage is high, resonance within the power system is indicated. If instead a large harmonic current flows in consumer A's load and leads the harmonic voltage at the busbar, resonance between the system inductance and the load capacitor is indicated.

Series resonance

Consider the system of Figure 5.2. At higher frequencies the load can be ignored as the capacitive impedance reduces. Under these conditions a series resonant condition will exist when

$$f_s = f\sqrt{\left(\frac{S_t}{S_c Z_t} - \frac{S_l^2}{S_c^2}\right)}, \tag{5.2.2}$$

where f_s is the series resonant frequency (Hz), S_t is the transformer rating, Z_t is the transformer per unit impedance, S_l is the load rating (resistive).

The concern with series resonance is that high capacitor currents can flow for

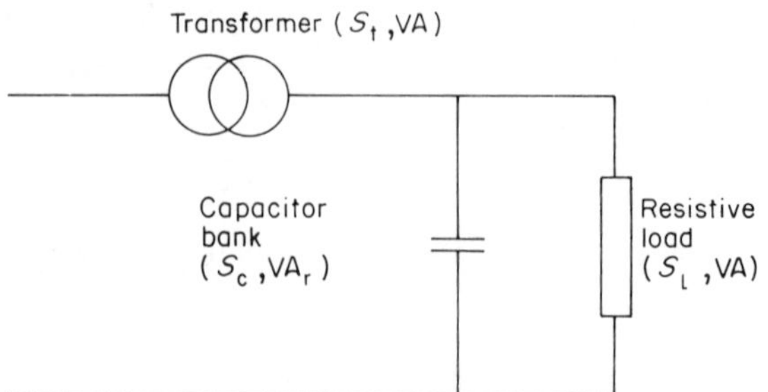

Figure 5.2. Series resonance circuit

relatively small harmonic voltages. The actual current that will flow will depend upon the quality factor, Q, of the circuit. This is typically of the order of 5 at 500 Hz.

Effects of resonance on system behaviour

Resonances have been considered in relation to capacitors and, in particular, power factor correction capacitors. These capacitors are made to a number of standards which specify varying levels of overload current capability.[2] Typically, overload capabilities range from 15% (UK), through 30% (Australia and Europe) to as high as 80% (USA). In many instances capacitors have been observed to operate well in excess of these levels with subsequent failure.

Another area where resonance effects may lead to component failure is associated with the application of power line signalling (ripple control) for load management. In such systems tuned stoppers (filters) are often used to prevent the signalling frequency from being absorbed in low impedance elements such as power factor correction capacitors. A typical installation is shown in Figure 5.3.

Where local resonances exist, excessive harmonic currents can flow, resulting in damage to the tuning capacitors. Figure 5.4 shows the harmonic currents recorded at one such installation where failure of this type occurred.

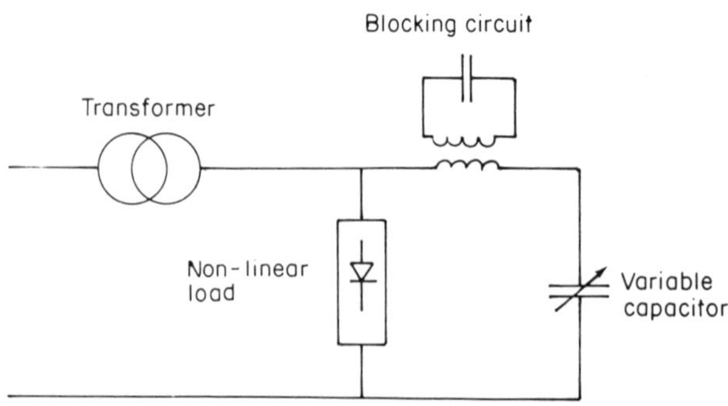

Figure 5.3. Tuned stopper circuit for ripple control signal

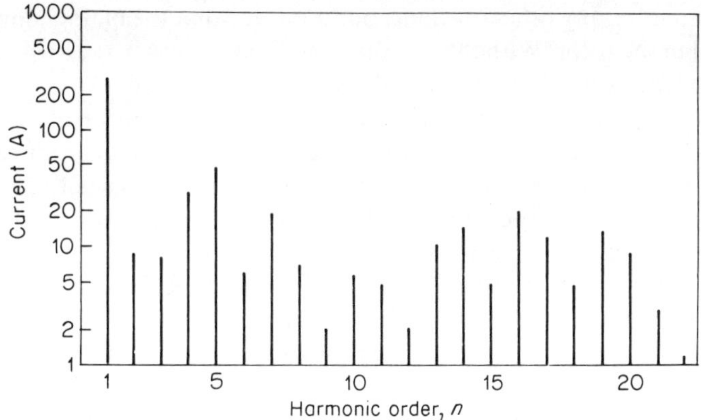

Figure 5.4. Harmonic currents measured through a blocking circuit

In another installation tuned stoppers (at 530 Hz) were fitted to 15 × 65 kVAr steps of power factor correction capacitance, each stopper rated at 100 A. Most of the stopper tuning capacitors failed within two days. The problem was eventually traced to a local power system harmonic at 350 Hz near to which frequency the tuned stoppers were found to series resonate with the power factor correction capacitors.

5.3 EFFECTS OF HARMONICS ON ROTATING MACHINES

Harmonic losses

Harmonic voltages or currents give rise to additional loss in the stator windings, rotor circuits, and stator and rotor laminations. The losses in the stator and rotor conductors are greater than those associated with the d.c. resistances because of eddy currents and skin effect.

Leakage fields set up by harmonic currents in the stator and rotor end-windings produce extra losses.

In the case of induction motors with skewed rotors the flux changes in both stator and rotor and high frequency can produce substantial iron loss. The magnitude of this loss depends upon the amount of skew, and the iron-loss characteristics of the laminations.

As an illustration of the effect of supply waveform distortion on the power loss, Klinghsirn and Jordan(3) considered the case of a 16 kW motor, operating at full output with 60 Hz supply, and rated fundamental voltage in each case. With a sinusoidal voltage supply the total loss is 1303 W, whereas with a quasi-square voltage supply the total loss is 1600 W.

The following distribution of losses caused by supply harmonics for the case of an invertor-fed machine is given by Chambers and Sarkar:(4) stator winding, 14.2%; rotor bars, 41.2%; end region, 18.8%; skew flux, 25.8%.

It would be inappropriate to apply the above loss breakdown for individual harmonics, or for any other machine, but it is clear that the major component of loss is that in the rotor. With the exception of the skew losses, the loss subdivision of a synchronous machine should follow a similar pattern.

When considering the harmonic heating losses in the rotor of synchronous machines it must be remembered that pairs of stator harmonics produce the same rotor frequency. For example the fifth and seventh harmonics both give induced rotor currents at frequency $6f_1$. Each of these currents takes the form of an approximately sinusoidal spatial distribution of damper bar currents travelling around the rotor at velocity $6\omega_1$, but in opposite directions. Thus for a linear system, the average rotor surface loss density around the periphery will be proportional to $(I_5^2 + I_7^2)$; however, because of their opposing rotations, at some point around the periphery the local surface loss density will be proportional to $(I_5 + I_7)^2$. If the fifth and seventh harmonic currents are of similar magnitude then the maximum local loss density would be about twice the average loss density caused by these two currents.

The additional power loss is probably the most serious effect of harmonics upon a.c. machines. The capability of a machine to cope with extra harmonic currents will depend on the total additional loss and its effect on the overall machine temperature rise and local overheating (probably in the rotor). Cage-rotor induction motors tolerate higher rotor losses and temperatures provided that these do not result in unacceptable stator temperatures, whereas machines with insulated rotor windings may be more limited. Some guidance as to the probably acceptable levels may be obtained from the fact that the level of continuous negative sequence current is limited to about 10% for generators, and that of negative sequence voltage to about 2% for induction motors. It is therefore reasonable to expect that if the harmonic content exceeds these negative sequence limits, then problems will occur.

Harmonic torques

The familiar equivalent circuit of an induction machine can be drawn for each harmonic as in Figure 5.5, where all the parameters correspond to actual frequencies of winding currents.

Harmonic currents present in the stator of an a.c. machine produce induction

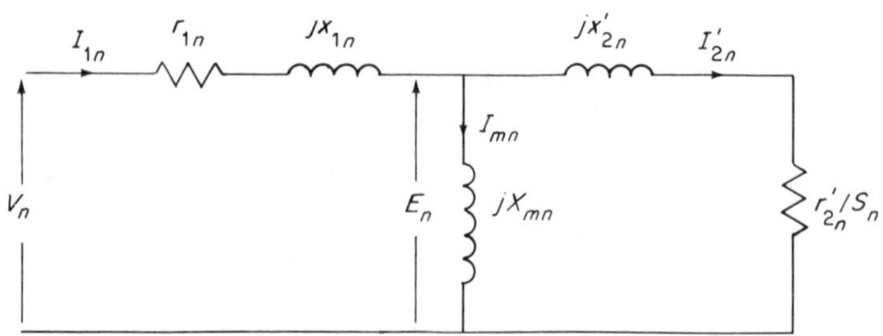

Figure 5.5. Equivalent circuit of induction machine per phase for harmonic n

motoring action (i.e. positive harmonic slips s_n). This motoring action gives rise to shaft torques in the same direction as the harmonic field velocities so that all positive sequence harmonics will develop shaft torques aiding shaft rotation whereas negative sequence harmonics will have the opposite effect.

For a harmonic current I_n, the torque per phase is given by $I_n^2(r'_{2n}/s_n)$ watts at harmonic velocity. Referred to fundamental velocity this becomes

$$T_n = (I_n^2/n)(r'_{2n}/s_n) \text{ synchronous watts,} \tag{5.3.1}$$

the sign of n giving the torque direction.

Since s_n is approximately 1.0 the above expression can be written as

$$T_n = (I_n^2/n)r'_{2n} \text{ per unit} \tag{5.3.2}$$

if I_n and r'_{2n} are per unit.

Using the relationship $V_n = I_n Z_n$ and $Z_n \sim nX_1$, the torque can be expressed in terms of the harmonic voltages, i.e.

$$T_n = (V_n^2/n^3)(r'_{2n}/X_1^2). \tag{5.3.3}$$

Because the slip to harmonic frequencies is almost unity, the torques produced by practical per unit values of harmonic currents is very small, and moreover the small torques occur in pairs which tend to cancel. This effect is illustrated in Figure 5.6. Therefore the effects of harmonics upon the mean torque may, in most cases, be neglected.

Although harmonics have little effect upon mean torque, they can produce significant torque pulsations.

Williamson[5] has developed the following approximate expression for the magnitudes of torque pulsations based on nominal voltage:

$$T_{3k} = [I_{n+}^2 + I_{n-}^2 - 2I_{n+}I_{n-}\cos(\phi_{n+} - \phi_{n-})]^{1/2} \text{ per unit,}$$

where I_{n+} and I_{n-} are per unit values, $n+$ represents the $1 + 3k$ harmonic orders

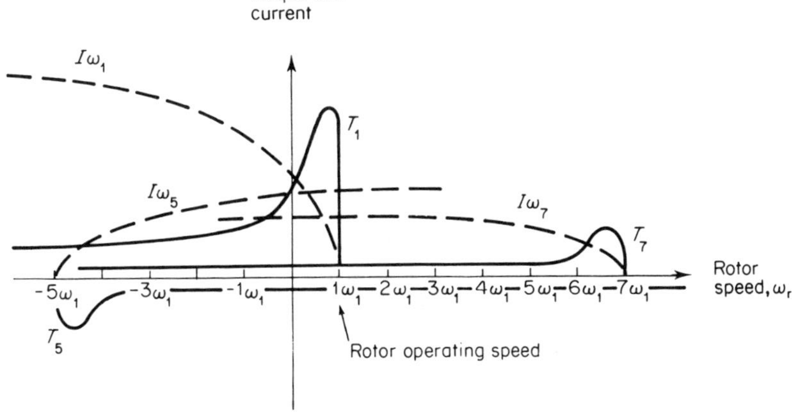

Figure 5.6. Rotor harmonic torques and currents

and $n-$ represents the $1-3k$ harmonic orders. This expression permits preliminary assessment of possible shaft torsional vibration problems.

As an example let us take the case of a supply voltage with total harmonic distortion of about 4%, resulting in machine currents of 0.03 and 0.02 per unit for the fifth and seventh harmonics respectively. If both harmonics have the same phase angle, then for a 50 Hz machine on full voltage the torque will have a varying component at 300 Hz with an amplitude of 0.01 per unit. If the harmonics have the most adverse phase relationship, the amplitude will be 0.05 per unit.

5.4 EFFECT OF HARMONICS ON STATIC POWER PLANT

Transmission system

The flow of harmonic currents in a network produces two main effects.

One is the additional transmission loss caused by the increased r.m.s. value of the current waveform, i.e.

$$\sum_{n=2}^{\infty} I_n^2 R_n,$$

where I_n is the nth harmonic current and R_n the system resistance at that harmonic frequency.

The second effect of the harmonic current flow is the creation of harmonic voltage drops across the various circuit impedances.

This means in effect that a 'weak' system (with a large amount of impedance and thus low fault level) will result in greater voltage disturbances than a 'stiff' system, with a high fault level and low impedances.

In the case of transmission by cable, harmonic voltages increase the dielectric stress in proportion to their crest voltages. This effect shortens the useful life of the cable. It also increases the number of faults and therefore the cost of repairs.

The effects of harmonics on Corona starting and extinction levels are a function of peak to peak voltage. The peak voltage depends on the phase relationship between the harmonics and the fundamental. It is thus possible for the peak voltage to be above the rating while the r.m.s. voltage is well within this limit.

Transformers

The presence of harmonic voltages increases the hysteresis and eddy current losses and stresses the insulation. The flow of harmonic currents increases the copper losses; this effect is more important in the case of convertor transformers because they do not benefit from the presence of filters, which are normally connected on the a.c. system side. Apart from the extra rating required, convertor transformers often develop unexpected hot spots in the tank.

An important effect particularly relevant to power transformers is the circulation of triplen zero sequence currents in the delta windings. The extra circulating currents can overrate the windings unless these are taken into account in the design.

Another important consideration exists for a transformer supplying an asymmetrical load. If the load current contains a d.c. component the resulting saturation of the transformer magnetic circuit (described in Section 4.2) greatly increases the level of all harmonic components of the a.c. excitation current.

Capacitor banks

The presence of voltage distortion produces an extra power loss in capacitors expressed by

$$\sum_{n=1}^{\infty} C(\tan\delta)\omega_n V_n^2,$$

where $\tan\delta = R/(1/\omega C)$ is the loss factor, $\omega_n = 2\pi f_n$ and V_n is the r.m.s. voltage of the nth harmonic.

Moreover, the total reactive power, including fundamental and harmonics, i.e.

$$Q = \sum_{n=1}^{\infty} Q_n,$$

should not exceed the rated reactive power.

Series and parallel resonances (Section 5.2) between the capacitors and the rest of the system can cause overvoltages and high currents thus increasing dramatically the losses and overheating of capacitors and often lead to their destruction.

5.5 HARMONIC INTERFERENCE WITH RIPPLE CONTROL SYSTEMS

Ripple control is often used for the remote control of street lighting circuits and for load reduction (such as domestic hot water heaters) during peak times of the day.

Supply authorities have experienced some practical difficulties with their ripple control equipment as a result of harmonic interference.

Since ripple relays are essentially voltage operated (high impedance) devices, harmonic interference can cause signal blocking or relay maloperation if present in sufficient amplitude. The exact amplitude at which the voltage harmonic will affect the relay is a function of the relay detection circuit (sensitivity and selectivity) and the proximity of the ripple injection frequency to the frequency of the interfering harmonic.

Signal blocking is the presence of sufficient interfering voltage to render the relay unable to detect the presence of the signal. Capacitors can produce the same effect by virtue of their capability to absorb the ripple signal. Relay maloperation is the presence of the harmonic voltage (usually in the absence of the signal) causing the relay to change state. The latter problem has effectively been solved by the use of suitably encoded switching signals in present generation of ripple relays.

Earlier ripple relays were electromechanical devices employing mechanical filter assemblies. Although their response was slow, they did achieve very good selectivity. However, these relays commonly suffered from maloperation, because they had inadequate signal encoding to cater for any harmonic interference which got past the filters.

Present generation ripple relays are basically electronic equivalents of their electromechanical forerunners. They commonly employ piezoelectric or active filter circuits, and a high degree of signal encoding to minimize maloperation. They have filter response curves defined by international standards,[6] and they require the use of precision electronic components (high stability and reliability) to ensure a satisfactory performance over the life of the relay.

In the future ripple relays will almost certainly employ digital filtering techniques which are ideally suited for the detection of signals of a specified frequency in the presence of harmonic voltages.

5.6 HARMONIC INTERFERENCE WITH POWER SYSTEM PROTECTION

Harmonics can distort or degrade the operating characteristics of protective relays depending on the design features and principles of operation. Digital relays and algorithms that rely on sampled data or zero crossings are particularly prone to error when harmonic distortion is present.

In most cases, the changes in operating characteristics are small and do not present a problem. Tests have been carried out[7] which indicate that for most types of relays operation is not affected significantly for harmonic voltage levels less than 20%. However, with the increase of large power convertors it is likely that the situation will change in the future.

The problems which arise are different during normal and fault operating conditions and these are considered separately.

Harmonic problems during fault conditions

Protective functions are usually developed in terms of fundamental voltages and/or currents and any harmonics present in the fault waveforms are either filtered out or ignored altogether. The latter is particularly true for electromagnetic relay applications, especially overcurrent protection. Electromechanical relays have significant inertia associated with them such that often they are inherently less sensitive to higher harmonics.

More important is the effect of harmonic frequencies on impedance measurement. Distance relay settings are based on fundamental impedances of transmission lines and the presence of harmonic current (particularly third harmonic) in a fault situation could cause considerable measurement errors relative to the fundamental based settings.

High harmonic content is common where fault current flow through high resistivity ground – i.e. ground impedance is dominant – so that the possibility of maloperation is great unless only the fundamental waveforms are captured.

In solid fault situations, the fundamental components of current and voltage are much more dominant (notwithstanding the d.c. asymmetry associated with fault waveforms). However, because of current transformer saturation, secondary induced distortion of current waveforms, particularly with large d.c. offsets in the primary waveforms, occurs. The presence of secondary harmonics in such instances can be a real problem, i.e. whenever current transformer saturation occurs it is very difficult to recover the fundamental current waveform.

Whenever high secondary e.m.f. exists during steady-state conditions, the non-linear current transformer exciting impedance only causes odd-harmonic distortion. During saturation under transient conditions, however, any harmonics can be produced with dominance of second and third harmonic components.[8]

Fortunately, these are often design problems. Correct choice of equipment in relation to the system requirements can eliminate many of the difficulties associated with instrument transformers.

Filtering of current and voltage waveforms, particularly in digital protection systems, is of special importance to distance protection schemes. Several papers have been written which contrast some of the digital filtering techniques available and it is clear that, although implementation is not always simple, recovery of fundamental data is not difficult using digital techniques.[9]

Harmonic problems outside fault conditions

The effective insensitivity of protective apparatus to normal system load conditions implies that, generally, the harmonic content of power system waveforms is not a problem during non-fault conditions.

The most notable exception is probably the problem encountered in energization of power transformers. In practice, constructive use of the high harmonic content of magnetizing inrush current prevents (most of the time!) tripping of the high voltage circuit breaker by the transformer protection due to the excessively high peaks experienced during energization.

The actual peak magnitude of inrush current depends on the air-core inductance of the transformer and the winding resistance plus the point on the voltage wave at which switching occurs.[10] Residual flux in the core prior to switching also increases the problem or alleviates it slightly depending on the polarity of flux with regard to the initial instantaneous voltage.

Since the secondary current is zero during energization, the heavy inrush current would inevitably cause the differential protection to operate unless it is rendered inoperative.

The simple approach is to use a time-delayed differential scheme, but this could result in serious damage to the transformer should a fault be present at energization.

In practice, the uncharacteristic second harmonic component present during inrush is used to restrain the protection, but protection is still active should an internal fault develop during energization.

5.7 EFFECT OF HARMONICS ON CONSUMER EQUIPMENT

(i) *Television receivers* Harmonics which affect the peak voltage can cause changes in TV picture size and brightness.

(ii) *Fluorescent and mercury arc lighting* Ballasts sometimes have capacitors which, with the inductance of the ballast and circuit, have a resonant frequency. If this corresponds to a generated harmonic, excessive heating and failure may result.

(iii) *Computers*[11] There are designer-imposed limits as to acceptable harmonic distortion in computer and data processing system supply circuits. Harmonic rate (geometric) measured in vacuum must be less than -3% (Honeywell, DEC) or 5% (IBM). CDC specifies that the ratio of peak to effective value of the supply voltage must equal 1.41 ± 0.1.

(iv) *Convertor equipment* Notches in the voltage wave resulting from current commutation may affect the synchronizing of other convertor equipment or any other apparatus controlled by voltage zeros.

(v) Harmonics could theoretically affect thyristor-controlled variable speed equipment of the same consumer in several ways: (a) voltage notching (causing brief voltage dips in the supply) cause maloperation through misfiring thyristors; (b) harmonic voltages can cause the firing of the gating circuits at other than the required instant; (c) resonance effects between various equipments can result in over-voltages and hunting.

(iv) The above described problems could also be experienced by other consumers if connected to the same (415 V or 11 kV) busbar. If a consumer has no problems with the simultaneous operation of his own thyristor-controlled equipment, then he is unlikely to interfere with other consumers. Consumers on different busbars could interfere with each other but 'electrical' remoteness (separation by impedances in the form of lines and transformers) will tend to reduce the problem.

5.8 EFFECT OF HARMONIC ON POWER MEASUREMENTS

Measuring instruments initially calibrated on pure sinusoidal alternating current and subsequently used on a distorted electricity supply can be prone to error.

The magnitude and direction of the harmonic power flow are important for revenue considerations as the sign of the meter error is decided by the direction of flow.

Studies have shown that errors due to harmonic content vary greatly as to the type of meter and that both positive and negative metering errors are possible.

The main energy measuring instrument is the Ferraris motor type kilowatt-hour meter. Its inherent design is electromagnetic, producing driving and braking fluxes which impinge on its rotor developing a torque. Secondary flux-producing elements are provided for compensation purposes to improve the accuracy and to compensate for friction in the register. These flux-producing elements, providing primary and secondary torques, are essentially non-linear in regard to amplitude and frequency. The non-linear elements include the voltage and current elements and overload magnetic shunts and the frequency sensitive elements include the disc, the quadrature and anti-friction loops.

The response of this meter is inefficient to frequencies outside the design parameter and large inaccuracies result. An expression for total power as seen by a meter is

$$\underset{(P_T)}{\text{Total power}} = \underset{(P_{dc})}{V_{dc}I_{dc}} + \underset{(P_F)}{V_F I_F \cos\phi_F} + \underset{(P_H)}{V_H I_H \cos\phi_H}. \qquad (5.8.1)$$

The meter will not measure P_{dc} but will be sensitive to its presence, it will measure P_F accurately and P_H inaccurately, the error being determined by the frequency. The total harmonic power P_H is obtained by adding all components derived from the products of voltage and currents of similar frequencies, both above and below the fundamental frequency.

Any d.c. power supplied to or generated by the customer will cause an error proportional to the power ratio P_{dc}/P_T, with the error sign related to the direction of power flow. Similarly any deficiency in measuring harmonic power P_H will cause an error represented by $\pm KP_H/P_T$, where the factor K is dependent on the frequency response characteristics of the meter, and the error sign again will be related to power flow direction.

D.C. power and harmonic voltages or currents alone should not produce torques, but will degrade the capability of a meter to measure fundamental frequency power. Direct currents distort the working fluxes and alter the incremental permeability of the magnetic elements. Fluxes produced by harmonic currents combine with spurious fluxes of the same frequency that may be present due to the imperfection in the meter element and produce secondary torques.

The normal kilowatt-hour meter, based on the Ferraris (eddy current) motor principle, has been found generally to read high to the extent of up to several per cent with a consumer generating harmonics through thyristor-controlled variable speed equipment (particularly if even harmonics and d.c. are involved) and notably if there is also a low power factor.

Convertor loads using the 'burst firing' principle can cause kilowatt-hour meters to read high by several per cent (cases in excess of 6% have been quoted) largely attributable to the lack of current damping during the no-load interval.

It appears that the consumer who is generating network harmonics is automatically penalised by a higher apparent electricity consumption, which might well offset the supply authority's additional losses.[12]

It is therefore in the consumer's own interest to reduce harmonics generation to the greatest possible extent.

Effect of harmonics on maximum demand meters

There is no evidence that the readings of 'kVA-demand' meters are affected by network harmonics. However, 'kW-demand' meters operating on the 'time interval Ferraris motor' principle will read possibly a few per cent high, as shown earlier for energy meters.

Solid state meters are available, at a higher cost, which can measure true power, irrespective of wave shape.

Harmonics present a problem to measurement of VAR values since this is a quantity defined with respect to a sinusoidal waveform.

A seldom mentioned effect, but nevertheless an important one, is that measurement and calibration laboratories, working to small tolerances of accuracy, may lose confidence in their results.

5.9 EFFECT OF HARMONIC DISTORTION ON POWER FACTOR[13]

In general, the instantaneous values of the voltage and current components can be expressed as

$$v = \sum_{1}^{n} \sqrt{2} V_n \sin(n\omega t + \alpha_n) + \sum^{m} \sqrt{2} V_m \sin(m\omega t + \alpha_m), \tag{5.9.1}$$

$$i = \sum_{1}^{n} \sqrt{2} I_n \sin(n\omega t + \alpha_n + \phi_n) + \sum^{p} \sqrt{2} I_p \sin(p\omega t + \alpha_p), \tag{5.9.2}$$

and the power factor is given by

$$\text{p.f.} = \frac{1/T \int_0^T vi\,dt}{V_{\text{rms}} I_{\text{rms}}} = \frac{\sum_1^n V_n I_n \cos\phi_n}{\left\{\left(\sum_1^n V_n^2 + \sum^m V_m^2\right)\left(\sum_1^n I_n^2 + \sum^p I_p^2\right)\right\}^{1/2}}. \tag{5.9.3}$$

This factor represents a figure of merit of the character of power consumption. Its low value indicates poor utilization of the source-power capacity needed by the load.

If the voltage waveform is sinusoidal equation (5.9.3) reduces to

$$\text{p.f.} = \frac{V_1 I_1 \cos\phi_1}{V_1 I_{\text{rms}}} = \frac{I_1}{I_{\text{rms}}} \cdot \cos\phi_1 = \mu \cos\phi_1, \tag{5.9.4}$$

where $\cos\phi_1$ is the displacement factor between the fundamental components of voltage and current and μ is a current distortion factor.

Unity p.f. can only be achieved when $\mu = 1$ since $\cos\phi_1$ (in equation (5.9.4)) cannot be greater than 1.

Power factor compensation is not straightforward with distorted waveforms.

As lossless devices are generally used for power factor compensation the minimization of the apparent power should lead directly to the optimum power factor. If, for example, a capacitance C is added in parallel to the load characterized by equations (5.9.1) and (5.9.2), the general expression for the apparent power in terms of C is

$$S=\left(\sum_1^n V_n^2+\sum^m V_m^2\right)^{1/2}$$

$$\cdot\left\{\sum_1^n (I_n^2+V_n^2 n^2\omega^2C^2+2V_nI_n n\omega C\sin\phi_n)+\sum^m V_m^2 m^2\omega^2C^2+\sum^p I_p^2\right\}^{1/2}. \tag{5.9.5}$$

The differentiation of this equation with respect to C and equating to zero leads to an optimum linear capacitance:

$$C_{\text{opt}}=-\frac{1/\omega\sum_1^n V_n n I_n\sin\phi_n}{\sum_1^n V_n^2n^2+\sum^m V_m^2m^2}. \tag{5.9.6}$$

An international committee for p.f. improvement[(14)] defined a component of apparent power called *reactive power* as

$$Q=\sum_1^n V_nI_n\sin\phi_n=\frac{1}{2\pi}\sum_1^n\frac{1}{n}\int_0^T V_n\mathrm{d}i_n. \tag{5.9.7}$$

Although no physical meaning for this expression was agreed, its adoption was due mainly to two considerations: (i) the property of conservation of this expression in linear sinusoidal systems; (ii) the possibility of expressing the apparent power in a general non-linear system by the relation

$$S=(P^2+Q^2+D^2)^{1/2}, \tag{5.9.8}$$

where p is the average power and D is an additional component designated as the distortion power.

The object of capacitor compensation is to improve the displacement factor if the voltage is sinusoidal. Improvement in the values of distortion factor can be done by filters, by higher pulse numbers, or by current waveform modification. These techniques are discussed in Chapter 10.

B Interference with communications

5.10 INTRODUCTION

Noise on communication circuits degrades the transmission quality and can interfere with signalling. At low levels noise causes annoyance; at higher levels the transmission quality is degraded and results in loss of information; in extreme cases noise can render a communication circuit unusable.

Continuous technological improvements in power and communication systems demands regular reconsideration of the interference problems in telephone lines located in the vicinity of power systems.

The signal to noise ratios, commonly used in communication circuits as a measure of the quality of transmission, must be used with caution when considering power system interference because of the relative power levels of power (in megawatts) and communication circuits (in milliwatts).

Due to the large difference in power levels, small unbalanced audiofrequency components within the power network may easily produce considerable noise voltage levels when coupled into a metallic communication circuit.

Moreover, the purpose of a power system is to transmit energy at high efficiency but with relatively low waveform purity; on the other hand, in a communication circuit the waveform must not be significantly distorted as the intelligence being conveyed may be destroyed, while the power efficiency is of secondary importance.

This chapter describes the means by which noise voltages may be induced into the telephone network and cause degradation of service. The main factors which influence the level of interference are discussed.

5.11 SIMPLE MODEL OF A TELEPHONE CIRCUIT

A physical telephone circuit consists of a twisted pair of wires with associated terminal equipment and a simplified model of such a circuit is illustrated in Figure 5.7.

For safety and practical reasons telephone circuits are referenced to earth and the equivalent circuit of the telephone system of Figure 5.7 includes the terminal impedances Z_{L1} and Z_{L2} to earth. An electromagnetic induced voltage is modelled as voltage source V_m, and an electrostatic induced voltage as V_s. The terminal impedances, Z_{L1} and Z_{L2}, are generally of high value and the telephone line self-impedance, being much smaller, may be neglected.

In the absence of an earth return conductor the earth return circuit is completed by the stray capacitances C_{S1} and C_{S2}.

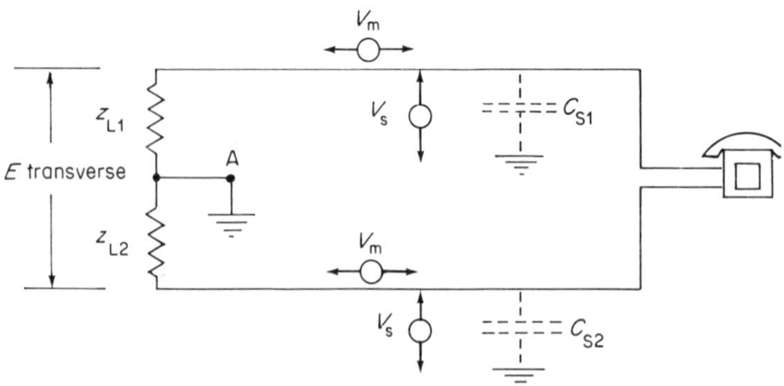

Figure 5.7. Simple model of a telephone circuit

5.12 FACTORS INFLUENCING INTERFERENCE

Three factors combine to produce a noise problem on a communication line, these are as follows:

(i) *Influence of power system* This depends on the source of audiofrequency components within the power system and the relative magnitude of unbalanced harmonic currents and voltages present in the power circuit in the vicinity of the communication circuit. This subject is discussed in other chapters.
(ii) *Coupling to communication circuits* This is coupling of interfering currents and voltages into a communication system.
(iii) *Effect on communication circuits* (*susceptiveness*) The effect of the noise interference on a communication circuit is dependent on the characteristics of the circuit and associated apparatus.

All three basic factors must be present in order for a noise problem to exist. The absence of any one of these factors eliminates the problem completely. A complete elimination is usually impractical and the degree of the problem will be a function primarily of the basic factors that have the highest influence.

5.13 COUPLING TO COMMUNICATION CIRCUITS

Noise voltages may be impressed on telephone circuits in several ways, i.e. by loop induction, by longitudinal electromagnetic induction, by longitudinal electrostatic induction, by conduction.

Loop induction

Loop induction occurs when a voltage is induced directly into the metallic loop formed by the two wires of a telephone circuit. This type of induction manifests itself directly as a transverse voltage across the terminations of the telephone circuit. It is cancelled out by regular transpositions of aerial wires or use of twisted pairs of cables. As these are standard practices in communication circuits, loop induction is not generally a problem.

In the case of crossings, telephone lines often come so near the power line that the loop effect may be important.[15]

Longitudinal electromagnetic induction

Longitudinal electromagnetic induction occurs when an e.m.f. is induced along the conductors of a telephone circuit. The residual current in a power line sets up a magnetic field which causes flux lines to intersect with any neighbouring telephone line and induces an e.m.f. longitudinally on it. This type of coupling, illustrated in Figure 5.8, constitutes the most common form of noise induction into communication lines.

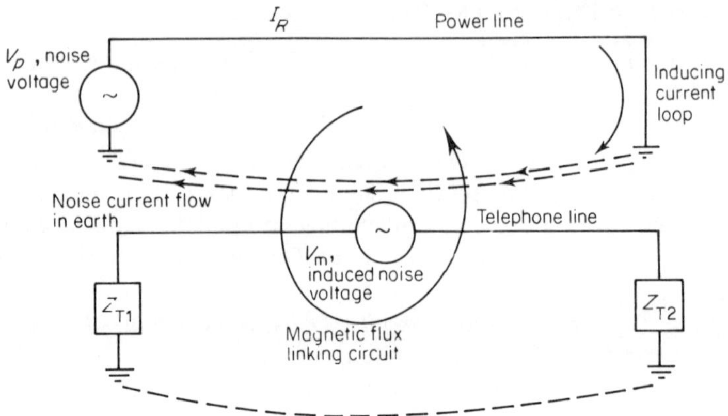

Figure 5.8. Electromagnetic induction

The residual current I_R in Figure 5.8 returns via earth to V_p, hence the current loop so formed has a large cross-sectional area for overhead transmission lines. Likewise, aerial openwire telephone circuits may have large cross-sectional areas. This leads to a longitudinal electromagnetic induction on telephone circuits given by

$$V_m = MI_R, \tag{5.13.1}$$

where M is the mutual impedance between the power and telephone systems.

The most widely accepted model for determining the mutual impedance between power and telephone systems was developed by Carson[16].

In general, electromagnetic propagation at power frequencies is not very sensitive to earth structure and resistivity and so Carson's model (which assumes a uniform flat earth) is valid for most cases.

Carson's equation for the mutual impedance of a power line at height h_1 and a telephone line at height h_2, as in Figure 5.9, is given by

$$M = \frac{j\omega\mu_0}{2\pi}\left[\ln\frac{d'_{12}}{d_{12}} - 2j\int_0^\infty [\sqrt{(u^2+j)} - u]e^{-u\alpha(h_1+h_2)}\cos(u\alpha x)\,du\right], \tag{5.13.2}$$

where M is the mutual impedance per unit length, x is the horizontal separation of power and telephone lines, h_1 is the height of the power line above ground (negative if below ground), h_2 is the height of the telephone line above ground (negative if below ground), $d_{12} = \sqrt{[(h_1 - h_2)^2 + x^2]}$ is the radial distance between lines, $d'_{12} = \sqrt{[(h_1 + h_2)^2 + x^2]}$ is the radial distance between one line and the underground image of the other, $\omega = 2\pi f$ is the angular frequency of the inducing current, μ_0 is the rationalized permeability of free space, $\alpha = \sqrt{2}/\delta$, $\delta = \sqrt{(2\rho/\mu_0\omega)}$ is the skin depth of uniform earth with resistivity ρ, and ρ is the resistivity of the earth in ohm-metres.

The first term of Carson's equation is the mutual impedance between two conductors as if they were above a perfectly conducting earth; while the second

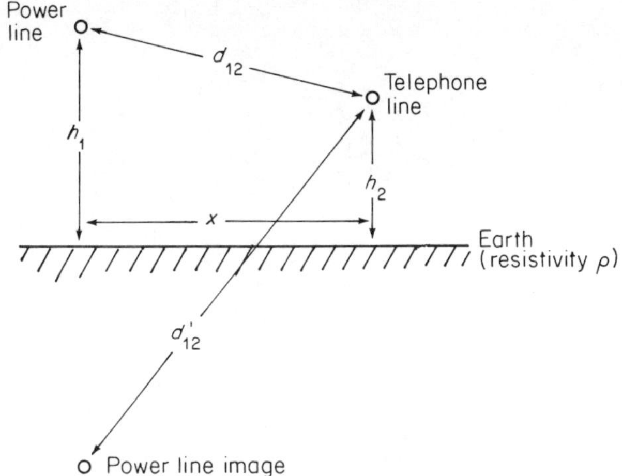

Figure 5.9. Power and telephone line configuration

term 'corrects' for finite resistivity of the earth. Unfortunately the second term usually dominates for typical parameters.

Carson's formula depends on measurements of the earth resistivity which can be difficult to obtain and simplified forms of Carson's equation are often used to obtain good approximations for the mutual impedance.

The factors influencing the mutual impedance between power and telecommunications lines can be summarized as follows: increases for increasing residual current loop area; increases for increasing common distance run; increases for increasing frequency; increases for increasing earth resistivity; decreases for increasing separation between circuits.

Longitudinal electrostatic induction

Longitudinal electrostatic induction occurs when an e.m.f. is induced between the conductors and earth.

The simplest way of visualizing electrostatic induction is by considering the capacitances in an exposure between a single power wire and a single telephone wire as illustrated in Figure 5.10. The voltage of the power wire to ground (residual voltage), V_r, divides over the capacitance between the power and telephone wire, C_{PT}, and the telephone wire and ground, C_{TC}, in the ratio of their impedances, i.e.

$$V_s = \frac{Z_T}{1/(j\omega C_{PT}) + Z_T} \cdot V_r, \tag{5.13.3}$$

where

$$Z_T = \frac{1}{j\omega C_{TG} + (1/Z_{T1}) + (1/Z_{T2})}.$$

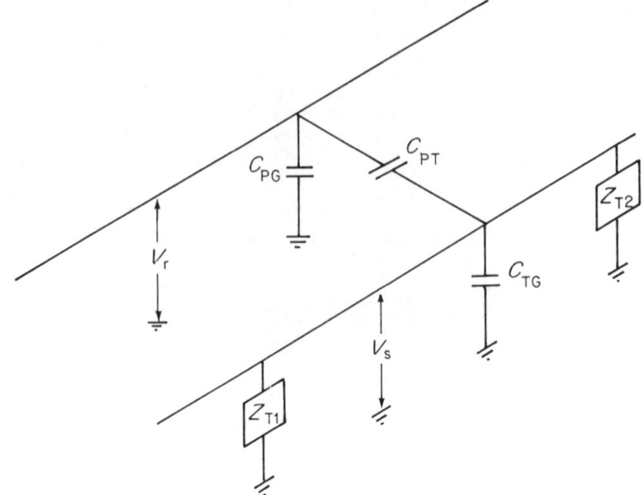

Figure 5.10. Electrostatic induction

Because of the loading effect of Z_{T1} and Z_{T2} and the relative separations, V_s is very small as compared with V_m and can be easily neutralized by cable screening.

Electrostatic induction is serious only when the residual voltage, V_r, is large (e.g. single wire power lines) or when C_{PT} is large (for example in joints construction and crossings where the two lines are very close together). Generally the telephone line is terminated in impedances which are small compared to the capacitive impedances and thus reduce the induced voltage, V_s. This form of induction is often a problem on long telephone lines in the neighbourhood of very high voltage transmission lines.

Conductive coupling

When a power system is in operation there is always some level of residual current flowing in the neutral due to out-of-balance components. In the multiple earthed neutral (MEN) system, some of this residual current will return to the transformer by the neutral wire, and some via earth. The earth currents will cause a local rise of earth potential at the earth electrode.

If one end of a telephone line is earth referenced in the area of influence of this earth potential rise, then a longitudinal voltage may be impressed on the line.

In the circuit of Figure 5.7, if a local earth potential rise occurred at A this would cause unequal currents to flow through Z_{T1} and Z_{T2} giving rise to a transverse noise voltage in the telephone circuit.

This source of interference is an increasing problem due to the following factors:

(i) MEN earth systems are carrying higher levels of noise currents.
(ii) The earth resistances of telephone exchange earth systems are increasing

as a result of the use of less lead armoured cable; it is very costly to achieve low earth resistances.

(iii) Despite (ii) the telephone exchange earth is often a relatively low impedance earth return circuit for a MEN earth system feeding the exchange and hence a considerable noise voltage can be impressed onto earth system from a MEN system.

5.14 EFFECT ON COMMUNICATION CIRCUITS (SUSCEPTIVENESS)

Telephone circuit susceptiveness

The effect that a 'noisy' power line will have on a communication line may be ascertained by considering the susceptiveness of the circuit to the effects of inductive interference. Three characteristics are of importance: (i) the relative interfering effects of different frequencies; (ii) the balance of the communication circuit; (iii) shielding effects of metallic cable sheaths and other buried metallic plant.

Harmonic weights

The effect of harmonic interference is not uniform over the audiofrequency spectrum. A 'standard' human ear in combination with a telephone set has a sensitivity to audiofrequencies that peaks at about 1 kHz.

To get a reasonable indication of the interference from each harmonic, various weighting systems are used to take into account the response of the telephone equipment and the sensitivity of the human ear. Two systems are in wide use: (i) psophometric weighting by the International Consultation Commission on Telephone and Telegraph Systems (CCITT)[17] – used in Europe; (ii) C-message weighting by Bell Telephone Systems (BTS) and Edison Electric Institute (EEI)[18] – used in the United States and Canada.

In both systems the weightings have been updated as telephone systems have evolved.

Psophometric frequency weighting

The psophometric weighting provides a means for evaluating the interfering effect that a power system will have on a communication circuit.

The sensitivity of the human ear in combination with a telephone set to audiofrequencies, determined by the CCITT, is shown in the psophometric weighting curve of Figure 5.11.

In practice the severity of the problem is assessed with the help of a psophometer, which is an r.m.s. voltmeter with a filter having the frequency characteristic of the psophometric telephone weighting curve.

The level of interference is described in terms of a telephone form factor (TFF),

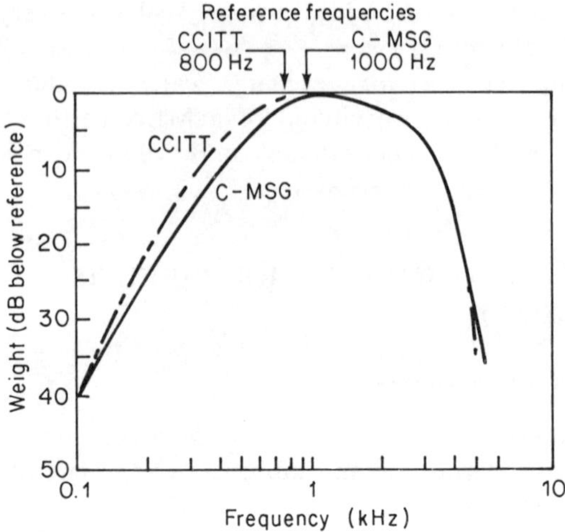

Figure 5.11. Comparison of BTS C-message and CCITT weights

which is a dimensionless value that ignores the geometrical configuration of the coupling. It is expressed as

$$\text{TFF} = \frac{\text{Equivalent 800 Hz disturbing voltage}}{\text{Service voltage of the power line}}. \tag{5.14.1}$$

The service voltage of a power line is the nominal voltage by which it is designated. The equivalent disturbing voltage is a voltage at 800 Hz which, if applied to the power line, would cause the same interfering effect to be experienced in a nearby telephone line as does the voltage on the power line and its harmonics.

Mathematically the TFF can be expressed as follows:

$$\text{TFF} = \frac{1}{U}\left[\sum_{n=1}^{\infty} (K_f p_f U_f)^2\right]^{1/2}, \tag{5.14.2}$$

where U is the r.m.s. voltage of the transmission line, U_f is the harmonic voltage of frequency $f, K_f = f/800$ is a coupling factor, and p_f is the psophometric weight divided by 1000.

The r.m.s. psophometric voltage is

$$V_\psi = \sqrt{[(p_f V_f)^2]}, \tag{5.14.3}$$

where V_f is the r.m.s. longitudinal or transverse voltage of frequency f on the telephone lines.

The CCITT directives recommend that the total psophometric weighted noise on a telephone circuit has an e.m.f. (i.e. open circuit voltage) of less than 1 mV. When measuring the noise voltage the telephone circuit is terminated with its characteristic impedance (which is a resistance of the order of 600 Ω) and the noise voltage is measured across such resistance. Therefore, the psophometrically weighted noise across the terminating resistor must be less than 0.5 mV.

As a general rule, if the TFF is greater than 0.5 mV it is likely to cause interference to telephone services. It must be stressed that the TFF is only a guideline measurement; it is not satisfactory as the sole measure of interference to a communication line as it takes no account of coupling and exposure factors.

C-message weighting

The C-message weighting system uses the telephone influence factor (TIF) instead of the TFF described in the previous section. Again TIF is a dimensionless value used to describe the interference of a power transmission line on a telephone line, and is expressed as

$$\mathrm{TIF} = \frac{1}{U}\left[\sum_{n=1}^{\infty} (K_f p_f U_f)^2\right]^{1/2}, \tag{5.14.4}$$

where U is the r.m.s. voltage of the transmission line, U_f is the harmonic voltage of frequency f, $K_f = 5000(f/1000) = 5f$ is the coupling coefficient, and p_f is the weight of the harmonic of frequency f, the maximum being 1 for $f = 1000$ Hz.

The psophometric and C-message weighting systems are only slightly different, as illustrated in Figure 5.11, their approximate ratio being

$$\frac{\mathrm{TIF}}{\mathrm{TFF}} = 4000 \tag{5.14.5}$$

(i.e. a TIF of 80 per unit corresponds to a TFF of 2%).

$I \cdot T$ and k$V \cdot T$ products

These are weighted currents or voltages in the power system. The voltages or currents of the power transmission line are represented by a single voltage or current obtained by weighting each harmonic voltage or current with the corresponding factor of the system (BTS–EEI or CCITT) used.

In the EEI–BTS system

$$\begin{aligned} I \cdot T &= \left[\sum_{n=1}^{\infty} (K_f p_f I_f)^2\right]^{1/2} \\ &= \left[\sum_{n=1}^{\infty} (T_f I_f)^2\right]^{1/2} \\ &= I \cdot (\mathrm{TIF}), \end{aligned} \tag{5.14.6}$$

where I_f is the r.m.s. current of frequency f and T_f is the corresponding single frequency TIF.

The k$V \cdot T$ product is derived similarly using voltages instead of currents.

Analogous quantities in the BTS–EEI system and the CCITT system are shown in Table 5.1.[19]

Table 5.1. Corresponding quantities in BTS–EEI and CCITT systems

BTS–EEI	CCITT
C-message weighting	Psophometric weighting
Telephone influence factor (TIF)	Telephone harmonic form factor (TFF)
C-message-weighted voltage:	
Longitudinal	Psophometrically weighted voltage
Transverse	Psophometric voltage
$I \cdot T$ product	Equivalent disturbing current
k$V \cdot T$ product	Equivalent disturbing voltage

Telephone circuit balance to earth

If the telephone line or terminal equivalent is not perfectly balanced with respect to earth, longitudinally induced voltage on that line can be transformed into transverse voltage and it is as a transverse voltage across the earpiece that we can hear noise on the telephone.

Consider the simple telephone circuit of Figure 5.7. If the impedances to earth of the two wires that form a telephone circuit are different, and assuming that the two wires are subjected to the same longitudinal induction, different currents will flow in each wire. Because the self-impedances of the wires will be very similar, the different current flows will give rise to a transverse voltage between the pair.

Factors which may affect the telephone circuit balance to earth are (i) any leakage paths to ground, e.g. across pole insulators or through cable insulation; (ii) any unbalanced terminal equipment, either subscriber's equipment (e.g. extension bells) or exchange equipment. Some of these factors may be corrected simply and quickly to reduce an induced noise problem, others rely on the proper design and manufacture of the equipment or plant.

Balance is simply defined as

$$20 \log_{10}\left(\frac{\text{longitudinal voltage}}{\text{transverse voltage}}\right)$$

and for the majority of relay sets it is of the order of 45–50 dB.

The most critical component balance-wise is generally any terminal relay set in a communication connection. As there are large numbers of such relay sets it is not practical to set balance objectives at a very high level as this would significantly increase the cost of the communication network.

Shielding

A metallic or conducting earthed screen such as a cable sheath which enclosed the telephone circuit over the length of an exposure is totally effective in eliminating electrostatic induction. Buried cables without a metallic screen or sheath are also immune to electrostatic induction due to conducting effect of the earth.

Metallic screens or sheaths are only partly effective in reducing the effects of electromagnetic induction.

The mechanism of electromagnetic shielding is as follows: The power line current causes longitudinal voltages to be induced in the wires of the cable and also in the shield. The resulting current flow in the shield is in thc opposite direction to the inducing current in the power line. This induced current in the shield generates in turn a voltage in the wires of the cable opposing the voltage induced by the power line current, thus tending to neutralize the latter.

If the shield current is large and/or well coupled to the inner conductors, reasonable shielding factors can be obtained.

The 'shielding factor' of cable is the fraction by which the shield reduces the voltage induced into the core; the shielding factor (K) of an installed shielded cable system is given by

$$K = \frac{\text{d.c. shield resistance} + \text{earth resistance}}{\text{a.c. shield resistance} + \text{earth resistance}}. \tag{5.14.7}$$

This simplistic formula shows that the shielding factor may be reduced by decreasing the resistance and/or increasing the inductance of the sheath and/or decreasing the earth resistance.

A good shielding factor for a cable can be achieved if the following set of conditions hold:

(i) At points where the shield is earthed, the earth resistance must be low (typically some 1–2 Ω).
(ii) There is a low resistance shield, i.e. plenty of metal in cable shield to keep resistance low.
(iii) Steel tapes are usually required.

All these factors mean that shielded cables for noise shielding are relatively expensive; typically they can be expected to cost some 30–50% more than standard unshielded types.

5.15 MITIGATION TECHNIQUES

The steps to be taken when a noise problem is known to exist are as follows:

(i) Check transverse noise voltage on telephone circuit. If the e.m.f. is less than 1 mV no further steps need be taken; if greater than 1 mV, then proceed as below.
(ii) Determine the mode of induction – probably electrostatic or electromagnetic induction.
(iii) Test telephone line and termination balance to earth.
(iv) Where considered useful, test telephone form factor in various parts of the power system to try to isolate the source of noise.

With the information gained from these tests, mitigation methods are considered.

Mitigation refers to the ways and means of reducing interference. Referring to

Section 5.4, there are three ways of reducing a noise problem on a communication line which results from the operation of a neighbouring power system, these are: power system influence, coupling, and susceptiveness.

(i) Reducing the influence of the power system can be achieved by (a) physical relocation of either system (usually an expensive exercise); (b) reducing the harmonics in the power system (the techniques are discussed in Chapter 10).

(ii) Reducing coupling is not usually practical except in cases of earth potential rise where a noisy incoming multiple earthed neutral can be excluded if required.

(iii) Reducing the susceptiveness of communications circuits can be achieved by the use of noise chokes, noise-neutralizing transformers, shielded cable and derived circuits.

NOISE CHOKES

Reduction factors upwards of 25 dB are achievable in certain circumstances. Generally noise chokes are only useful for improving the balance of substandard terminal relay sets. They work by increasing the a.c. longitudinal line impedance, thereby reducing noise current and thus reducing the transverse noise voltage level.

NOISE-NEUTRALIZING TRANSFORMERS

These work by inducing an equal but opposite (in-phase) noise voltage in affected cable pairs, thus reducing the induced longitudinal noise. Reduction factors of 15–20 dB can be achieved.

SHIELDED CABLE

Reduction factors upward of 60 dB can easily be achieved, but it is costly. Generally the use of shielded cable is only applicable for new work.

DERIVED CIRCUITS

By providing circuits via a PCM, FDM or equivalent system can be made relatively immune to noise induction. The degree of improvement depends on the circumstances.

It is emphasized that the actual reduction factors depend on the particular circumstances and the factors mentioned above are not achievable in all cases.

5.16 REFERENCES

1. Ross, N. W. (1982). 'Harmonic and ripple control carrier series resonances with P. F. correction capacitors'. *Trans. Electr. Supply Authority* (*N. Z*), **52**, 48–62.
2. Standard specifications for capacitors for connection to power frequency systems: (i) British Standard BS6650 (1971); (ii) Australian Standard AS1013 (1971); (iii) IEC Publication 70 (1967); (iv) ANSI/IEEE Standard 18 (1980).

3. Klinghsirn, E. A., and Jordan, H. E. (1968). 'Polyphase induction motor performance and losses on non-sinusoidal voltage sources'. *IEEE Trans.*, **PAS-87**, 624–631.
4. Chalmers, B. J., and Sarkar, B. R. (1968). 'Induction motor losses due to non-sinusoidal waveforms'. *Proc. IEE*, **115**, 1777–1782.
5. Williamson, A. C. (1981). 'The effects of system harmonics upon machines'. Paper presented at an international conference on Harmonics in Power Systems, UMIST, Manchester.
6. 'Harmonisation document for ripple control receivers', CENELEC Standard TC102: 1978.
7. Jost, F. A., Menzies, D. F., and Sachdev, M. S. (1974). 'Effect of system harmonics on power system relays'. Paper presented at a Power System Committee Meeting, Canadian Electrical Association.
8. *GEC Protection Applications Guide*, GEC, London Chap. 5, pp. 76–78.
9. McClaren, P. G., and Redfern, M. A. (1975). 'Fourier series techniques applied to distance protection', *Proc. IEE*, **122**, 1301–1305.
10. Van Warrington, A. R. (1971). *Protective Relays: Their Theory and Practice*, Vol. 1, Chapman and Hall, London.
11. Goldberg, G. (1975). 'Behaviour of apparatus under the influence of voltage and current harmonics'. *Bull. Soc. R. Belg. Electr.*, **91**, 225–235.
12. Baggott, A. G. (1974). 'The effect of waveshape distortion on the measurement of energy tariff meters'. *IEE Conf. Publ.*, **110**.
13. Shepherd, W., and Zand, P. (1979). *Energy Flow and Power Factor in Non-sinusoidal Circuits*, Cambridge University Press, New York.
14. CIGRE (1929). 'Rapports et discussions sur la puissance reactive', Pt. III, pp. 117–218, Paris.
15. Kuussaari, M., and Pesonen, A. J. (1976). 'Measured power-line harmonic currents and induced telephone noise interference with special reference to statistical approach'. Paper 36–05, CIGRE, Paris.
16. Carson, J. R. (1926). 'Wave propagation in overhead wires with ground return'. *Bell Sys. Tech. J.*, **5**, 539–554.
17. *Directives concerning the Protection of Telecommunication Lines against Harmful Effects from Electricity Lines*, International Telegraph and Telephone Consultative Committee (CCITT) published by the International Communications Union, Geneva, 1963.
18. *Engineering Reports of the Joint Subcommittee on Development and Research of the Edison Electric Institute and the Bell Telephone System*, New York, 5 volumes, July 1926 to January 1943.
19. Kimbark, E. W. (1971). *Direct Current Transmission*, Vol. I, Wiley–Interscience, New York, p. 331.

6

Power system harmonic measurements

6.1 INTRODUCTION

In order to maintain an efficient and effective electricity supply it is necessary to define the levels of distortion that can be permitted on the power systems. This need has resulted in the development of a series of measuring instruments and techniques of steadily increasing sophistication leading to the present-day generation of microprocessor-based measuring equipment.

Any measuring system is formed by the appropriate connection of a number of component parts as illustrated by Figure 6.1. The actual combination required for any particular measurement will vary with the system on which the measurement is to be made and its purpose. The responsibility of determining the specific combination to be used lies with the operator/interpreter controlling the overall measurement function.

The role of this operator/interpreter is central to the measurement function. Once the need for information has been established, the measurements required to provide this information must be identified by determining whether the measurement includes full or partial harmonic spectra, simultaneous measurements of current and voltage, harmonic power and power flow, single or multiphase measurements and other related factors.

Once the choice of measurement parameters has been made then the measuring system can be defined. At lower voltage levels a direct connection of the instrument may be possible, dispensing with the need for transformers or communications links. As system current and voltage levels increase so the need for a proper choice of transformer becomes important to the measurement

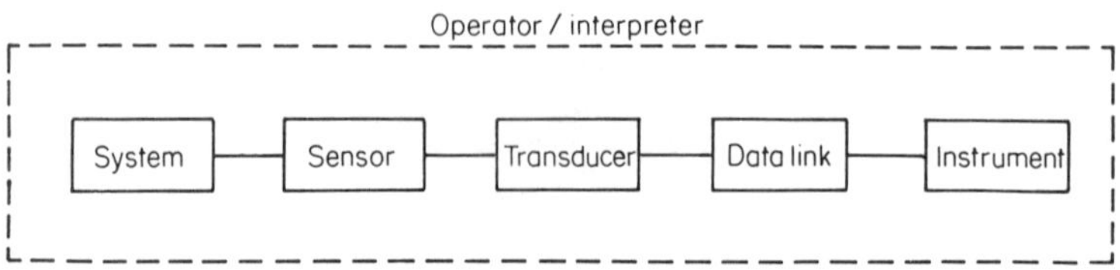

Figure 6.1. Measurement system

function. Similarly a requirement for remote siting of the measuring instrument and the need to transmit data through an electrically noisy environment such as a switchyard will influence the choice of communication link used. For the measuring instrument the choice lies between analogue and digital equipment, each of which is capable of providing a particular range of functions. In choosing the instrument, consideration must be given to the purpose of the measurement and the use to which the results are to be put. The selection of a more sophisticated instrument than actually required can impose cost penalties and could lead to data-handling problems. Similarly, an attempt to use an inadequate instrument will result in the failure of the whole measurement process.

Finally, the way in which the information provided is to be used must be examined and here it is necessary to reconsider the purpose of the measurement.

(i) Is it to check system harmonic levels against a standard?
(ii) Is there some problem of which harmonics are suspected as the cause?
(iii) Is it to provide background information on system harmonic levels?

Each of these and other questions will affect the choice of instrument, the way in which the measurement is made and the interpretation of the data.

6.2 THE DEVELOPMENT OF POWER SYSTEM HARMONIC MEASUREMENTS

Initially, harmonic measurements were based on the manual calculation of harmonic levels using some form of recorded data. Typical in this approach was Thompson who in 1905 used 18 ordinates taken from an oscillogram recording to obtain the first nine terms of the Fourier series (Section 2.3) by means of the 'selected ordinate method'. The analysis took some 70 min.

Russell, in 1916, performed a similar analysis using the amplitude and time information from a single cycle of a waveform recorded on an oscillograph and obtaining approximations to the integrations in the Fourier coefficient equations. The harmonic amplitudes and phase angles were then evaluated.(1)

The use of oscillograph recordings continued to provide the basis for the harmonic analysis of power system waveforms for some time. The method of integration was, however, modified to improve the process and make it easier to perform. Nevertheless, the analysis procedure remained tedious and for any individual harmonic the accuracy achievable was only about half a percent of the fundamental amplitude. This led to the development of instruments capable of providing a direct measurement of the harmonic components.

In 1925 Cockroft *et al.*(2) proposed an instrument using a dynamometer movement of which an improved version was produced by Coe(3) in 1929. In this instrument the fixed coil of the dynamometer movement was energized with a sinusoidal 'analysing current' at the desired harmonic frequency whilst the current or voltage to be analysed was passed to the moving coil. As a non-zero mean torque is produced only by components at the same frequency the deflection of the moving coil provided a measure of the harmonic amplitude. The analysing current was derived from a synchronously driven disc carrying a series of concentric rings, one for each harmonic frequency. Each ring consisted of an alternating series of conducting and insulating segments from which brushes picked up a series of pulses at an exact multiple of the supply frequency. From these pulses a constant amplitude sinusoidal current of the desired frequency was produced which was fed to the fixed coil. The phase angle was measured by rotating the brushes, to achieve a maximum deflection on the meter, and then noting the angle through which the brushes had been rotated relative to the fundamental frequency position. The instrument was capable of measuring harmonic amplitudes to an accuracy of '1/20th of 1 percent of the fundamental amplitude' with a measurement time per harmonic of 4 min.

The development of valve technology and the increasing availability of associated components led to the development by Prescott in 1939 of an electrostatic wave analyser.(4) This used a similar principle to the Coe instrument but with an electrometer rather than a dynamometer movement.

The analysing sinewave was produced electronically using a dynatron oscillator and, as before, the frequency and phase angle could be adjusted relative to the fundamental frequency. The reference point for phase angle measurements was obtained by using a thyratron valve which was triggered at a fixed point on the waveform to be analysed. The analysis time was similar to that of the Coe instrument.

By 1946 the derivation of harmonic content was entirely based on Fourier analysis, either of the actual waveform or of recorded waveshapes, with an increasing tendency to try to avoid calculations either by direct measurement or through the use of 'electric calculating bridges'.(5)

As the quality and availability of electronic components improved still further it became possible to produce stable, variable frequency oscillators and this in its turn led to the development of the techniques of harmonic measurement which are still used by the current generation of analogue wave, frequency and spectrum analysers.(6 – 9) The development of integrated circuit technology and microprocessors has resulted in a range of instruments employing digital techniques, based on the fast Fourier transform,(9 – 11) to provide the spectral information. Such instruments enable the system to be continually monitored to provide information at previously unavailable levels of detail, bringing with them problems of data analysis and interpretation.

Instruments currently available for the measurement of power system harmonics fall into two broad categories, harmonic analysers and spectrum analysers; this latter category also encompassing wave and frequency analysers. The separation into these categories is expressed by the following outlines of the two instrument types, which may employ either analogue or digital techniques.

SPECTRUM ANALYSERS

These scan a range of frequencies to provide a measurement of signal amplitude at all frequencies within that range. For harmonic measurements the harmonic frequencies must be identified by reference to the fundamental to enable them to be isolated from the overall spectrum. This is usually possible with the provision of some additional external circuitry.

HARMONIC ANALYSERS

These measure signal amplitudes at harmonic frequencies only. In order to achieve this the measurement is referenced to the fundamental frequency and allowance made for any variation in that frequency. The harmonic analyser provides an output spectrum which is a specific subset of the spectrum of signal amplitudes that would be produced by a spectrum analyser covering only the frequency range containing the harmonics.

6.3 FILTERS

Mathematically, the Fourier analysis described in Chapter 2 generates a spectrum on an infinitely narrow bandwidth. This is obviously not a realizable condition for any practical filter, either analogue or digital, and the concept of bandwidth must be introduced.

An ideal bandpass filter, shown in Figure 6.2, has the property of only allowing the passage of that portion of the total power of the signal which is carried by frequencies within its passband. All frequency components which lie outside the passband, that is for $f < f_l$ (the lower filter frequency) and $f > f_u$ (the upper filter frequency) are completely attenuated. The bandwidth B of the filter is then

$$B = f_u - f_l. \tag{6.3.1}$$

In practice, the characteristic of a filter is more like that shown in Figure 6.3, with amplitude variations in the passband, and small but finite amplitude outside the passband. Both these characteristics introduce error into the analysis of the input signal.

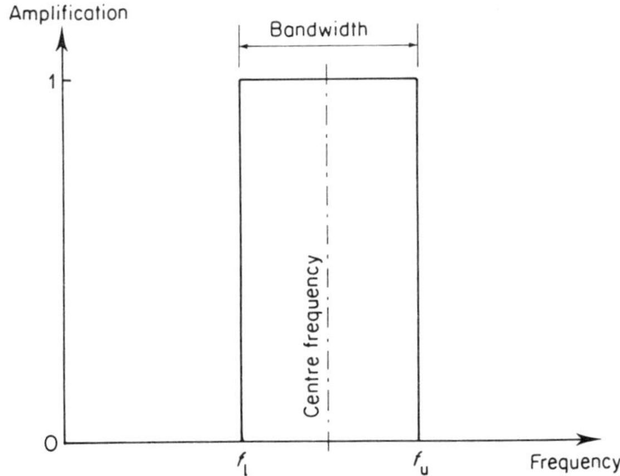

Figure 6.2. Ideal bandpass filter

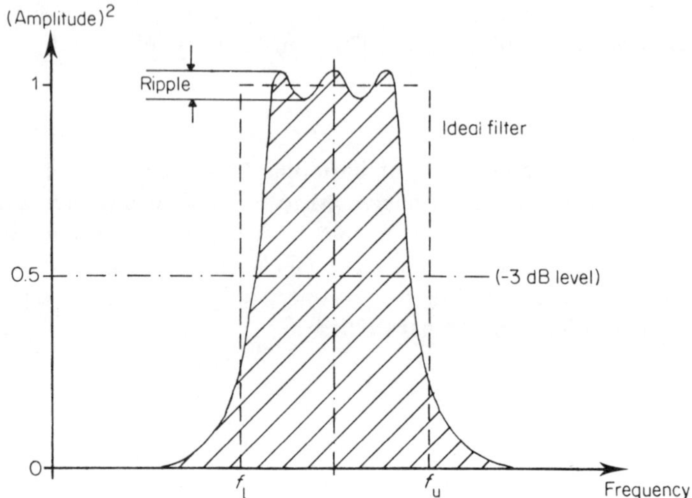

Figure 6.3. Practical filter characteristic. Noise bandwidth equals bandwidth of ideal filter

The concept of bandwidth that is used to relate the performance of the ideal, non-realizable filter of Figure 6.2 to that of the practical filter of Figure 6.3 is the 'effective noise bandwidth'. This is the bandwidth of an ideal filter with the same amplitude, which transmits the same power as the actual filter for a signal which has a constant power spectral density with frequency, i.e. a white noise source.

The effective noise bandwidth can be obtained by dividing the area under the practical filter characteristic of Figure 6.3 by the $|\text{amplitude}|^2$.

Bandwidth is also influenced by the length of the record on which the measurement is based. Consider a frequency measuring system in which the frequency is obtained by counting the number of positive going transitions through zero in an interval of T seconds. Division of the number of transitions by T is then used to give a value for frequency. The conditions illustrated by Figure 6.4(a) and (b) will both produce a result in a measured frequency of

$$f = 3/T.$$

The lowest possible frequency which would produce this result is

$$f_1 = 2/T$$

whilst the higher possible frequency is

$$f_2 = 4/T.$$

Such a system is seen only to be sensitive to frequency differences of $1/T$, which is then the system bandwidth, i.e.

$$B = 1/T.$$

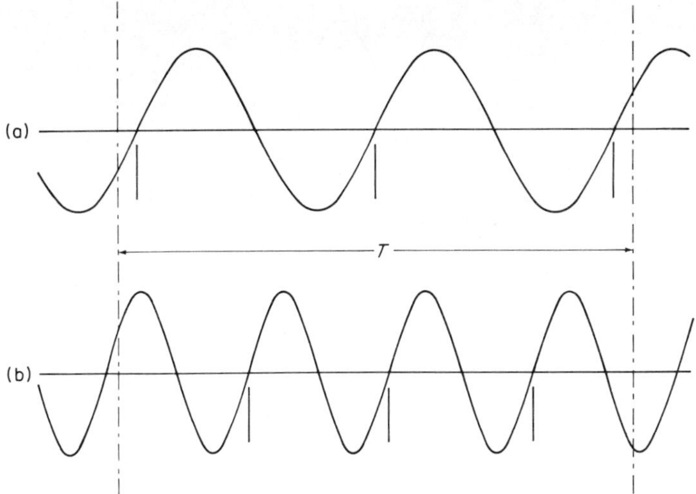

Figure 6.4. Frequency measurement using zero crossing

The characteristic of a filter operating on a record of length T seconds can be obtained by considering the situation where T does not correspond to an integer number of periods. An equation of the form of (6.3.2) has been shown to yield a zero solution only when T corresponds to an integer number of periods; namely frequencies of $1/T$, $2/T$, etc. (Section 2.6):

$$S(f_n) = \int_0^T x(t) e^{-j2\pi f_n t}\, dt. \tag{6.3.2}$$

Consider a component of frequency f_x where

$$f_x = f + \Delta f \quad \text{and} \quad f = 1/T.$$

After multiplication by the unit vector $e^{-j2\pi f_n t}$ and integration, there will remain a component at a frequency Δf which will have a non-zero value. Plotting this response gives the curve shown in Figure 6.5. This is the $\sin x/x$ (sinc) curve in the frequency domain with $x = \pi f T$.

Using the definition given earlier, the effective noise bandwidth can be found from the power transmitted by integrating the power spectrum with respect to frequency, i.e.

$$\text{Transmitted power} = \int_{-\infty}^{\infty} \sin^2(\pi f T)/(\pi f T)^2\, df = \frac{1}{T}. \tag{6.3.3}$$

The power transmitted by the ideal filter (from Figure 6.2) is

$$\int_{f_l}^{f_u} df = f_u - f_l = B. \tag{6.3.4}$$

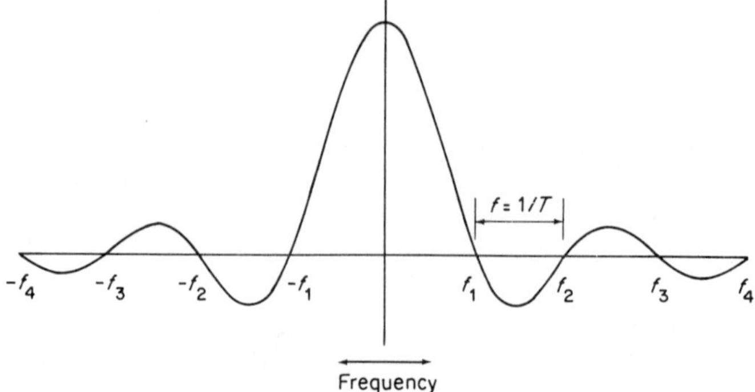

Figure 6.5. Sin x/x characteristic for a record of length T. $f_1 = 1/T$, etc.

Hence

$$\text{Effective bandwidth} = 1/T.$$

More commonly used than the effective noise bandwidth is the 3 dB bandwidth, which is taken as the width of the attenuation or power transmission characteristic between points 3 dB below the peak level. For many practical filters the 3 dB bandwidth is found to correspond closely to the effective noise bandwidth, the exceptions being filters of poor selectivity. The 3 dB bandwidth is particularly useful when dealing with a signal comprised of constant amplitude harmonics as it allows a ready assessment of how the individual harmonics will be separated by the filter.

Filters for spectrum analysis

Two types of filter are generally used for spectrum analysis, namely constant absolute bandwidth and constant percentage bandwidth.

The centre frequency f_c, of a constant absolute bandwidth filter is given by

$$f_c = \frac{(f_1 + f_u)}{2}, \tag{6.3.5.}$$

where f_1 is the lower filter frequency and f_u is the upper filter frequency.

The constant absolute bandwidth filter presents a bandwidth of B_a to the incoming signal irrespective of the centre frequency. This type of filter is to be preferred for the measurement of harmonics as it provides the same resolution and separation at each frequency, though at the expense of a reduced frequency range.

For the constant percentage bandwidth filter, the centre frequency is given by

$$f_c = \sqrt{f_1 f_u}. \tag{6.3.6}$$

The constant percentage bandwidth filter is a constant Q filter and presents a

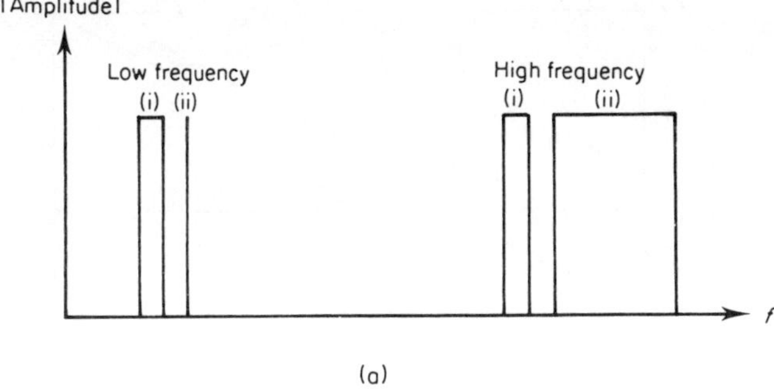

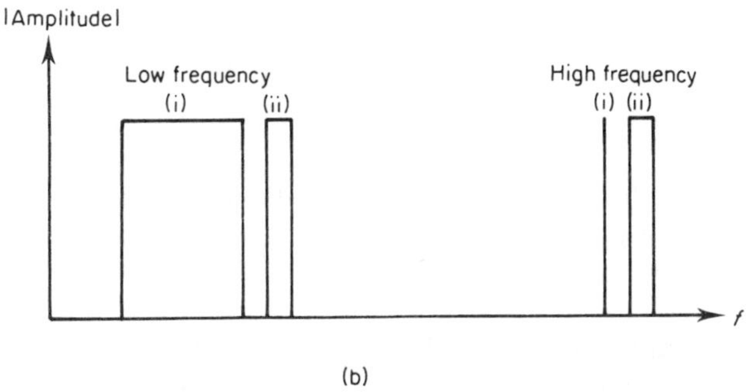

Figure 6.6. Constant absolute (i) and constant percentage (ii) bandwidth filter characteristics on (a) a linear frequency scale and (b) a logarithmic frequency scale

bandwidth of B_p of the incoming signal such that

$$B_p/f_c = \text{constant.} \tag{6.3.7}$$

The relationship of the bandwidth for both the constant and constant percentage bandwidth filters, with respect to frequency on a linear and logarithmic scale, is shown in Figure 6.6.

With the constant bandwidth filter, a uniform resolution (separation between spectral lines) is obtained on a linear frequency scale. This is useful for the separation of harmonic components of a waveform. At frequencies above a few kilohertz this type of filter has limited use, since the bandwidth becomes too small a percentage of the waveform for easy detection of a particular frequency. At these frequencies the constant percentage bandwidth filter, where the bandwidth is a fixed percentage of the centre frequency, is used.

The performance of the two types of filter is explained in Figure 6.7 and their relative suitability for harmonic measurement is further illustrated by the following example:

A system with a fundamental frequency of 50 Hz is analysed for its harmonic content, using both constant absolute bandwidth and constant percentage

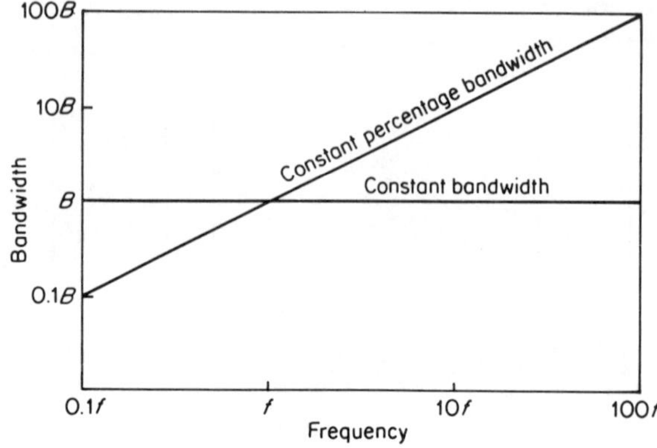

Figure 6.7. Comparison of constant bandwidth and constant percentage bandwidth filters

bandwidth filters. Both filters have a bandwidth of 10 Hz (± 5 Hz on the centre frequency) when tuned to the fundamental frequency. At the 20th harmonic (1000 Hz) the filter bandwidths are as follows:

Constant absolute bandwidth filter	10 Hz,
Constant percentage bandwidth filter	200 Hz.

Thus the bandwidth of the constant percentage bandwidth filter will now encompass all the harmonics from the 18th (900 Ha) to the 22nd (1100 Ha).

SELECTIVITY

In addition to bandwidth a filter is characterized by its selectivity. This is a measure of its ability to separate components of different frequency and is expressed by the 'shape factor'. This is defined for constant bandwidth filters, as the ratio of the width of the filter characteristic at an attenuation of 60 dB from its maximum to the 3 dB bandwidth.

Where the dynamic range is less than 60 dB, a '40 dB shape factor' is sometimes used based upon the width of the characteristic at 40 dB attenuation (Figure 6.8).

For constant percentage bandwidth filters an 'octave selectivity' is used. This is the attenuation of the filter characteristic at one octave either side of the centre frequency.

RESPONSE TIME

A filter cannot respond instantaneously to a change in the input signal level, the actual time required for output signal to reach its final steady state level being of the order of $1/B$, where B is the filter bandwidth. This is another manifestation of the physical requirement that a measurement with a bandwidth B, requires a measurement time of at least $1/B$.

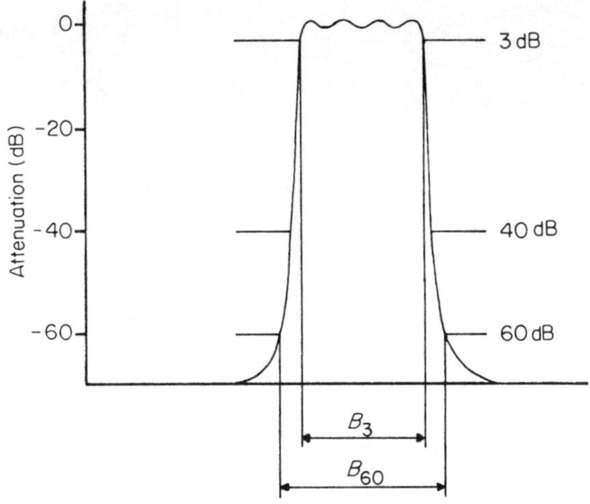

Figure 6.8. Shape factor ($= B_{60}/B_3$)

The response may be expressed as follows:

$$\text{Bandwidth} \times \text{Filter response time} = BT_f \approx 1, \qquad (6.3.8)$$
$$\text{Relative bandwidth} = b_r = B/f_c,$$

where f_c is the centre frequency of the filter.

The number of cycles in response time is

$$n_f = f_c T_f;$$

therefore

$$b_r n_f \approx 1 \qquad (6.3.9)$$

Analogue and digital filters

Typically, analogue filters are constructed by combining conventional circuit elements to provide the desired response. Such filters are liable to variation in performance brought about by factors such as component ageing, drift, long term stability and thermal effects. They therefore require careful setting up and regular checking and recalibration to maintain full performance.

A recursive digital filter receives continuous sampled data at its input which is then subject to some sequence of digital processing to produce an output which has been filtered in some way with respect to the input. As the data is continuous with an output value obtained for each input data point, the operation of the digital filter is comparable with the action of an analogue filter.[12]

Indeed it is possible to design digital filters with properties equivalent to those of virtually any practical analogue filter and even to produce filters with characteristics which are not realizable in analogue form.

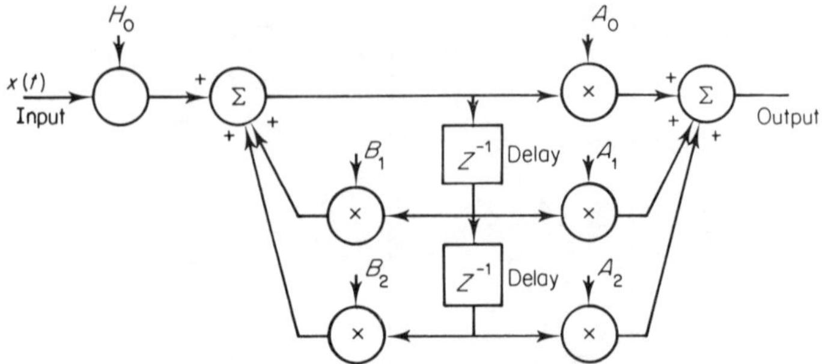

Figure 6.9. Two-pole recursive digital filter

Digital filtering is based on the Z transform, which is defined as

$$F(z) = \sum_{n=-\infty}^{\infty} f(n\Delta t)z^{-n}, \tag{6.3.10}$$

where z is a complex variable. This is of the same form as the sampled time function transform when

$$z = e^{j2\pi f\Delta t}.$$

Here, a delay corresponds to an integration and consequently by inserting multiple delay circuits, multiple pole filters can be constructed. A simple two-pole recursive digital filter is shown in Figure 6.9.

One of the main advantages of a digital filter is that the same hardware can be used to generate virtually any filter shape just by changing the filter coefficients used in the calculations. Once established, the filter coefficients completely determine the filter properties, and as these coefficients are numeric in form they do not change with time, with the result that the filter never requires trimming.

6.4 SIGNAL AVERAGING

Signals derived from any real source are unlikely to be 'steady' but will contain both random and systematic variations. Reference to, and use of, instantaneous spectra may be required when considering the effect of specific transient or short term variations in the amplitude of individual frequency components. However, any characteristic spectrum present may be masked by the random content of the instantaneous spectra.

Further, the Fourier transform itself effectively assumes a continuous signal, with averaging taking place over infinite time. In practice this is not possible, and a finite measuring time must be used, the effect of which is to introduce a ripple into the frequency domain solution. Averaging is used to provide low pass filtration to eliminate or remove this ripple component.

Consider a running average of the function covering a period of T seconds. This running average corresponds to viewing the signal through a moving

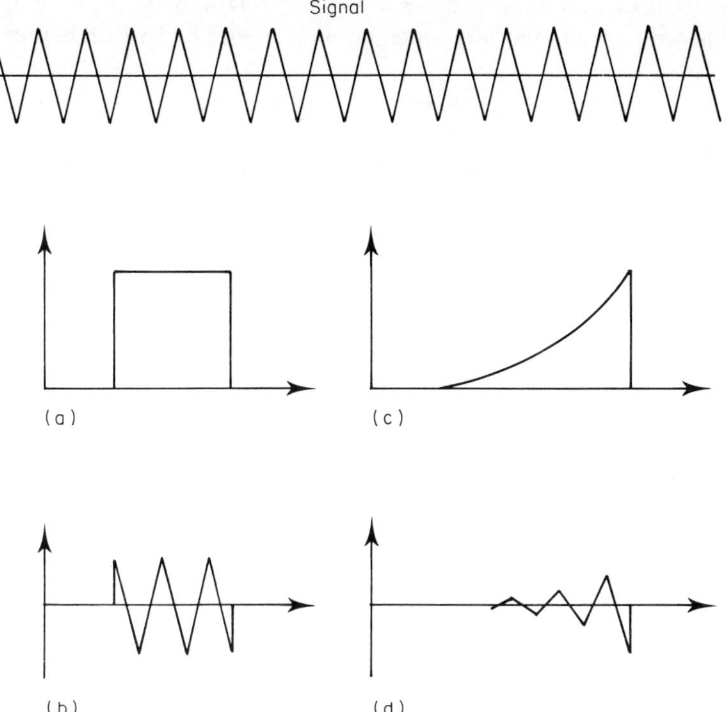

Figure 6.10. Linear and exponential signal averaging: (a) linear running average; (b) linear averager output; (c) exponential averaging; (d) exponential averager output

rectangular window as in Figure 6.10(a) and is expressed by

$$[x]_t = \frac{1}{T}\int_{t-\tau}^{t} x(\tau)\mathrm{d}\tau \tag{6.4.1}$$

which can be written as the convolution equation

$$[x]_t = \frac{1}{T}\int_{-\infty}^{\infty} x(\tau)y(t-\tau)\mathrm{d}\tau, \tag{6.4.2}$$

where $y(\tau) = 1$ for $0 < \tau < T$ and $y(\tau) = 0$ otherwise, and is illustrated in Figure 6.10(b).

A convolution in the time domain has been shown (Section 2.6) to correspond to a multiplication in the frequency domain of the respective Fourier transforms. The Fourier transform of the rectangular function has been shown to produce an amplitude spectrum in the form of the sin x/x function. This corresponds to a low pass filter with ripple decaying outside its pass-band at the rate of 20 dB per decade.

Instead of linear averaging, represented by the use of a rectangular time window, exponential averaging can be used to emphasize the more recent trends in the signal; this situation is shown in Figure 6.10(c) and (d). Such averaging may be achieved by using the 'leaky integrator' of Figure 6.11 which has an impulse response of

$$x(t) = 0, \qquad -\infty < t < 0$$

and

$$x(t) = A_c e^{-t/RC}, \qquad 0 < t < \infty.$$

The Fourier transform of this function is

$$X(f) = A_c \int_{-\infty}^{\infty} e^{-t/RC} e^{-j2\pi ft}\, dt$$

$$= A_c RC/(1 + j2\pi f RC). \tag{6.4.3}$$

By scaling such that

$$A_c = 2A_r \quad \text{and} \quad T = 2RC$$

where A_r is the amplitude of the sinc function at $x = 0$ (cf equation 2.6.10).

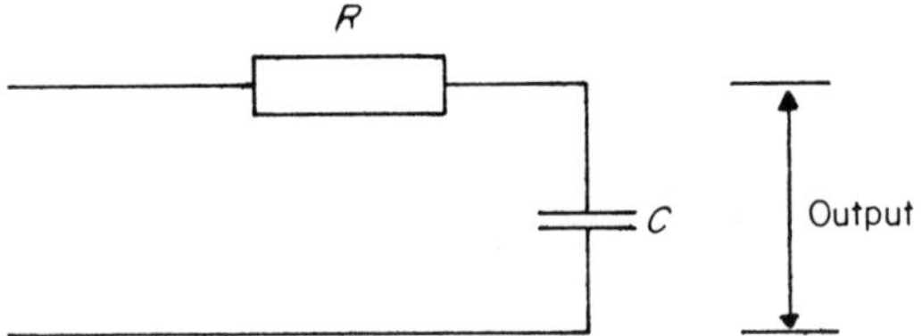

Figure 6.11. Leaky integrator

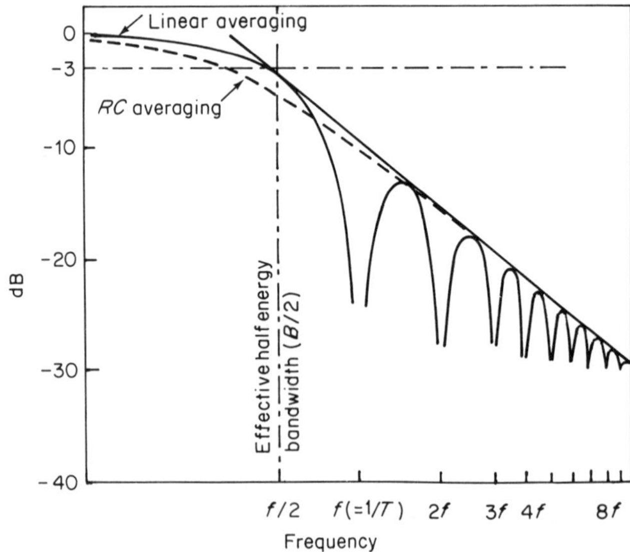

Figure 6.12. Comparison of linear and RC averaging in the frequency domain. $B = 1/T$

the two characteristics are found to have the same amplitude at $f=0$, the same asymptotic lines and the same effective bandwidths, as is shown by Figure 6.12. Further, for random signals they will have the same standard deviation.

If digital filters are used then linear averaging may be achieved by adding the squares of the individual filter output samples and then dividing by the number being averaged. Exponential averaging weights each individual output sample to emphasize the most recent data. Typically a period of $T=2RC$, where RC is the time constant of the equivalent leaky integrator, would be used.

6.5 ANALOGUE SPECTRUM AND HARMONIC MEASUREMENTS

Analogue spectrum and harmonic measurements are based on the use of analogue filtration to isolate and identify the individual frequency components of the input signal. The nature and form of the resulting spectrum is thus controlled by the characteristics of the filter used and the way the instrument performs the measurement.

The output from the analogue filter is obtained by the convolution of the incoming signal with the impulse response of the filter. This corresponds to a multiplication of the two frequency response functions in the frequency domain. The resulting signal from the filter will therefore be formed from the product of the individual amplitude spectra. Although the filter modifies the phase relationships of the individual frequency components, this has no effect on the power spectrum (a function of the square of the amplitude) which is the information displayed by the analysers.

The output power spectrum is formed as the product of the individual power spectra of the incoming signal and the filter. The output from a filter, which may be varying in time, must then be processed to produce a measure of the signal at a particular frequency. Usually the power of the signal is required, and this can be achieved mathematically by squaring the instantaneous value of the signal and integrating this over a specified time interval. By this means, a root mean square estimate of the value of the signal is obtained according to

$$y_{\mathrm{rms}} = \left[\frac{1}{T}\int_{t-\tau}^{t} [y(\tau)]^2 \,\mathrm{d}\tau.\right]^{1/2}. \tag{6.5.1}$$

Alternatively, the filter output signal can be rectified and an average of the signal obtained by integrating over a finite period according to

$$y_{\mathrm{av}} = \frac{1}{T}\int_{t-\tau}^{t} |y(\tau)|\,\mathrm{d}\tau. \tag{6.5.2}$$

This is similar to the averaging process previously discussed.

The analogue spectrum and harmonic analysers available can be divided into a number of broad categories dependent upon their mode of operation. These are (i) discrete filter analysers, (ii) parallel analysers, (iii) swept frequency analysers, (iv) time compression analysers.

Discrete filter analysers

Figure 6.13(a) shows a block diagram for a discrete filter, or bank-of-filters, analyser. The signal, after conditioning by the input amplifier stage, is applied to the inputs of a number of filters, each tuned to a different centre frequency to cover the whole of the frequency range. The spectrum is then obtained by stepping the detector and its associated amplifier across the outputs of the filters at a controlled rate.

Such an approach requires a large number of accurately tuned and stable filters. In practice therefore only two, or possibly three, such filters are used and the system designed to step the filters up in frequency in sequence, so that the signal is being applied to the next filter(s) in the series whilst any particular filter is being used for measurement. Provided that an adequate time is allowed for each filter to settle at its new frequency, this will give the same results as the multiple, bank-of-filters approach.

If the filters were ideal, having the perfectly rectangular characteristic of Figure 6.2, then a discrete power spectrum of the signal, based on the centre frequencies as in Figure 6.13(b) would be obtained.

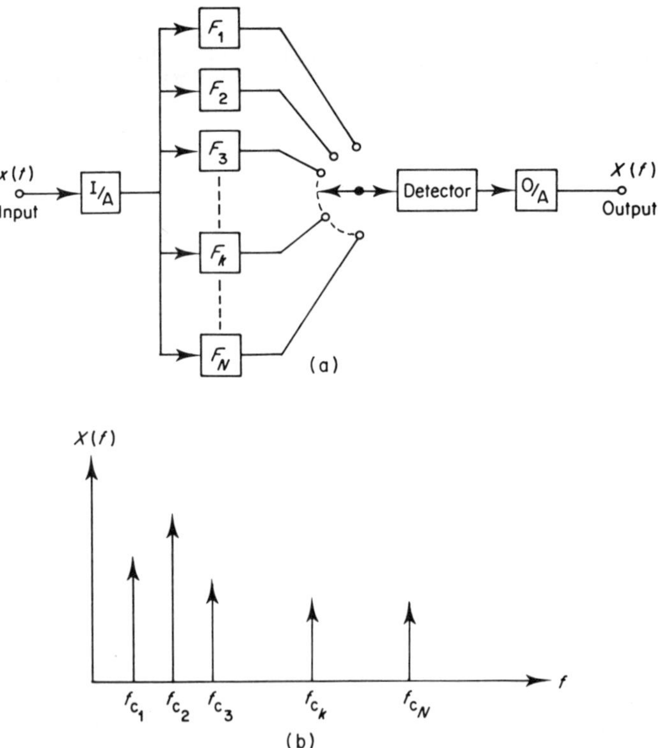

Figure 6.13. Discrete filter or bank of filter technique for spectral analysis: (a) signal processing hardware; (b) output power spectrum. f_{c_k} = centre frequency of the kth filter

Because of the sequential nature of obtaining the output of each filter this method is inaccurate for non-stationary signals with continuously changing spectra.

Parallel analysers

To overcome the problem of analysing practical signals which may be non-stationary, the sequential detection of the discrete filter analyser can be replaced by a multiple filter-detection circuit, incorporating a detector with each filter as shown in Figure 6.14. This effectively provides for 'real-time' analysis of the input signal, which can be updated at regular intervals if required. However, due to the multiplicity of filters required, this technique tends to be limited in narrowness of bandwidth.

The analogue filter form of the parallel analyser has now been superseded by analysers based on digital filtering and the fast Fourier transform.

Swept frequency analysers

For narrow bandwidth analysis a single filter with a tunable centre frequency is used, as shown by the block diagram of Figure 6.15. By sweeping the centre

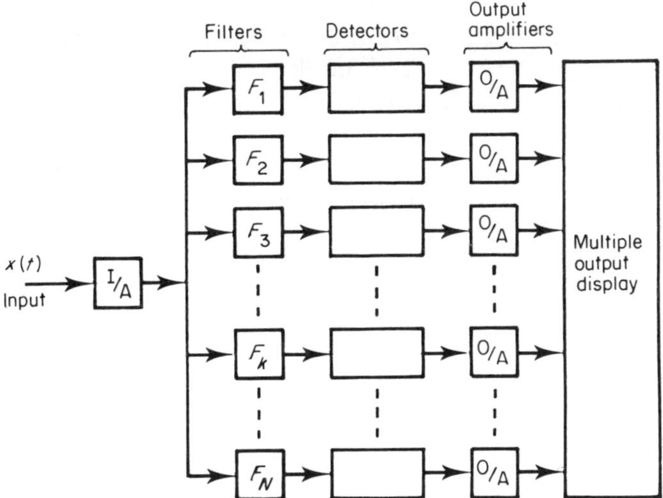

Figure 6.14. Parallel analyser for 'real time' spectrum analysis of analogue signals

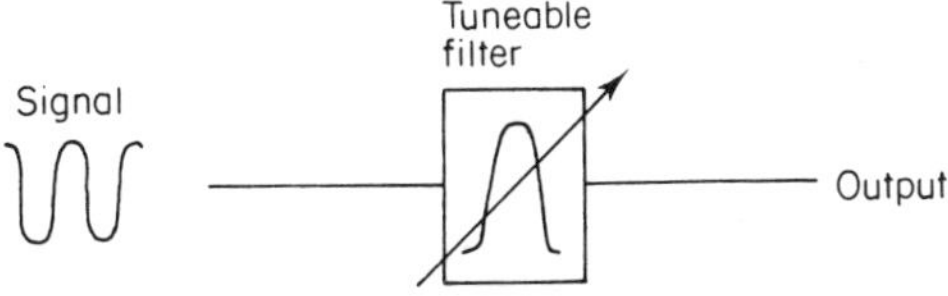

Figure 6.15. Tuned filter analyser

frequency through a range of frequencies, either continuously or in discrete steps, a spectrum is produced with each point representing an integration of the true spectrum over a frequency response corresponding to the filter characteristic.

A major limitation of the swept frequency analyser is the time taken to cover the frequency range, which results in the time correspondence between the individual frequency components being lost. The sweep time is influenced by the response time of the filter and the averaging time used. The time taken for the filter to be swept across its bandwidth (B) is referred to as the 'filter dwell time' (T_d). Hence the sweep speed (S, in hertz per second) is given by

$$S = B/T_d. \tag{6.5.3}$$

If T_d is made equal to the filter response time (i.e. $T_f = 1/B$), this will have the effect of shifting the resulting spectrum in the frequency domain by an amount equal to the bandwidth. For this reason the relationship

$$T_d = K_f T_f, \tag{6.5.4}$$

in which K_f is a numerical constant, is used to establish the dwell time.

Using the relationship between response time and bandwidth ($T_f = 1/B$) then gives

$$BT_d = K_f. \qquad (6.5.4)$$

A similar expression may be used to express the relationship between the dwell time and the averaging time. However, for deterministic signals, such as those dominated by harmonics, the filter response time will largely determine the dwell time used.

The frequency range of interest for harmonic measurement extends from d.c. to around 3 kHz. For a filter of bandwidth 10 Hz the response time is 0.1 s. Choosing $K_f = 4$, then

$$S = 10/0.4 = 25\ \text{Hz/s},$$

giving a sweep time of 120 s for the frequency range to 3 kHz. The duration of the sweep means that this method is only accurate for stationary (time invariant) signals.

Heterodyne analysers

A common form of swept frequency analyser is the heterodyne analyser, shown in block diagram form in Figure 6.16. The input signal and the frequency from the local oscillator are combined in a balanced modulator (multiplier) to produce an output consisting of the sum and difference frequencies, i.e.

$$\text{Input signal} = v_i = V_i \sin(\omega_i t + \phi),$$

$$\text{Reference signal} = v_r = V_r \sin(\omega_r t),$$

$$\begin{aligned} v_o = v_i v_r &= V_i V_r \sin(\omega t + \phi)\sin(\omega_r t) \\ &= \frac{V_i V_r}{2}\{\cos[\omega_r - \omega_i)t - \phi] - \cos[(\omega_r + \omega_i)t + \phi]\}. \end{aligned} \tag{6.5.5}$$

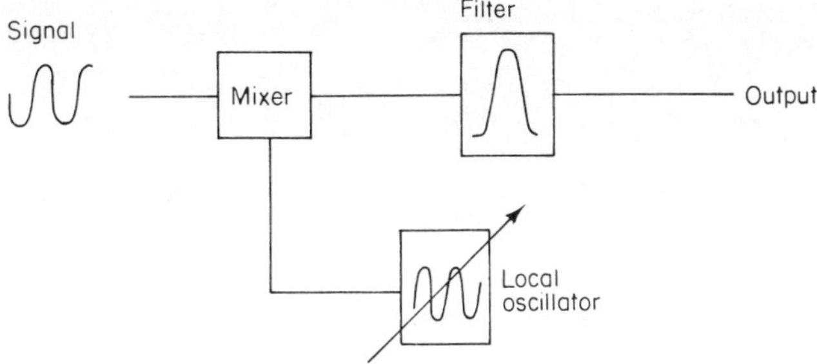

Figure 6.16. Heterodyne analyser

A narrow bandpass filter is then used to select the difference frequency, the r.m.s. value of which is then presented at the output.

As the difference frequency detected by the narrow bandpass filter is constant the local oscillator can be calibrated directly in terms of the frequency selected.

By matching the local oscillator frequency to the required frequency ($\omega_i = \omega_r = \omega$) the difference frequency becomes zero and the output after filtering (filter centre frequency $= 0$ Hz) becomes

$$v_o = (V_i V_r/2)\cos\phi. \tag{6.5.6}$$

Similarly, a second modulator can be used with a reference signal shifted by 90° to produce a second output (v_{oo}) such that after filtering

$$v_{oo} = (V_i V_{rr}/2)\sin\phi. \tag{6.5.7}$$

In this way the heterodyne system can be used to provide information on both the amplitude and the phase relationship of the frequency components.

Time compression analysers

Time compression analysis provides a means of improving the performance of a swept frequency analyser by allowing it to operate at a higher frequency range than the actual analysis frequency range.

Referring back to equation (6.5.4), the only way in which T_d can be reduced without adjusting K_f is by increasing the bandwidth B. This can be achieved by reconstituting the time domain signal over a shorter time-scale, hence time compression. This has the effect of multiplying each of the component frequencies by some constant.

Time compression can be achieved by first recording the signal at one speed and then replaying it R times faster. All the frequency components in the original signal will then be multiplied by R, enabling the same discrimination to be achieved using a filter of R times greater bandwidth and reducing the dwell time by the same factor R.

In its simplest form this could mean making an analogue recording of the time domain and then replaying it to the analyser at an increased tape speed.

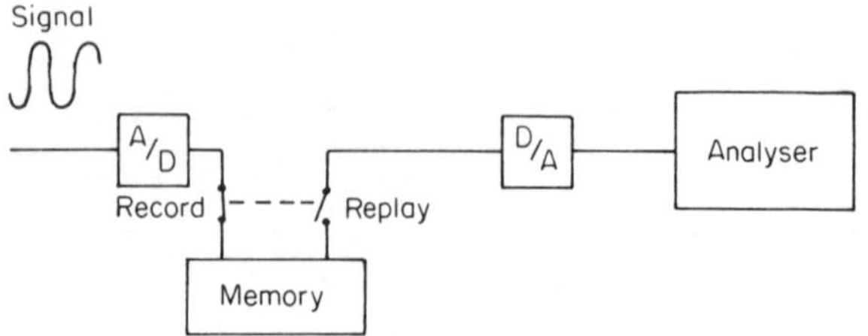

Figure 6.17. Time compression analyser using digital memory

Multiplication factors of the order of 10 can be achieved using this technique.

For higher multiplication factors a digital system, shown in simplified block diagram form in Figure 6.17, is used. With this arrangement the signal to be analysed is recorded into a block of digital memory and then replayed at high speed, through a digital to analogue convertor and anti-aliasing filter, to the analyser.

By using parallel memory areas and recording data into one area, whilst playing back the other, it can be arranged that none of the incoming data is lost, and the analyser will be operating in real time.

The rate at which data is entered into memory will be controlled by the Nyquist frequency (Section 2.11) for the measurement and the spectral response by the type of window function used.

Analogue harmonic analysers

The harmonic spectrum is a subset of a frequency spectrum covering the frequency range of interest. The principle of operation of an analogue harmonic analyser follows closely that of the spectrum analyser but with the addition of some means of isolating and identifying the harmonic frequencies.

In making measurements of power system harmonics two further points need to be considered. Firstly, the amplitudes of the individual harmonic components are usually small in relation to the amplitude of the fundamental and, secondly, the frequency of the fundamental may drift within prescribed limits about its nominal frequency.

In an analogue harmonic analyser a notch filter is often incorporated, centred on the nominal fundamental frequency and with a bandwidth sufficient to encompass the normal frequency drift, to remove the fundamental component and pass only the harmonic components to the measuring circuit.

The effect of the variation in the fundamental frequency is to alter the frequencies of the individual harmonics. Consider a system with a nominal fundamental frequency of 50 Hz and a permitted variation of ± 1 Hz. The frequency of the 20th harmonic will therefore lie anywhere between 980 and 1020 Hz and that of the 50th harmonic anywhere between 2450 and 2550 Hz. If an analyser operating on the heterodyne principle with a filter centred on 0 Hz is used, then the reference voltage must be continually adjusted to maintain the system in tune as the fundamental frequency varies. This is usually achieved by reference to a phase locked loop.

THE PHASE-LOCKED LOOP

The phase-locked loop (PLL) is a feedback system comprising a phase comparator (multiplier), a low pass filter and an amplifier in the forward path together with a voltage controlled oscillator (VCO) in the feedback path. A block diagram for the basic phase locked loop is given in Figure 6.18.

As the phase-locked loop is a feedback system it is characterized mathematically by equations similar to those for other, conventional feedback systems. The error signal is, however, a measure of phase difference between the input signal and the output of the voltage controlled oscillator.

The operation of the phase locked loop may be described qualitatively as follows. With no input signal the error voltage v_e will be zero and the VCO operates at a set frequency known as the free running frequency. When an input signal is applied, the phase comparator compares the phase and frequency of the input signal with the VCO frequency to generate an error voltage, related to the difference in phase and frequency between the two signals. This error voltage is filtered and amplified to generate the control voltage v_d, which is applied to the control terminal of the VCO. This control voltage acts to adjust the output frequency of the VCO in such a way as to reduce the frequency difference between the VCO output signal and the reference signal to zero.

If this frequency difference is within the 'capture range' of the phase-locked loop (PLL), defined as the frequency range centred on the VCO free running frequency over which the loop can acquire lock with the input signal, then this feedback action will synchronize the VCO with the input signal. Once in lock, the VCO frequency is identical with that of the input signal. However, a small phase difference $\Delta\phi$ is needed to generate the control voltage v_d and keep the PLL in lock.

Once in lock the PLL will track the input frequency over a range of frequencies known as the 'lock range', which is never less than the capture range. The lock range is defined as the frequency range, usually centred on the VCO free running frequency, over which the loop can track the input signal once lock has been

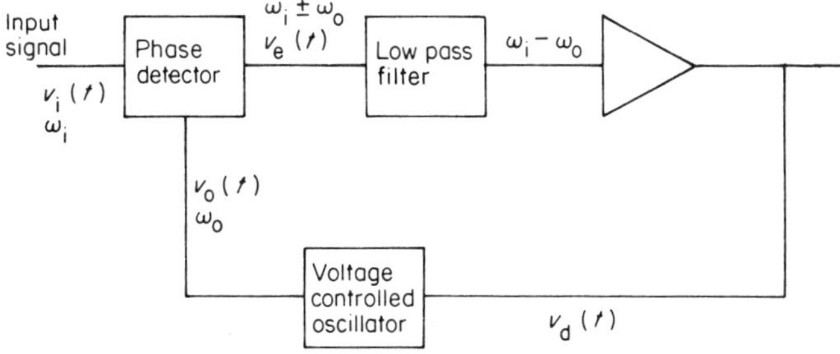

Figure 6.18. Phase-locked loop

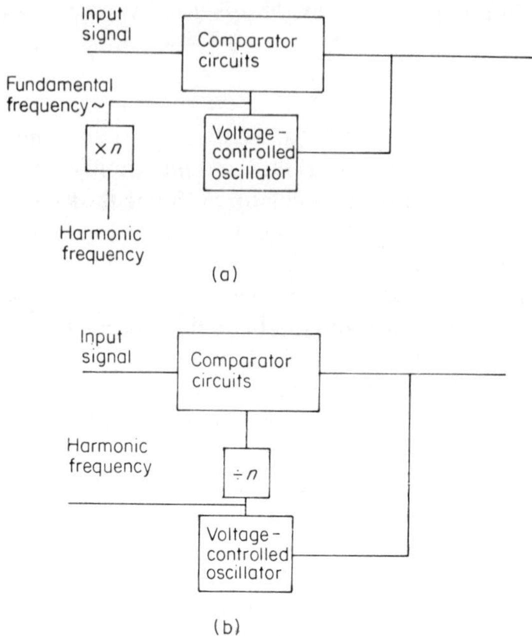

Figure 6.19. The phase locked loop for harmonic measurement

achieved. The lock range is limited by the range of variation that can be achieved in the output frequency of the VCO.

The PLL can be arranged to lock on to any component of the input signal by adjustment of the free running frequency to place the desired harmonic within the capture range. Figure 6.19 shows two possible ways of using a PLL to obtain a signal at harmonic frequencies by locking to the fundamental and a harmonic respectively.

6.6 DIGITAL METHODS FOR SPECTRAL ANALYSIS

Analysers using digital filters

Digital filter analysers can be constructed to provide responses equivalent to those of the range of analogue analysers already considered. The most usual form is that of a parallel, bank-of-filters type of analyser.

Discrete Fourier transform analysers

The general form of the Fourier transform for digital spectral analysis, the discrete Fourier transform, can be written as

$$X(f) = \sum_{n=0}^{N-1} W(n)x(n)\mathrm{e}^{-j2\pi kn/N}, \qquad 0 \leqslant k \leqslant N-1, \tag{6.6.1}$$

where $W(n)$ are uniformly spaced samples of the time window (see section 2.8), $x(n)$ are uniformly spaced samples of the waveform and N is the number of samples of both the window and the signal within the window length (Section 2.8).

Equation (6.6.1) can be computed using the fast Fourier transform (FFT) algorithm (Section 2.9) provided that N is equal to 2^m, where m is a positive integer. Typically, a figure of $N = 1024$ ($m = 10$) is used.

The inverse transform, which includes the time window, is given by

$$W(n)x(n) = \frac{1}{N}\sum_{k=0}^{N-1} X(k)e^{j2\pi kn/N}, \qquad 0 \leqslant n \leqslant N-1. \tag{6.6.2}$$

If $W(n) = 1$ for all n, then equation (6.6.1) gives the Fourier series coefficients of the sampled signal, $x(t)$. In this case, the window function is just the rectangular window. Alternatively any of the window functions of Figure 2.26 can be used in their discrete form.

Resolution and bandwidth

The spectral result of an FFT analysis is distributed linearly with frequency. The resolution (β) of the FFT is defined as the frequency increment between successive lines in the spectrum. This is related to the number of samples (N) on which the analysis is based and the sampling frequency (f_s), i.e.

$$\beta = f_s/N. \tag{6.6.3}$$

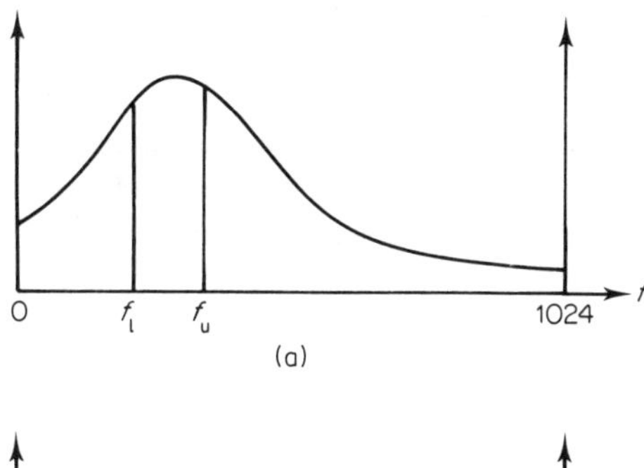

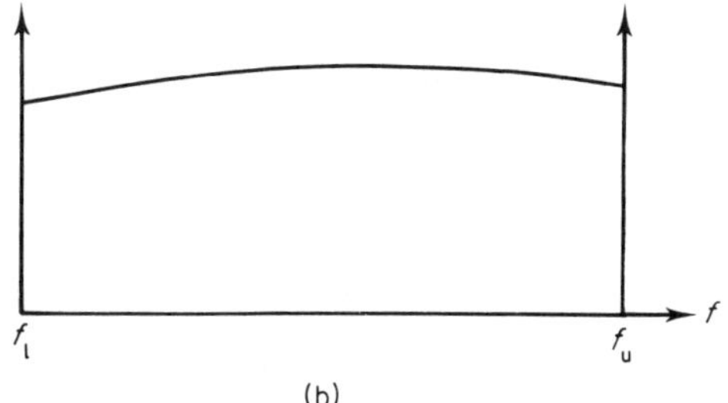

Figure 6.20. Illustration of 'zoom FFT' facility: (a) original signal of 1024 samples; (b) portion of original signal expanded to 1024 samples

The bandwidth is in turn determined by the resolution and the effect of any window function applied to the data. For a rectangular window function the effective noise bandwidth (B) equals the resolution:

$$B = \beta. \tag{6.6.4}$$

If any other time window is used, then the bandwidth of the result is the bandwidth of that window function.

It may be desirable to obtain finer resolution over a limited portion of the spectrum. The 'zoom FFT' procedure is used for this and is often incorporated on spectrum analysers. It can be considered as 'zooming in' on a limited portion of the spectrum as in Figure 6.20 with a resolution power corresponding to the number of lines normally used for the whole spectrum. This is achieved by multiplying a time domain signal by a rotating unit vector of frequency $-f_k$. This effectively changes the frequency origin to f_k. If the resulting signal is digitally low pass filtered, aliasing effects are removed.

The zoom FFT facility is useful for signals which have closely spaced resonances or a large number of harmonics. Also a spectrum with steep changes in phase with frequency can be viewed in finer detail.

Time domain averaging

Both linear and exponential (RC) averaging can be employed on the results of the FFT analysis in exactly the same manner as previously considered.

In addition, for harmonic measurements, where the sampling frequency and sampling period are continuously controlled with respect to the fundamental frequency, time domain averaging may be used.

Consider a harmonic analysis system using the FFT technique and for which a bandwidth of at least 6.25 Hz is required. Using a rectangular window this implies a sampling period of no less than 0.16 s, or eight cycles of a 50 Hz fundamental.

If the highest frequency of interest is 3 kHz, then a suitable sampling rate would be 6400 Hz, corresponding to 128 samples per cycle of the fundamental. Over the eight cycles this would yield a total of 1024 samples for the FFT. As explained in Section 2.9 the number of multiplications required is $(N/2)\log_2 N$. In this case this would mean a total of 5120 multiplications for solution.

As only harmonic frequencies are of interest, an analysis based on one cycle of data would give the information for these frequencies using only 448 multiplications, based on a sample rate of 128 samples per cycle but at a bandwidth of 50 Hz. By sampling over the eight cycles required to achieve the required bandwidth and averaging on a point by point basis, a single, composite cycle can be formed which is then analysed to yield the harmonic information.

Dynamic range and quantizing error

As part of the sampling process the time domain signal is converted to a series of digital data by means of an analogue to digital (A/D) converter. The number of

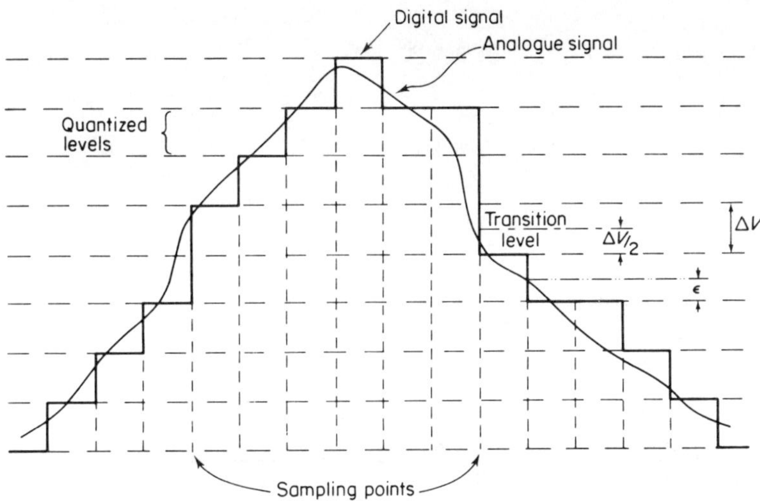

Figure 6.21. Digitized waveform

bits used in the conversion will control the dynamic range of the analysis.

Consider a 10-bit analogue to digital conversion. This uses 1024 levels to represent a signal capable of varying from 0 to V. The smallest variation in signal amplitude, capable of being registered at the output of the converter, is therefore one of $(V/1024)$. This represents a dynamic range of about 60 dB.

In practice an A/D conversion with 12 bits is commonly used to give a dynamic range of around 72 dB. By using 16-bit arithmetic this dynamic range can then be maintained in the calculation.

The conversion of the analogue signal into a digitized series of levels, each ΔV apart (Figure 6.21), itself introduces an error referred to as the quantizing error, which can vary between $\pm\frac{1}{2}$ least significant bit.

To minimize the quantizing error, the input signal should be normalized to the full range of the A/D convertors. Alternatively, the bit resolution of the A/D convertors should be increased.

Quantizing error is also introduced by the finite precision arithmetic available in the computing device performing the FFT. By performing the arithmetic with a greater number of bits than the A/D convertor this contribution can usually be ignored.

Real time measurements using FFT instruments

Spectrum and harmonic analysers using analogue or digital filtering to obtain information about the individual harmonic frequencies can be arranged to provide a continuous output signal with no loss of information.

For the analysis using FFT techniques to be considered as real time, the collection and digitizing of the data must be continuous at the defined sampling rate and all data must be used in the analysis. In order that this requirement is

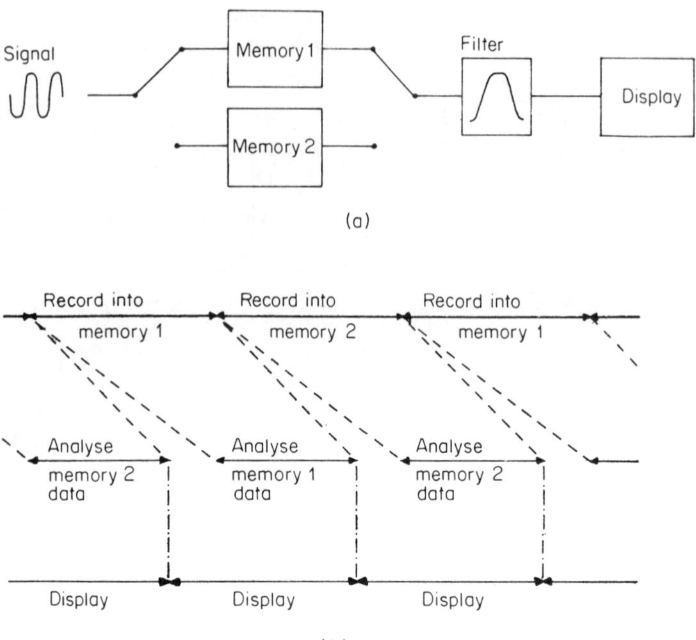

Figure 6.22. (a) Schematic diagram and (b) timing diagram for a real time analyser using dual memory

met, any individual analysis must be carried out in a time less than the data collection time. In practice this can be achieved by using two memories with the data from one memory being analysed whilst the second continues recording the incoming data. Figure 6.22(a) shows this arrangement whilst Figure 6.22(b) gives the timing diagram. Current analysers are limited to a few kilohertz range when operating in real time.

6.7 OFF-LINE MEASUREMENTS

The instruments described so far are all intended for on-line measurement of power system harmonic levels. Alternatively, by recording the waveform in either analogue or digital form, analysis can be performed off-line using any suitable computing facility.

Analogue recording

Various systems have been developed based on the analogue recording of system voltage using a range of recorder types from cheap, simple cassette recorders to precision FM tape decks.[13,14]

In order to provide adequate dynamic range on the recorders used, some form of signal conditioning has normally to be introduced prior to the recording stage.

This usually takes the form of some type of a pre-emphasis filter, used in conjuction with a fundamental rejection filter (Figure 6.23(a)). The signal is then played back to the analyser through an inverse (de-emphasis) filter (Figure 6.23(b)).

This filtering introduces an additional phase shift into the harmonic components whilst retaining the amplitude characteristics. It is also possible for the recording system to introduce distortion (Figure 6.24), which must be compensated for if waveshape information is required.

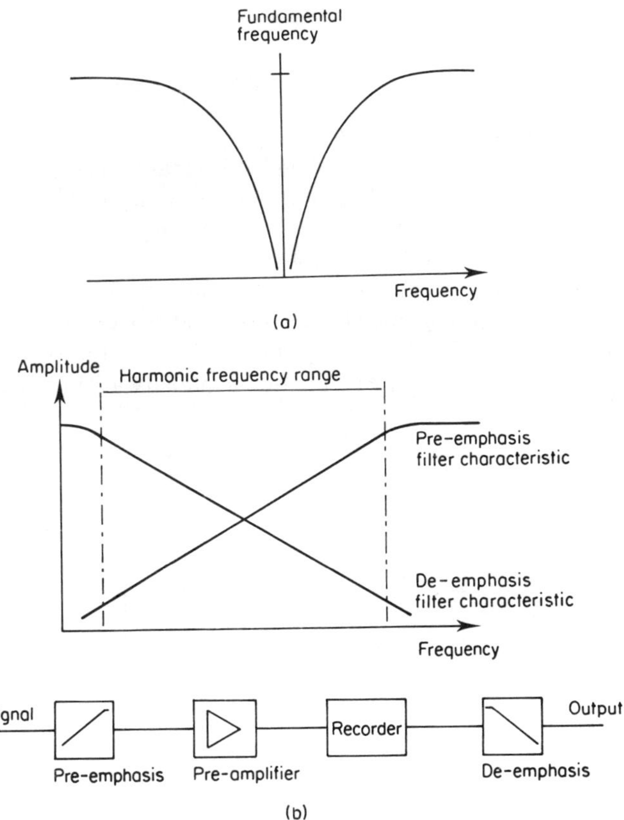

Figure 6.23. A recording system with (a) a fundamental rejection filter and (b) pre- and de-emphasis filtering

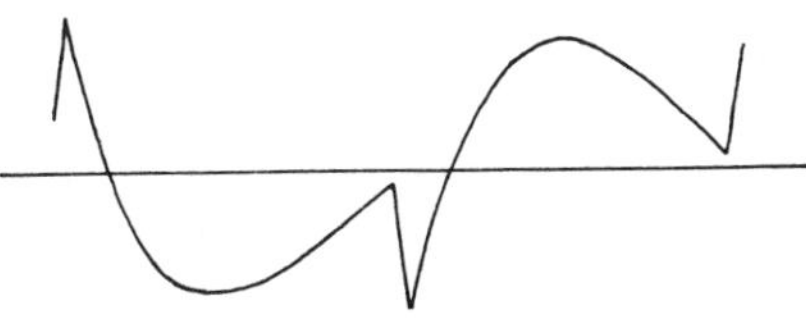

Figure 6.24. Distortion produced on a square wave input by recorder characteristics

Digital recording

The use of digital recording techniques eliminates the problems of recorder dynamic range and frequency response encountered on analogue systems. The high rate of data acquisition involved does, however, require high tape speeds. Such recording systems tend to be expensive and limit the amount of data that can be obtained in a single recording.[15]

For this reason such systems tend to be used more for the investigation of short duration or transient distortions of the supply system waveforms than for harmonic monitoring.

6.8 THE PRESENTATION OF HARMONIC DATA

Analogue harmonic analysers of the swept or stepped filter type take several minutes to produce a set of harmonic data with repetitions taking place at fixed intervals. The resulting data may be presented in the form of a table of harmonic amplitudes or by using a suitable chart recorder in graphical form. These forms of output are shown in Table 6.1 and Figure 6.25.

In either case the data produced lacks any time correspondence between the individual harmonic components. This means that all waveshape data is lost. The fact that data is collected at widely separated points in time can also lead to problems of interpretation. Figure 6.26 shows a set of data for a single harmonic.

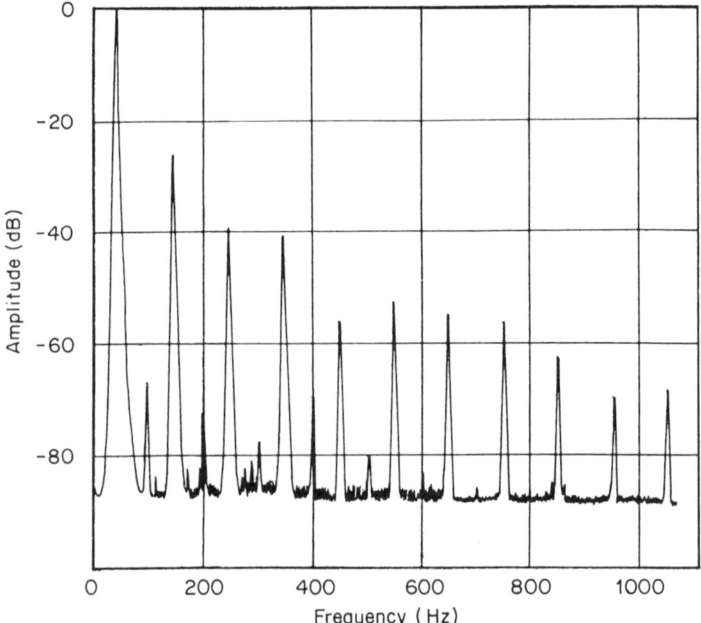

Figure 6.25. Output from an analyser (see Table 6.1 for comparison). Fundamental, 240 V (0 dB); filter bandwidth, 3 Hz; sweep rate, 5 Hz/s

Table 6.1. Tabulated data indicating output from an analogue analyser (cf. Figure 6.25)

Frequency (Hz)	Amplitude (V)
50	240.0
100	0.1
150	12.0
200	0.1
250	2.7
300	0.0
350	2.1
400	0.0
450	0.3
500	0.0
550	0.6
600	0.0
650	0.4
700	0.0
750	0.3
800	0.0
850	0.2
900	0.0
950	0.1
1000	0.0

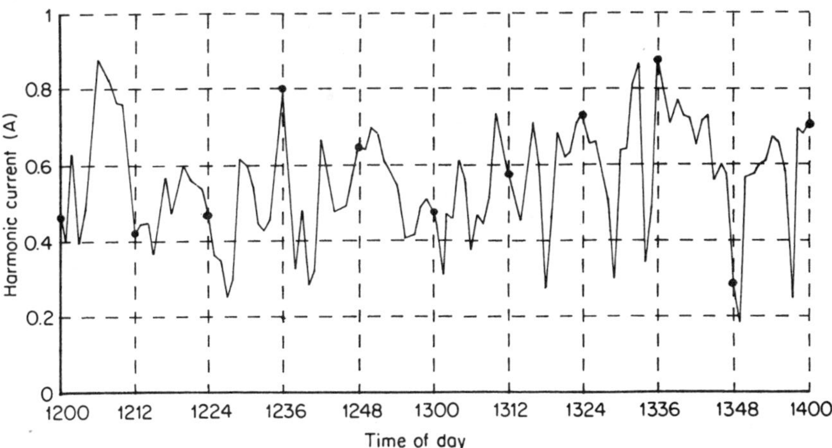

Figure 6.26. Seventh harmonic current showing effect of sampling at 12 min intervals. ●, analogue measurement points. From Bradley, Graphical presentation of power system harmonic data, Copyright © 1984, *International Conference on Harmonics in Power Systems*

If this harmonic was recorded every 12 min, then it is apparent that a large volume of information is omitted and that the results could easily be misleading. This serves to illustrate a basic limitation of such systems where harmonic levels are varying.

Where digital measuring systems are used, the problem changes from one of a

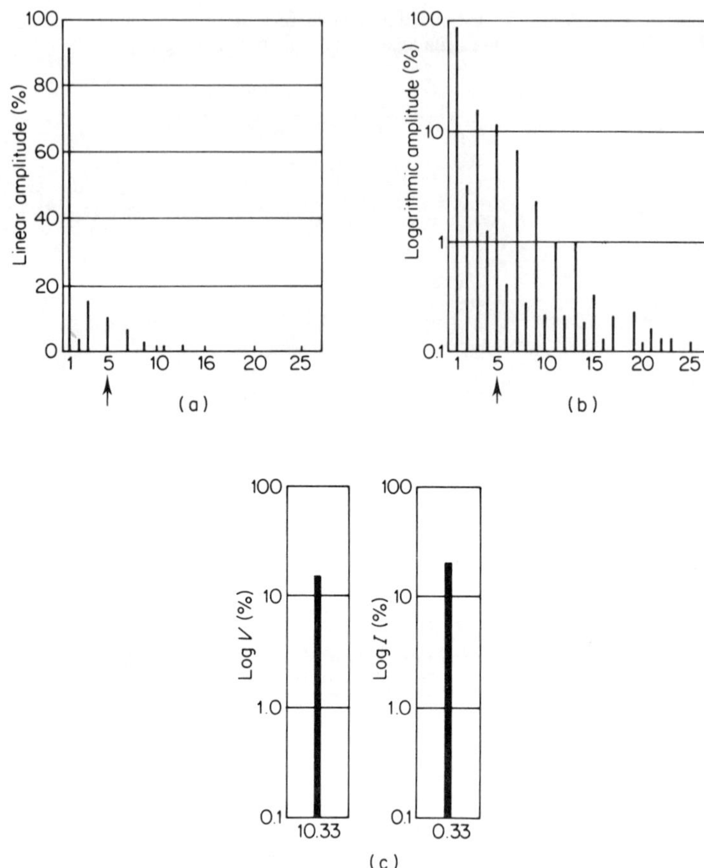

Figure 6.27. Display options from digital measuring systems: (a) linear amplitude; (b) logarithmic amplitude. Full scale = 110 V. *Cursor details*: harmonic number, 5; frequency, 250 Hz; amplitude, 10.33 V. (c) Logarithmic voltage and current. Volts scale = 110 V; current scale = 2 A; frequency, 250 Hz; current, 0.33 A; harmonic power, 2.83 W; power factor, 0.83.

sparsity of data to that of extracting from the volume of data available that which is relevant.[16,17] These instruments provide a variety of display options, some of which are illustrated by Figure 6.27. By combining the instrument with a printer, graph plotter or computer logger, a range of output forms is made available to the user. For many purposes, however, the local display is inadequate for detailed examination of conditions. More detailed examination therefore requires some form of off-line processing thus introducing further, and possibly significant delays between the collection of the original data and it becoming available to the user.

Based on stored harmonic data, various graphical displays are considered as a means of extracting the relevant from the mass of irrelevant information.

In order to examine these various forms of data presentation it is assumed that the measuring system only records a single amplitude of information for each harmonic at one minute intervals. This 'minute value' for the harmonic may itself be formed from a single sample over a group of cycles or as the average of a number of such samples. The limitations imposed by adopting this arbitrary measuring system configuration are recognized, but are considered to be acceptable in the context for which they are applied.

On the basis of the display system described above, data presentation falls into two broad categories, i.e. those relating to individual harmonics and those which provide information over a range of harmonics.

The first category includes the following: (i) the presentation of the individual harmonic amplitudes on a minute by minute basis; (ii) trend lines providing information over a longer time span than can readily be obtained from (i); (iii) cumulative probability plots; (iv) histograms.

The second category includes: (v) some form of display which provides information about the range of harmonics in a form which allows for a ready comparison between harmonics; (vi) a bar chart form representing the harmonic spectrum; (vii) the total harmonic distortion of the signal.

Displays involving individual harmonics

MINUTE VALUES

This form of presentation is shown in Figure 6.26, in which harmonic amplitude is plotted directly against time. Presentation may be in absolute units or as a normalized value, relative to some level. This latter would usually be some particular limiting value such as the level imposed by a standard. The limitation of this form of presentation is the amount of data to be presented, and the extent to which the time scale can be compressed, whilst retaining the legibility of the display.

TREND LINES

In many instances the detail provided by plotting the minute values is not required. Instead, what is needed is information about the general form of variation in harmonic amplitude over an extended period. This has been achieved by the use of the trend line presentation of Figure 6.28. These trend lines are formed from a series of overlapping averages which results in a smoothing of the data. By choosing to display the means of the upper and lower quartiles of the data in the averaging period, instead of the maximum and minimum in this period, the prospect of undue distortion by extremes is reduced.

The trend line form of presentation has been found to be of particular use in conducting long term surveys and in identifying the specific periods for which more detail examination is required.

CUMULATIVE PROBABILITY FUNCTION

Where time information is not required then the cumulative probability function of Figure 6.29(a) could be used. This provides information on the proportion of

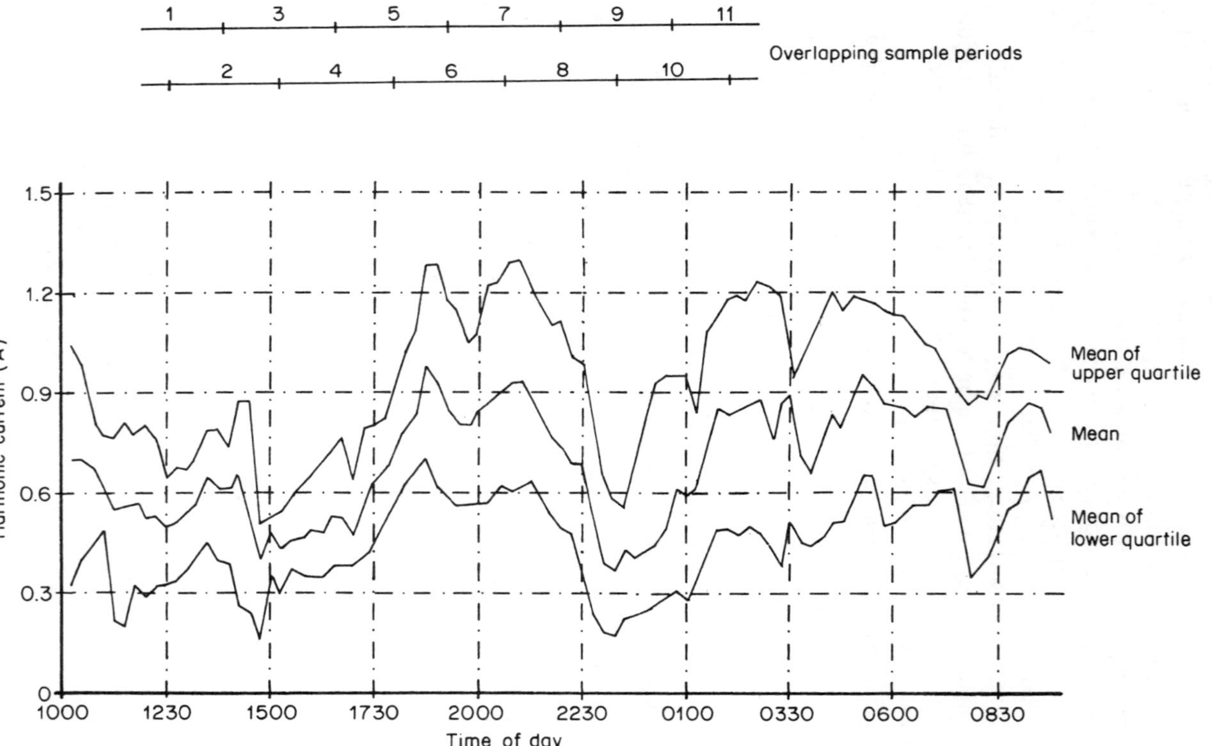

Figure 6.28. Trend line harmonic display. From Bradley, Graphical presentation of power system harmonic data, Copyright © 1984, *International Conference on Harmonic is Power Systems*

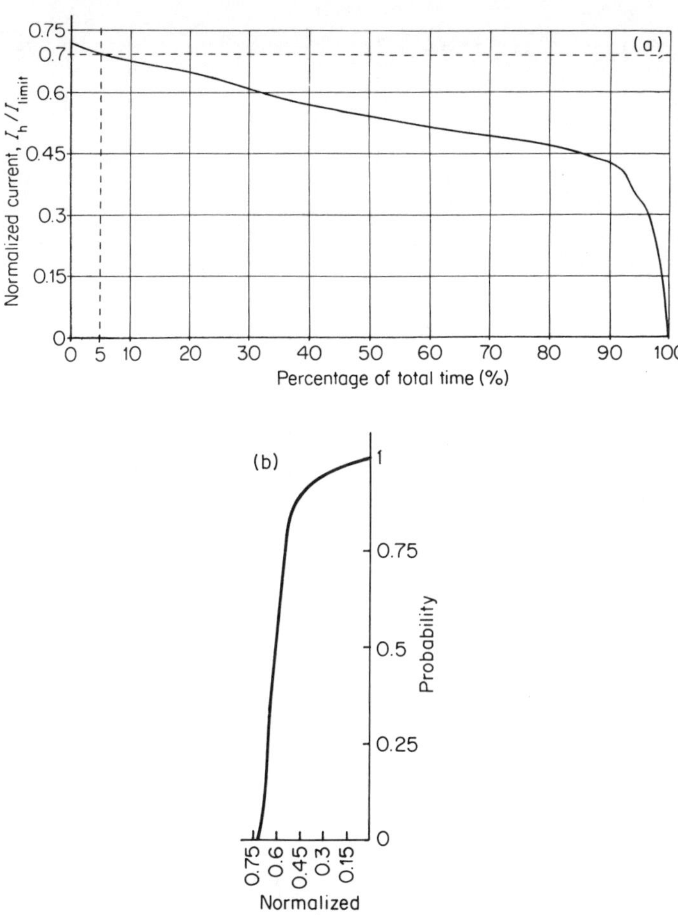

Figure 6.29. (a) Cumulative probability function and (b) probability curve. In part (a), seventh harmonic current limit is 2.5 A at 220 kV, the measurement start time was 1000 hours and the measurement stop time was 1200 hours. Part (a) from Bradley, Graphical presentation of power system harmonic data, Copyright © 1984, *International Conference on Harmonics in Power Systems*

the specified time interval that the harmonic amplitude exceeds any particular level.

By altering the orientation of the axes to those of Figure 6.29(b), the function becomes an expression of the probability that the harmonic amplitude will exceed a particular level.

HISTOGRAMS

As an alternative to the cumulative probability function, histograms can be used to provide information on the variation of harmonic amplitudes. Such a histogram is shown in Figure 6.30.

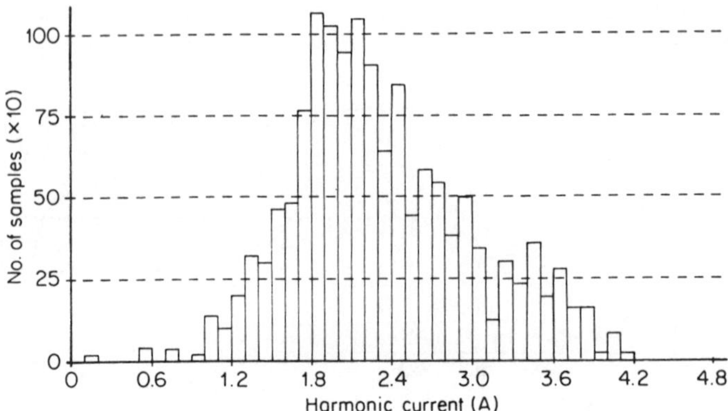

Figure 6.30. Histogram of harmonic amplitudes. The fifth harmonic current limit was 3.4 A at 220 kV, the measurement start time was 1200 hours and the measurement stop time was 2400 hours. From Bradley, Graphical presentation of power system harmonic data, Copyright © 1984, *International Conference on Harmonics in Power Systems*

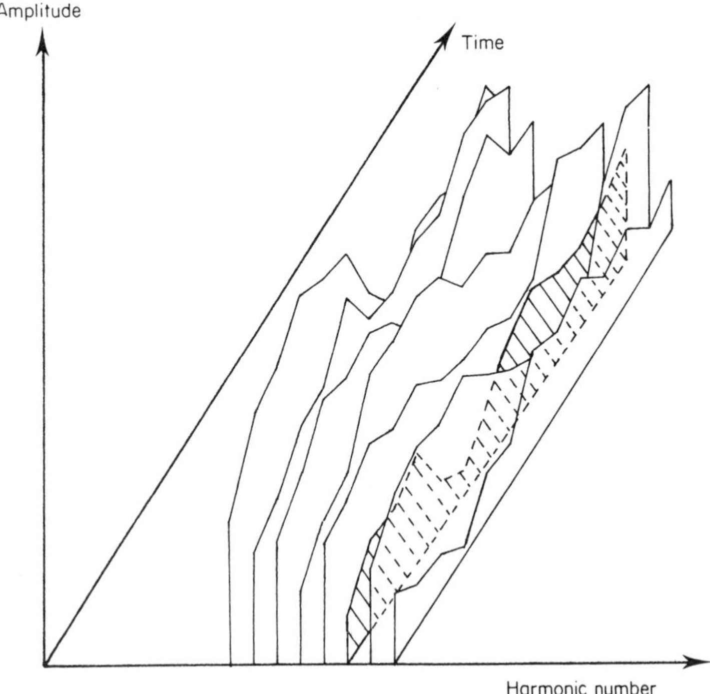

Figure 6.31. Three-dimensional trend line plot with one harmonic shaded. From Bradley, Graphical presentation of power system harmonic data, Copyright © 1984, *International Conference on Harmonic in Power Systems*

Displays involving groups of harmonics

The ability to provide information about the inter-relationships between groups of harmonics is obviously an essential feature of any display and data presentation system. A number of graphical forms have been examined as means of achieving this.

THREE-DIMENSIONAL PLOTS

A three-dimensional plot of individual harmonics on a minute by minute basis has been found to be impractical because of the quantity of data involved and the resulting tendency to mask significant portions of the plot.

If instead, trend lines are used then, although masking still occurs, useful information about general variation and the relationships between individual harmonics is obtained. With some form of enhancement, such as colour, to highlight the particular harmonics of interest the problem of masking is reduced in its significance. Figure 6.31 shows a section of a three-dimensional trend line plot with a single harmonic isolated, in this case by shading.

LEVEL PLOT

As an alternative to the three-dimensional plot, the form of presentation shown in Figure 6.32(a) and (b) can be used, where information is required about the amplitude of each harmonic relative to a defined level.

The result is a presentation which retains time information and can be used to identify intervals of harmonic activity across the full range of harmonics.

SPECTRUM

Most harmonic analysers provide their output as some form of spectral plot. This same approach can be adopted for display, as shown by Figure 6.33. The difference between this form of spectral plot and that provided by most instruments is that it can be made to carry information from a single minute to any time interval. Typically, for each harmonic the maximum, mean and minimum in the period could be given. This could, however, be readily extended, particularly through the use of colour graphics to provide other statistical information about the particular harmonic. This information could be made selectable via a cursor used to identify the particular harmonic.

Total harmonic distortion

The total harmonic distortion can be calculated from the recorded harmonic data and then plotted directly against time. Similar curves could be produced for other computed values, such as the psophometrically weighted currents and voltages required by some standards.

Interpretation of harmonic data

All the above forms of presentation do not in themselves provide a solution to the problem of extracting the required information from the mass of data available.

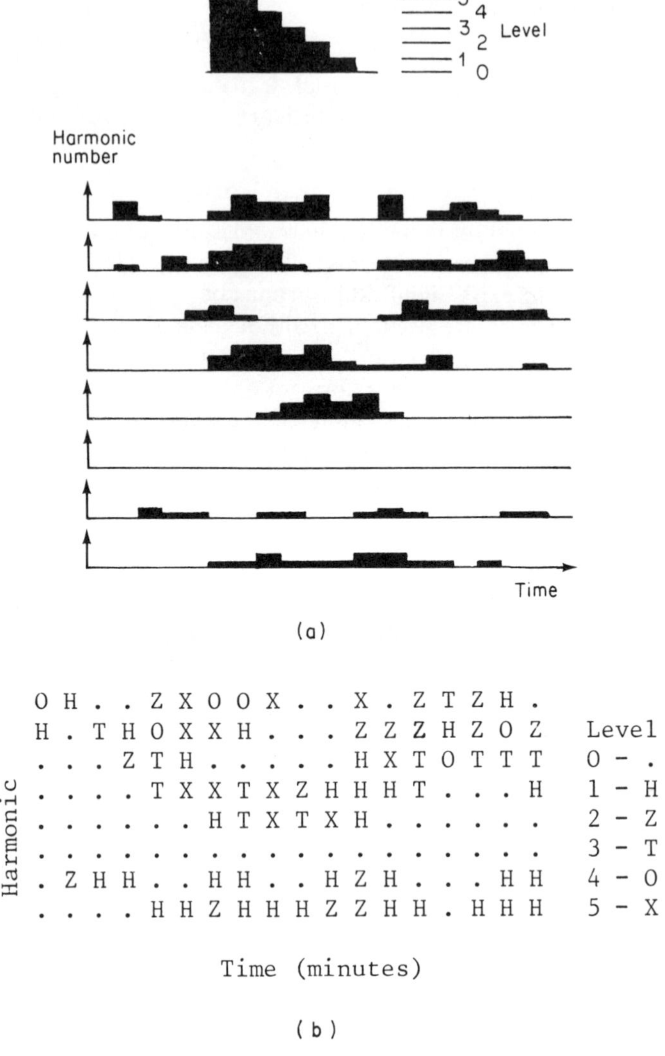

Figure 6.32. (a) Amplitude of each harmonic relative to a defined level and (b) the same level plot using symbols. From Bradley, Graphical presentation of power system harmonic data, Copyright © 1984, *International Conference on Harmonics in Power Systems*

What is of significance is their use in combination with large volumes of data. By using the various forms of data presentation it is possible for the user first to isolate and then to focus upon the data relating to the problem.

For example, the use of trend lines enables general information about system conditions to be provided over an extended period. Where the data set permitted, the superposition of trend lines for corresponding time periods, such as days of the week, enable variations to be observed and recorded.

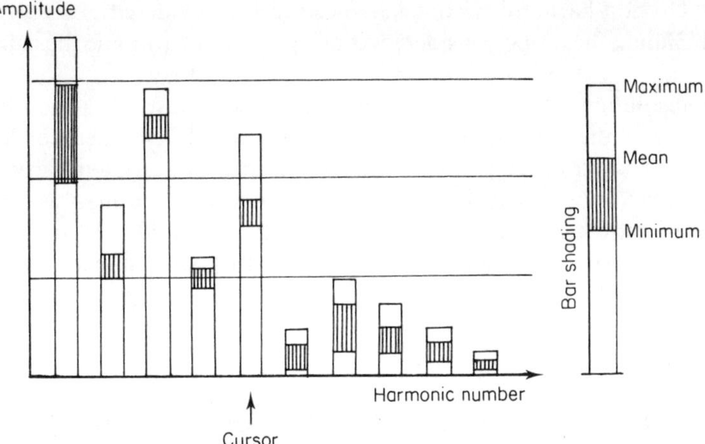

Figure 6.33. Spectral plot. The following information was provided by use of a harmonic select cursor: the harmonic number, n; the mean; the median; the mode; the maximum and minimum; the standard deviation; the upper and lower quartiles; readings in excess of the limit, Copyright © 1984, *International Conference on Harmonics in Power Systems*

The level plots, by providing information over the full range of harmonics, enable periods of high harmonic activity to be isolated and identified. A more detailed examination can then be made using, for example, the spectrum plot and the plot of individual minute values.

The present tendency is to measure and record as much data as possible and then try to identify that of significance by techniques of the type discussed above.

In the longer term a number of questions must be resolved to improve the techniques for analysing and presenting power system harmonic data:

(i) To what extent can data analysis and the sorting of data be incorporated at the point of measurement?
(ii) What sampling and averaging periods must be associated with any particular measurement function?
(iii) To obtain an adequate background picture of system harmonic behaviour and conditions for comparison purposes, how often and in what way should measurements be made?
(iv) How should measurements taken on analogue and digital instruments be compared?
(v) Is real time measurement necessary?
(vi) At what points in the system should measurements be made for greatest effect?

In addition to the above, the following, non-harmonic, conditions need to be considered as part of the measuring system:

(vii) How should bursts of harmonics and other transient forms of waveform distortion such as notching be monitored and recorded?

(viii) Do non-harmonic frequencies need to be considered?
(ix) Should flicker be considered as an element of harmonic measurements?

In the resolution of these questions importance must be attached to the integration of the measurement function with the development of harmonic analysis and harmonic penetration studies. The measuring system can be used to provide data for the analysis programs and the results of the analysis can help to identify measurement requirements. This framework could lead to a simplification of measurement procedures and assist in the establishment of harmonic standards.

6.9 REFERENCES

1. Lipscombe, F. W. (1971). *The Wise Man of the Wires – The History of Faraday House*, Hutchinson, London.
2. Cockroft, J. D., Coe, R. D., Tyacke, J. A., and Walker, M. (1952). 'An electric harmonic analyser', *J. IEE*, (63), 69.
3. Coe, R. T. (1929). 'A portable electric harmonic analyser'. *J. IEE*, (69), 1249.
4. Prescott, J. C. (1939). 'An electrostatic analyser for complex waves of small amplitude', *J. IEE*, (85), 302.
5. Gall, D. C. (1946). 'Scientific instruments'. *J. IEE*, (93), 353.
6. Barnes, H. (1974). 'An automatic harmonic analyser for the supply industry'. *IEE Conf. Publ.*, **110**, 203–208.
7. Kearley, S. J.(1982). 'A low cost harmonic analyser for the distribution network'. *IEE Conf. Publ.*, **210**.
8. Edward, L. N. M., Arrillaga, J., Ross, N. W., and Baird, J. F. (1981). 'Harmonic measurement and the selective AF power analyser'. *IEE Conf. Publ.*, **197**, 81–85.
9. Randall, R. B. (1977). *Application of B & K Equipment to Frequency Analysis*, Bruel and Kjaer.
10. Roberts, J. B. G., Moule, G. L., and Parry, G. (1980). 'Design and application of real time spectrum analyser systems'. *IEE Proc., F.* **127**, 76–91.
11. Wilson, L. A., and Morfee, P. J. (1978). 'Harmonics beware!' *New Zealand Engineering*, **33** (7), 164.
12. Hamming, R. W. (1977). *Digital filters*, Prentice Hall, Englewood Cliffs, N. J.
13. Defty, J. W., and Maples, G. C. (1970). 'Harmonic measurement facility for power systems'. *Proc. IEE*, **117**, 1993.
14. Bradley, D. A., and Grindrod, S. J. (1982). 'The influence and measurement of harmonics on an industrial system'. *IEE Conf. Publ.*, **210**, 136–141.
15. Hensman, G. O. (1978). 'A digital recording and measuring (DREAM) equipment for power system data acquisition'. Paper presented at the 13th Universities Power Engineering Conference, Heriot Watt University.
16. CIGRE, 'Harmonics, characteristic parameters, methods of study, estimating existing values in the network'. *Electra*, (27), 35–54.
17. Kendall, P. G. (1981). 'Harmonics in power systems'. Paper presented at a seminar on harmonics, UMIST, Manchester.

7 Transducers and data transmission

7.1 INTRODUCTION

The function of a current or voltage transformer is to provide a replica of power system current or voltage, at a level compatible with the operation of the instrumentation, in circumstances where direct connection is not possible.

For harmonic measurements the primary requirement is that the transformer should have a defined frequency response. Whilst this would ideally be of constant ratio and phase shift (Figure 7.1), any stable and definable response could be used. In this latter case a correction would need to be applied to allow for the transformer characteristic.

The majority of transformers used on power systems are of the following types:

(i) Toroidally wound current transformers using a ferromagnetic core. These usually have a single turn (bar) primary. An air-gap may be incorporated in the core to reduce remnance and d.c. current effects.
(ii) Magnetic voltage transformers using a ferromagnetic core.
(iii) Capacitive voltage transformers in which a capacitive potential divider is combined with a magnetic voltage transformer.

In addition a range of non-conventional current transformers have been proposed, based upon the Hall and Faraday effects and gyromagnetic materials. For voltage measurements a number of divider forms have been produced.

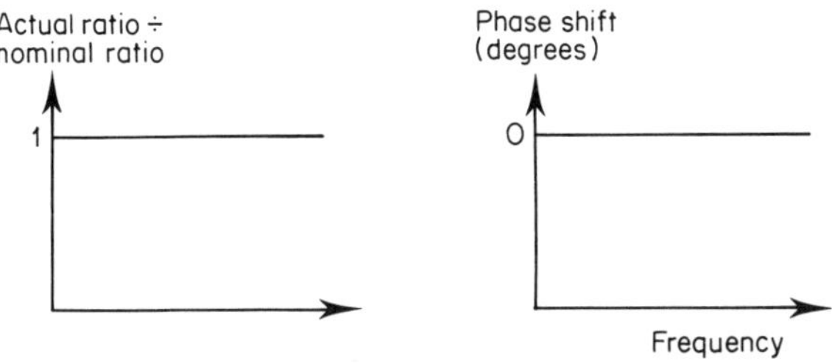

Figure 7.1. Ideal transformer frequency response

Whilst the behaviour of the conventional current and voltage transformers at fundamental frequency is well understood and defined, the behaviour at higher frequencies has not been as fully examined. With the need to measure power system harmonic content, their performance in transforming current and voltage signals containing harmonic components is essential to the measurement process.

7.2 CURRENT MEASUREMENT

The most common type of current transformer is the toroidally wound transformer with a ferromagnetic core. This has, by virtue of its construction, low values of primary and secondary leakage inductance and primary winding resistance. Under normal operating conditions the transformer primary current will be substantially less than that required for saturation of the core, and operation will be on the nominally linear portion of the magnetization characteristic. Thus the simple linear equivalent circuit of Figure 7.2 can be used as the basis for the current transformer model.

The frequency response of current transformers is effectively determined by the capacitance present in the transformer and its relationship with the transformer inductance. This capacitance may be present as inter-turn, inter-winding or stray capacitance. The effect of these various capacitances can then be modelled in the

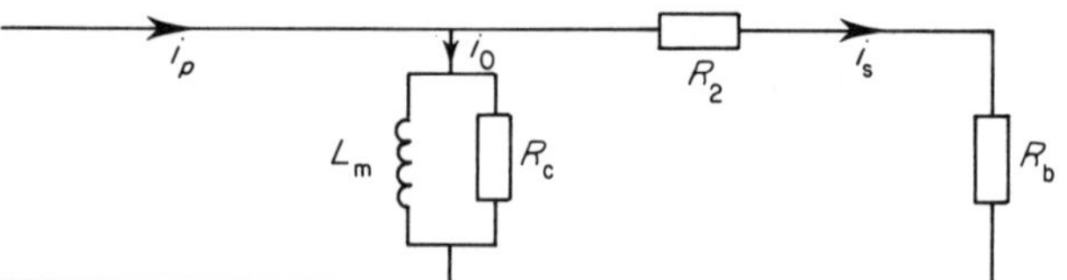

Figure 7.2. Basic current transformer equivalent circuit

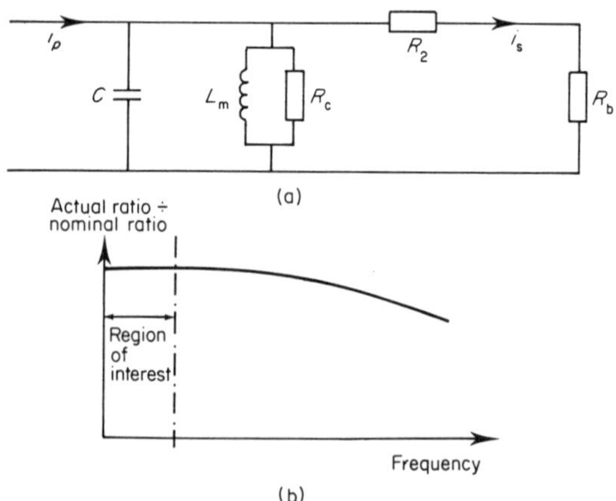

Figure 7.3. Frequency response characteristics of a current transformer: (a) current transformer equivalent circuit including capacitance; (b) frequency response

equivalent circuit by including an appropriate capacitance in parallel with the magnetizing branch as in Figure 7.3(a).

Tests have shown [1–3] that, while this capacitance can have a significant effect on the high frequency response, the effect on frequencies to the 50th harmonic is negligible, as its impedance at these frequencies is much greater than that of the magnetizing branch (Figure 7.3(b)).

In addition to harmonic frequencies it is also possible that the primary current will contain a d.c. component. If present, this d.c. component will not be transformed but will cause the core flux of the transformer to become offset (Section 4.2). If this offset is sufficiently large the transformer will be forced into saturation with a subsequent loss of information (Figure 7.4). A similar condition could arise from remnant flux present in the transformer core as a result of switching.

For this reason, where the presence of a d.c. component is suspected, or remnance a possibility, a current transformer with an air-gap in the core would be used. This air-gap reduces the effect of the d.c. component by increasing core reluctance and enables linearity to be maintained. Because the current transformer burden tends to increase with frequency, the associated power factor reduces with increasing frequency and the transformer will produce a higher harmonic output voltage than it would for a purely resistive load. The resulting increased magnetizing current will cause further error.

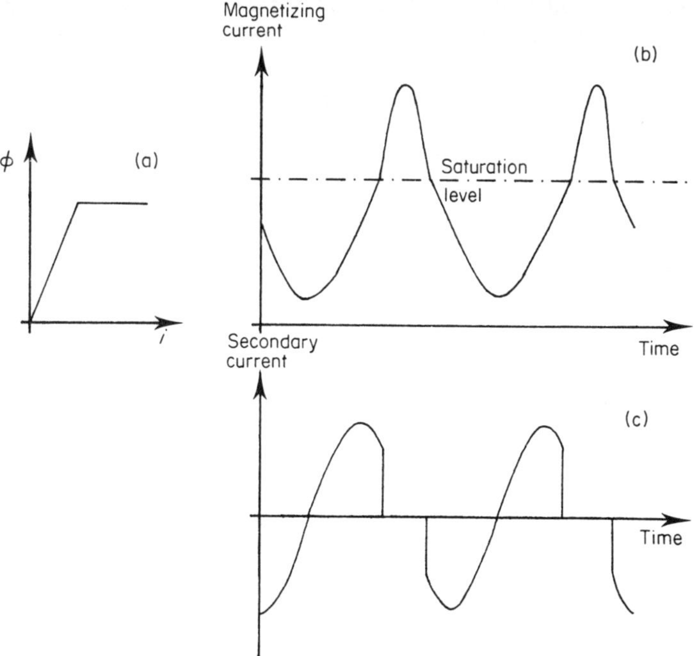

Figure 7.4. Effect of core saturation on current transformer secondary current for a resistive burden: (a) idealized magnetizing characteristic; (b) magnetizing current; (c) secondary current

The following practical recommendations are worth observing whenever possible:

(i) If the current transformer is a multi-secondary type, the highest ratio should be used. Higher ratios require lower magnetizing current and tend to be more accurate.
(ii) The current transformer burden should be of very low impedance, to reduce the required current transformer voltage and, consequently, the magnetizing current.
(iii) The burden power-factor should be maximizing to prevent its impedance from rising with frequency and causing increased magnetizing current errors.
(iv) Whenever possible, it is suggested that the secondary of the measuring current transformer is short-circuited and the secondary current monitored with a precision clamp-on current transformer.

7.3 VOLTAGE MEASUREMENT

The magnetic voltage transformer

The magnetic voltage transformer is generally modelled by the equivalent circuit of Figure 7.5(a). Whilst this circuit adequately models the fundamental frequency it takes no account of higher frequencies. To allow for these the equivalent circuit is modified to that of Figure 7.5(b), with the capacitors C_1 to C_4 included to represent the inter-turn, inter-winding and stray capacitances of the transformer.

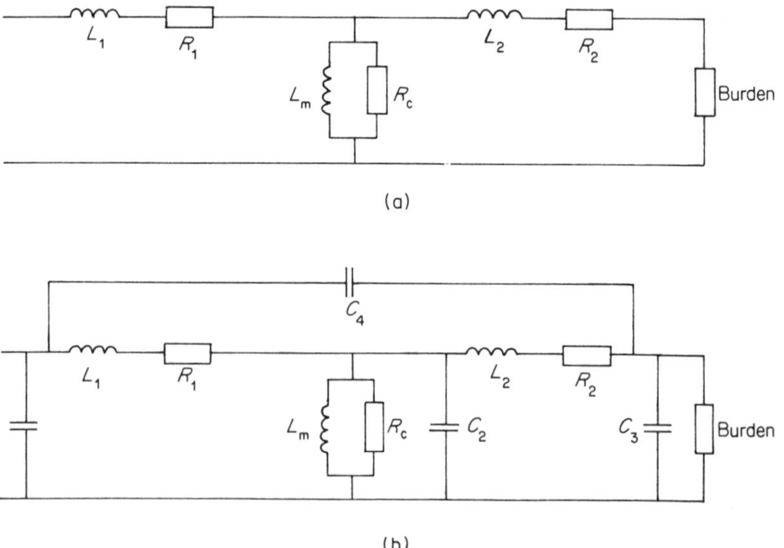

Figure 7.5. Magnetic voltage transformer equivalent circuits: (a) basic equivalent circuit; (b) with added capacitance

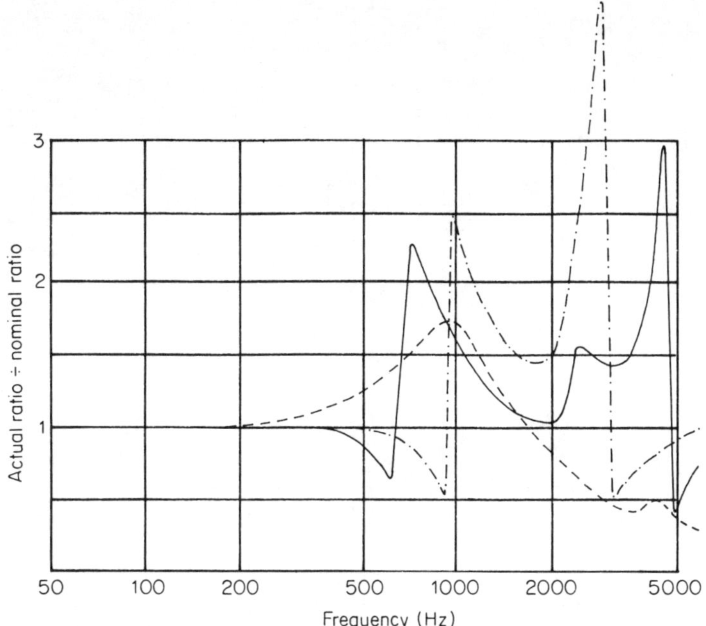

Figure 7.6. Frequency response of magnetic voltage transformers (CIGRE). ——, 400 kV primary voltage; -·-·-·, 200 kV primary voltage; - - - -, 20 kV primary voltage. Reproduced by permission of CIGRÉ[5].

Tests have shown that for transformers operating on voltages to about 11 kV, a linear response is obtained up to 1 kHz and possibly 2 or 3 kHz. The precise nature of the response has been found to be dependent upon the burden used with the transformer.[(4)]

At higher voltage levels the transformer tends to exhibit resonances at lower frequencies, as the internal capacitance and inductance values vary with insulation requirements and construction. The results of Figure 7.6 are illustrative of the type of response obtained. The precise response for a particular unit will be a function of its construction.[(5)]

Capacitive voltage transformer

The capacitive voltage transformer (CVT) combines a capacitive potential divider with a magnetic voltage transformer as shown in Figure 7.7. This combination enables the insulation requirements of the magnetic unit to be reduced with an associated saving in cost.

The additional capacitance provided by the capacitive divider will influence the frequency response of the CVT and Figure 7.8 is illustrative of the responses obtained.

It has been found that the form of frequency response obtained is dependent upon the magnitude of the fundamental component and its relationship to any

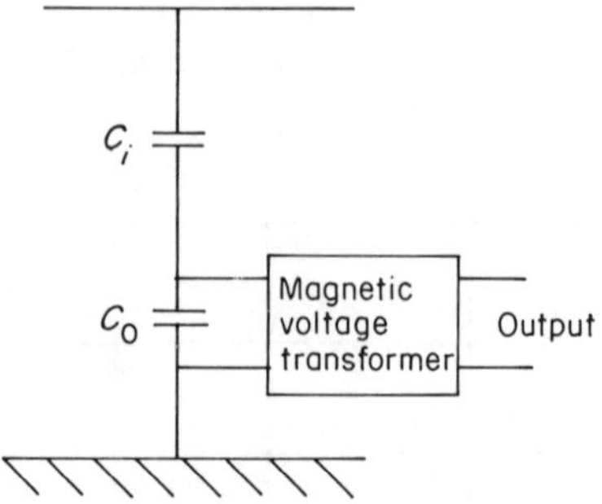

Figure 7.7. Capacitive voltage transformer

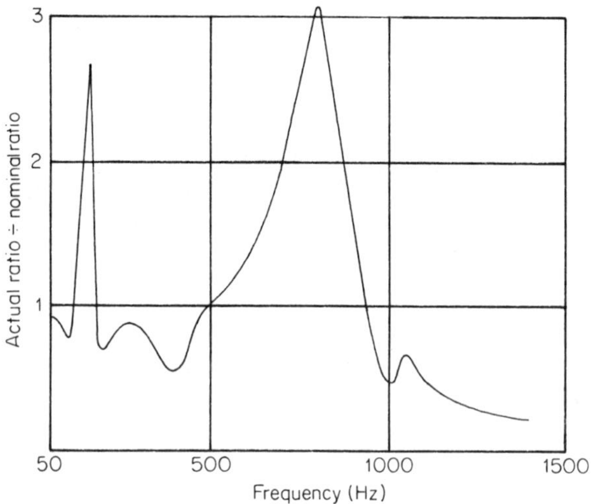

Figure 7.8. Frequency response of a capacitive voltage transformer. Reproduced by permission of CIGRÉ[5]

transition point in the magnetization characteristic of the transformer steel. A large change in the form of the frequency response can result from small changes in fundamental magnitude. Figure 7.9 shows this effect.(5 – 7)

Cascade voltage transformer

The cascade voltage transformer is formed by connecting two magnetic transformer units together in the manner of Figure 7.10. By using this arrangement two low voltage units may be used to perform high voltage measurements.

The frequency response of this combination is illustrated by Figure 7.11.(8)

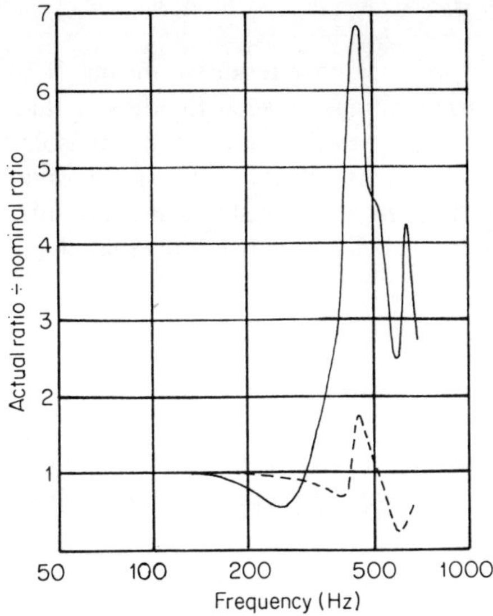

Figure 7.9. Frequency response of a capacitive voltage transformer showing the effect of a small variation in fundamental voltage. ——, V_1; – – – –, V_2; $V_2 > 99V_1/100$. From Bradley, Bodger and Hyland, Harmonic response tests on voltage transducers for the New Zealand power system, *IEEE Summer Power Meeting*, Paper No. 84 SM 687–0. Copyright © 1984 IEEE

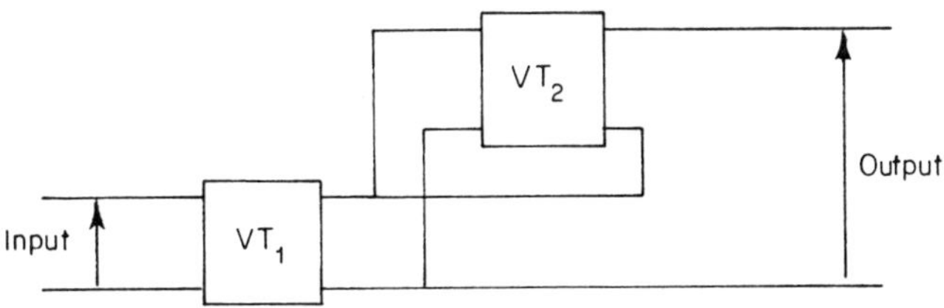

Figure 7.10. Cascade connection of voltage transformers

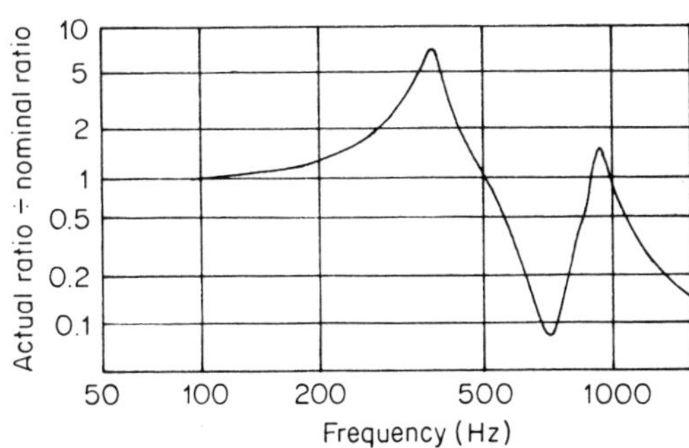

Figure 7.11. Frequency response of cascade voltage transformers

Voltage dividers[9]

The voltage divider provides an alternative means of obtaining the required replica voltage. A disadvantage of such dividers is that, unlike transformers employing magnetic units, they do not provide any isolation between the high voltage and measuring systems. In a practical system some isolation is needed for the protection of both operator and instrument, particularly in an environment where high common mode voltages may be anticipated due to faults or switching surges.

RESISTIVE DIVIDERS

Ideally, a resistive divider should have a linear frequency response. In practice the presence of stray capacitance to ground, any inductive component in the resistor construction and skin effects can cause this ideal response to be modified.

Stray capacitance varies with distance to ground as shown in Figure 7.12(a). By

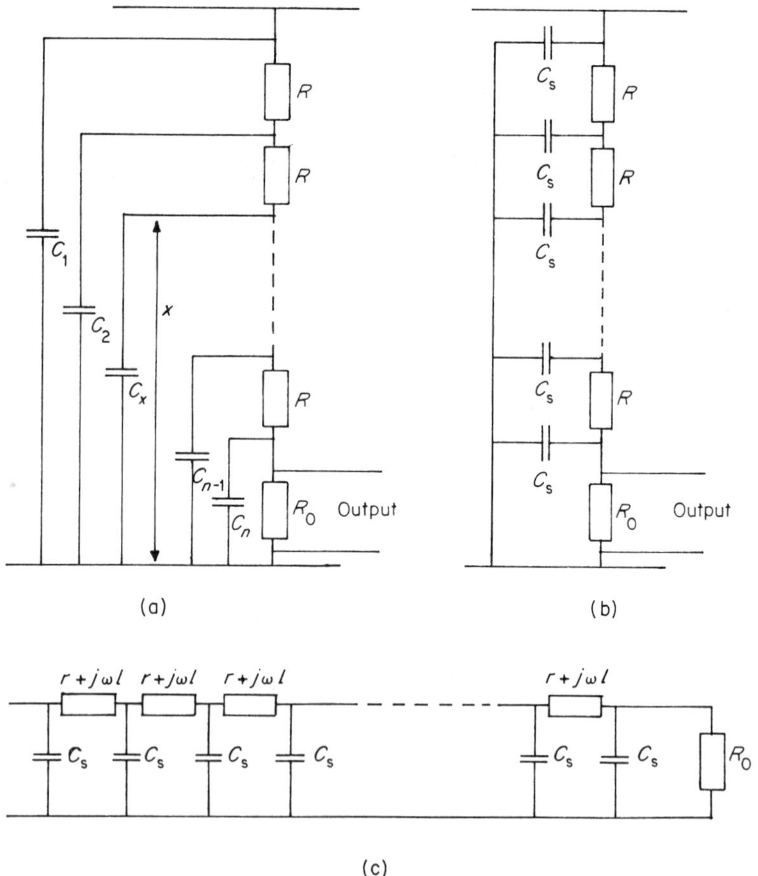

Figure 7.12. Resistive dividers: (a) unscreened resistive dividers; (b) screened resistive dividers; (c) screened resistive divider as a transmission line including imperfect resistors

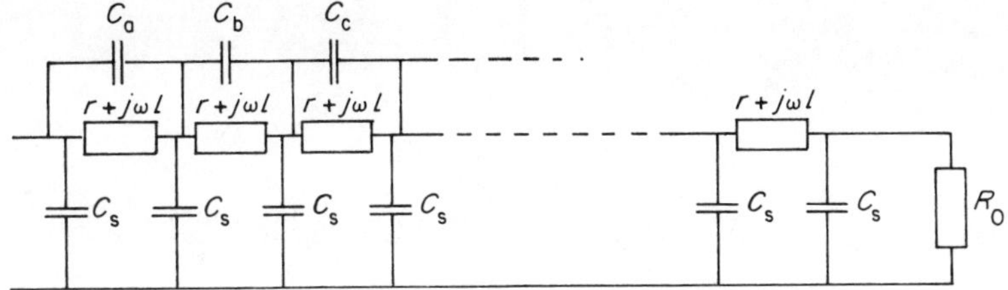

Figure 7.13. Compensated screened resistive divider: $C_a > C_b > C_c >$ etc.

providing an earthed screen (Figure 7.12(b)) a constant value of stray capacitance can be obtained, enabling the divider to be analysed as a transmission line terminated in a lumped resistance R_0 (Figure 7.12(c)).

The frequency response can be further improved by the addition of capacitance in parallel with the resistive elements, as in Figure 7.13. The values of added capacitance are a function of the stray capacitance and position along the divider chain.[(10)]

For measurements at typical harmonic frequencies, however, the influence of these factors is unlikely to be significant.

CAPACITIVE DIVIDER

The basic capacitive divider of Figure 7.14 provides an alternative to the resistive divider for use on a.c. systems. For measurement of harmonics either a purpose built divider could be assembled or, alternatively, use could be made of the divider unit of a capacitive voltage transformer, with the magnetic unit disconnected, or the loss tangent tap on an insulating bushing.

When subject to an impulse such as might arise from local switching, the capacitive divider is subject to 'ringing' due to the interaction between the divider capacitors and their internal inductances. This can lead to high common mode voltages, particularly in areas of high earth impedances. To minimize ringing, the capacitors forming the divider circuit should have a low inductance and the low voltage capacitor should be screened. Ringing can be further reduced by using a

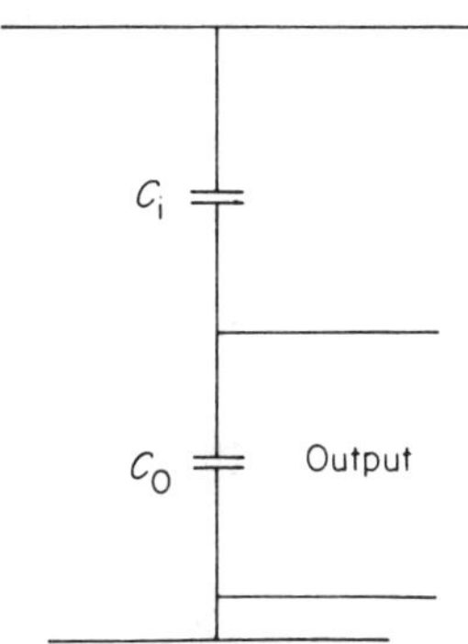

Figure 7.14. Capacitive divider

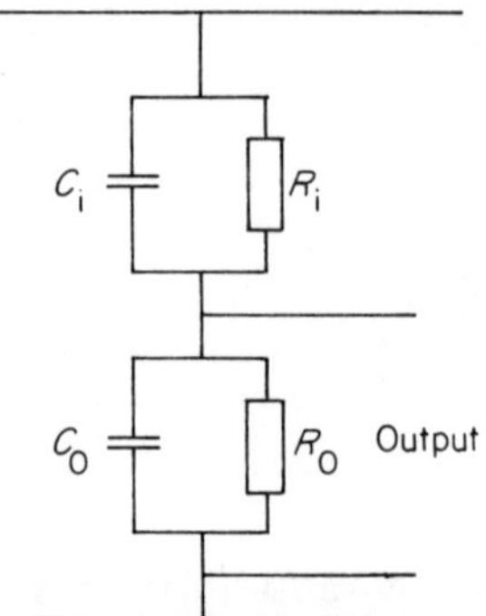

Figure 7.15. Parallel RC divider

mixed *RC* divider in which resistance is included in the divider circuit to damp out the oscillations.

PARALLEL RC DIVIDER

The parallel *RC* divider is shown in Figure 7.15. In this divider the resistance acts to damp out ringing, whilst the parallel capacitance reduces the effect of any stray capacitance.

By using the same time constant for each element, a frequency-independent divider is obtained. For a practical divider this time constant should be less than a half-cycle of the fundamental frequency.

SERIES RC DIVIDER

The series *RC* divider incorporates a series resistance element as part of the divider stack. An output can be obtained either across this resistor or the low voltage capacitor. Figure 7.16 shows both these arrangements.

With the output taken across the resistor the divider acts as a high pass filter

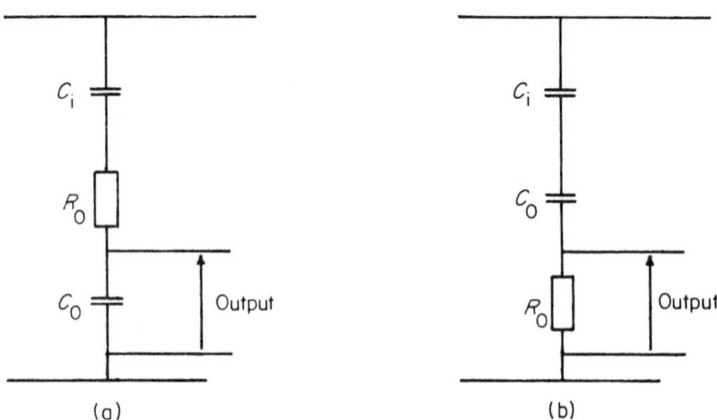

Figure 7.16. Series RC dividers: (a) low pass form; (b) high pass form

with the frequency response of Figure 7.17, defined by equation (7.1):

$$\frac{V_o}{V_i} = \frac{1}{1 + \omega_o/j\omega}, \tag{7.1}$$

where $\omega_o = 1/R_oC$ and $C = C_iC_o/(C_i + C_o)$. By choosing ω_o to be several times greater than ω for the highest frequency to be measured, the divider can be made to operate on the linear asymptotic region of the response curve. In this way the divider can be used to reduce the system dynamic range requirement, by reducing the amplitude of the lower order harmonics with respect to the higher orders.

When used in this way a correction must be applied, in the form of the inverse filter characteristic, to obtain the system harmonic amplitudes.

If instead the output is taken across the low voltage capacitor, the frequency response of Figure 7.18 and equation (7.2) is obtained.

$$\frac{V_o}{V_i} = \frac{C_i}{(C_i + C_o)} \cdot \frac{1}{(1 + j\omega/\omega_o)}. \tag{7.2}$$

If ω_o is again chosen to be several times greater than the largest value of ω,

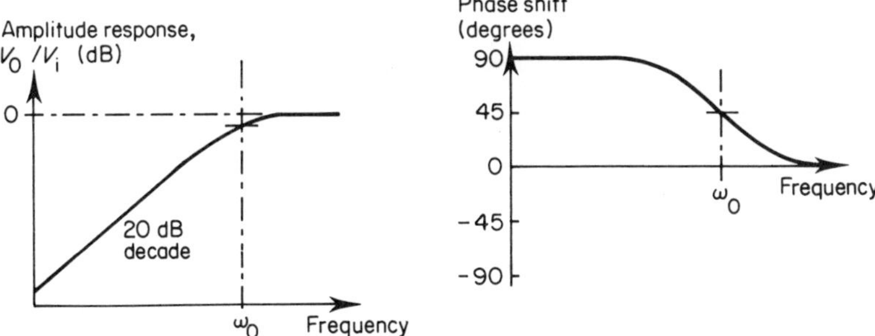

Figure 7.17. Series RC divider, frequency response of high pass form

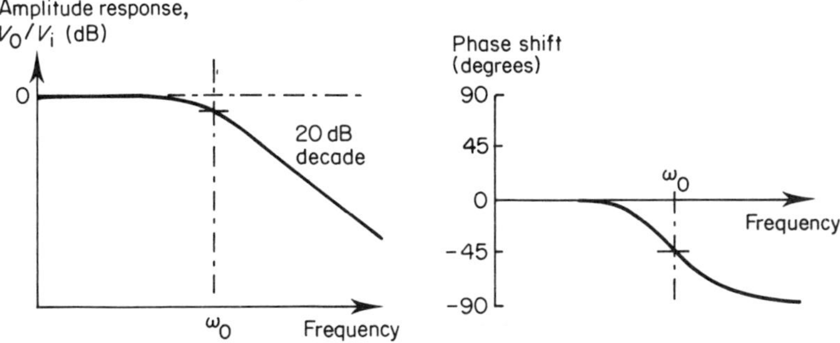

Figure 7.18. Series RC divider, frequency response of low pass form

operation will now be in the constant, linear region, of the frequency response curve. Depending on the value of ω_0 some correction may still be required at the higher harmonic frequencies.

Amplifier-based systems with capacitive dividers[11,12]

In recent years a number of amplifier-based capacitive divider systems have been developed. Intended primarily for use with high speed protection schemes, they also have obvious application for harmonic measurements.

The basic arrangement for a single phase unit is shown in Figure 7.19. Ideally the amplifier should be sited as close as possible to the divider to minimize the effects of cable capacitance. This is particularly so where an insulating bushing is used as divider, as its capacitance will be several orders of magnitude less than that of a purpose-built divider. Where it is not possible to position the amplifier close to the divider, the arrangement of Figure 7.20 can be used, with the C_o capacitor incorporated into the feedback loop of the amplifier. With this arrangement the capacitive current flowing in C_i is forced to flow through C_o by

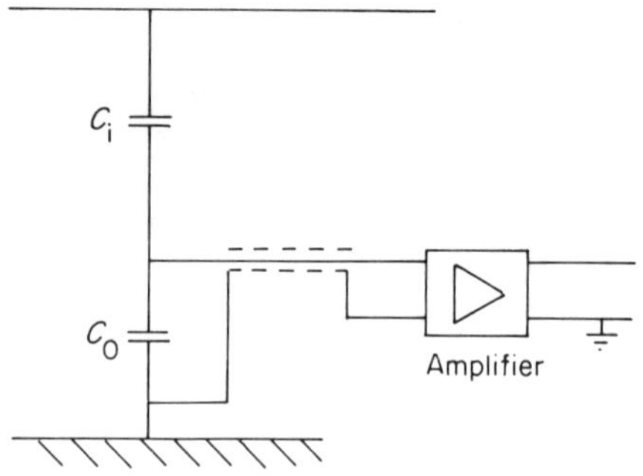

Figure 7.19. Capacitive divider incorporating amplifier circuit

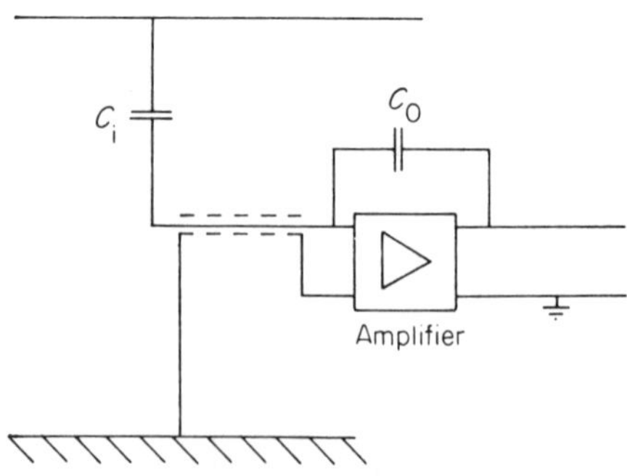

Figure 7.20. Capacitive divider with capacitor in feedback loop of amplifier: current driven system

the high input impedance of the amplifier. The amplifier input is therefore a virtual earth and the circuit acts as a capacitive divider with the intermediate point at earth potential. All stray capacitance in the connection between the divider and the amplifier can therefore be neglected and the output voltage will be dependent entirely on C_i and C_o.

The amplifier itself should provide a high input impedance, to avoid loading the divider, together with a low impedance output for good voltage regulation. In addition it must be adequately screened to prevent problems caused by interference.

To prevent damage to the amplifier system and any associated equipment connected, due to an excessive difference in earth potential between the transformer and the measuring point, the inputs and outputs of the arrangement of Figure 7.19 should be isolated. This isolation may be achieved by a variety of means, including isolating amplifiers, transformers or fibre-optic links. If the circuit of Figure 7.20 is used, then surge protection must be fitted at the capacitive divider base.

The amplifier must also be provided with its own power supply isolated from other voltage sources. This will require the provision of a suitable screened and isolated supply derived from some external source.

Also, where carrier injection is required at the measuring point, the capacitances in the amplifier input connections and capacitor divider earth must be small to prevent the carrier signals from being bypassed and to avoid any detuning.

The performance requirements of a divider/amplifier arrangement for harmonic measurements are generally the same as those for protection purposes, if a little less stringent in certain areas such as transient response and the reproduction of overvoltages.

Hybrid dividers

By introducing an additional capacitor in the base of the divider stack the conventional CVT output can be maintained whilst providing the additional amplifier driven signal. Care must be taken when using this arrangement for harmonic measurement, as the presence of the magnetic circuit can modify the frequency response.

7.4 HIGH VOLTAGE PROBES AND CLAMP-ON CURRENT TRANSFORMERS

While at points of harmonic injection it is possible to use permanent transducers, at remote locations, high voltage probes and clamp-on current transformers will often be needed. However, their voltage limitation to under 15 kV restricts their application to low voltage distribution systems. A typical probe[13] used to measure phase to ground voltages is constructed using a glass-epoxy strut to which one end of a current-limiting power fuse is clamped; the other end of the fuse carries the clamp for the phase conductor. A compensated coaxial cable is used to achieve an acceptable frequency response to several kilohertz.

Clamp-on current transformers have been used[13] to monitor the three-phase currents and the neutral current at test sites on the distribution feeders at voltages to 8700 V to ground. Again, the frequency response of the devices with long instrument leads has been checked up to several kilohertz; the magnitude transfer is accurate but the phase-angle measurement is not reliable.

7.5 UNCONVENTIONAL CURRENT AND VOLTAGE TRANSFORMERS[14]

Various alternatives to the conventional current transformer have been investigated. These include (i) passive optical or microwave systems using ground-based transmitting and receiving stations; (ii) active systems using a conductor-mounted transducer and transmitter and a ground receiving station; (iii) hall effect systems; (iv) hybrid current transformers.

Passive systems

In a passive system a transmitted signal is modulated by a transducer mounted at the conductor. No power source is required at the conductor.

Optical systems use the Faraday magneto-optic effect by which the plane of polarization of a beam of linear polarized light is rotated by a magnetic field along its axis.

Designs for Faraday effect current transformers use either the open path or the closed path optical system[15] shown in Figure 7.21(a) and (b).

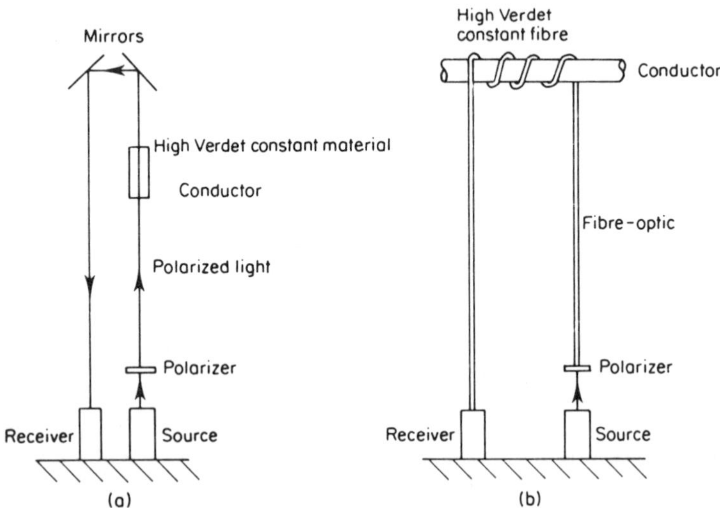

Figure 7.21. Faraday effect current transformer: (a) open path Faraday magneto-optic effect current transformer; (b) closed path Faraday magneto-optic effect current transformer

Microwave systems make use of gyromagnetic materials to modulate a microwave carrier by a magnetic field. The form of modulation is controlled by the arrangement of the gyromagnetic material and the form of polarization of the microwave signal.

Active systems

An active system uses a conductor-mounted transducer to provide a modulating signal for a carrier generated at the conductor. Transmission of the carrier to the receiving station is then achieved via a radio or fibre-optic link. The power for the transmitter is usually line derived using a magnetic current transformer together with some battery back-up (Figure 7.22).

Hall effect transducers

The Hall effect (Figure 7.23) is used in a variety of probes and transducers covering a range of current levels. For current transformer applications a major

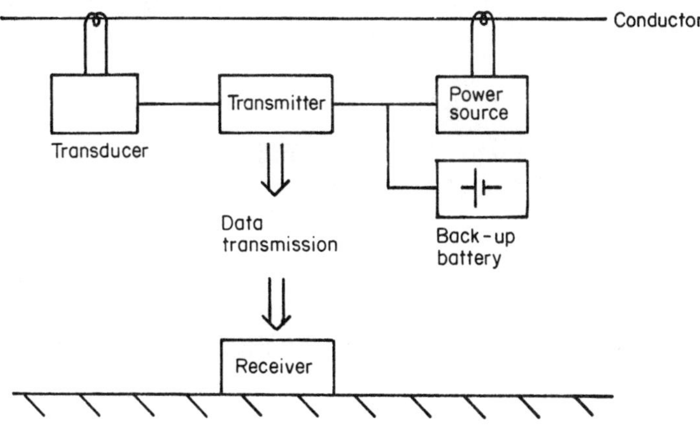

Figure 7.22. Active system with line derived power supply

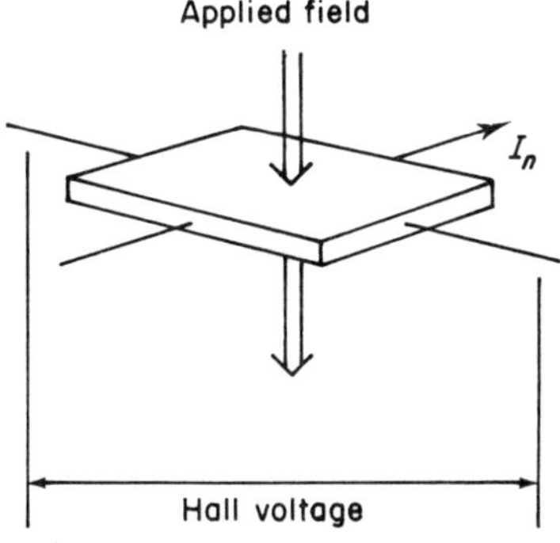

Figure 7.23. Hall effect transducer

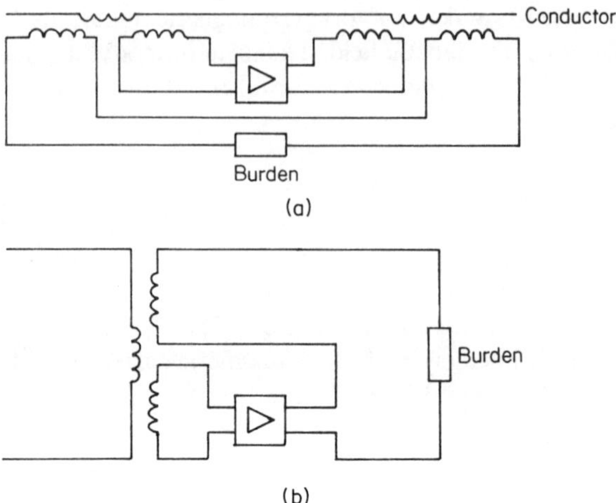

Figure 7.24. Forms of hybrid current transformer

problem is that of maintaining calibration over long periods, particularly where a wide range of operating temperatures is likely.

Hybrid current transformer[16]

In its original form as the zero flux current transformer, an amplifier was used to supply the load and reduce the transformer secondary voltage. In later versions a sensing winding is provided, the output of which is taken to a high gain amplifier which supplies the load via the secondary winding. Any change in flux levels will alter the amplifier output to compensate for the change. As an alternative to the sensing winding, a Hall probe could be used to detect any change in flux. Figure 7.24 shows some of the hybrid current transformer forms.

Unconventional voltage transformers[14]

Electro-optical and electrogyration effects can be used to measure voltage in a manner akin to the Faraday effect used for current measurement.

The electro-optical effect causes linearly polarized light passing through the material to become elliptically polarized. As the two mutually perpendicular components propagate in the crystal at different velocities, they have a difference in phase as they emerge from the material. This phase difference will be proportional to the path length in the material and to either the field (the Pockels effect) or the square of the field (the Kerr effect). By measuring this phase difference the electric field strength, and hence the votage, can be obtained.[15]

If a linearly polarized beam is propagated through an electrogyrational material in an electric field, the effect is to rotate the plane of polarization in a manner analogous to that occurring in magneto-optic materials. The electrogyration effect can therefore be used to measure voltage in a manner similar to the Faraday effect current transformers.

7.6 DATA TRANSMISSION

A high proportion of power system harmonic measurements will be made using instrumentation sited remotely from the transformers providing the replicas of system current and voltage. Some means of communication must therefore be provided between the transformers and the instruments.

This communication link may pass wholly or partially through a high voltage switchyard, in which case particular attention must be given to the effects of electrostatic and electromagnetic interference as well as the necessary screening.

The provision of increased noise immunity may require the use of other forms of data transmission such as current loop systems, modulated data or digitally encoded data.

Where high common mode voltages can occur, the communications link may be required to provide insulation up to several kilovolts to protect both users and equipment. This may require the use of isolation amplifiers or, where higher levels of isolation are required, fibre-optic links.

The principal transmission medium used is, however, likely to be some form of cable. This may be of the non-coaxial type, with data rates and bandwidths of the order of 10 MHz over distances of a few hundred metres. Coaxial cables will provide an order of magnitude increase in these figures. In either case they may be used for the direct connection of the instrument to the transformer, or as the link along which a processed signal is passed. A fibre-optic link, though providing higher data rates and bandwidths, as well as isolating the instrument from the source, has the disadvantage of a lower signal to noise ratio than a coaxial cable for similar operating conditions.

Information may be transmitted either as an analogue signal for direct connection of the instrument, or in a modulated or encoded form using both analogue and digital data systems. If direct analogue transmission is used, then a system of sufficiently high signal to noise ratio is obviously required. For certain harmonic measurements a dynamic range of the order of 70 dB may well be required and hence the achievable signal to noise ratio must be in excess of this figure.

It must also be borne in mind that most harmonic measuring instruments are designed to receive an analogue signal at their input terminals. This suggests that an analogue form of data transmission would be preferred where considerations of noise immunity, common mode behaviour and isolation permit. If a digital system is used then care must be taken to ensure that it is of adequate dynamic range, and that the resulting signal at the instrument is compatible with that instrument.

Reduction of electrostatic interference[17,18]

The capacitive coupling effect discussed in Section 5.13 can be a source of significant interference on a measuring system. To shield a conductor from electrostatic fields, a conducting plane at earth potential must be introduced to enclose the conductor.

For a measuring system this screen must be extended to enclose all the components of the system together with their interconnections, as in Figure 7.25. Such an ideal arrangement is unlikely to be fully realizable in practice.

A practical screen will have some resistance and inductance as well as capacitance to ground. Any currents flowing in the screen will cause a potential gradient along the screen. As the screen itself can be capacitively coupled to the components it encloses, this can result in some interference. The return path for any screen currents will be through the system earth. Where multiple earth points are used, as in Figure 7.26(a), then the effect of any earth impedance will be to introduce a potential difference between the earth points, which will appear as the common mode voltage in the measuring circuit (Figure 7.26(b)).

During normal operation this common mode voltage is likely to be small. However, during transient conditions such as faults or switching operations

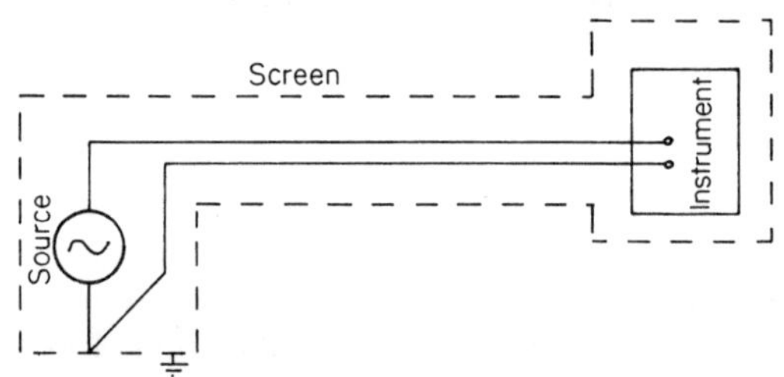

Figure 7.25. Fully screened measuring system

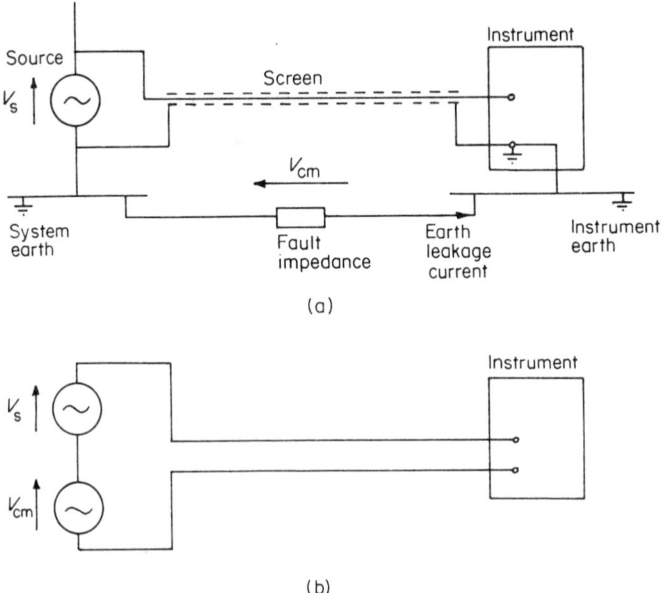

Figure 7.26. Effect of multiple earths on measurement: (a) multiply earthed measuring system; (b) common mode voltage

could, depending upon the resistance of the earth return and the type of transformer used, reach several kilovolts. In such cases it will be necessary to break the group loop in some way, to ensure safe operation and prevent damage to equipment.

For an instrument on which one of the input terminals is also the earth terminal, then only partial screening can be achieved (Figure 7.27(a)). In addition, only signals from sources which are themselves isolated from earth can be measured without introducing errors due to common mode voltages.

Improved screening and common mode rejection can be achieved by using an instrument with an isolated, differential input stage. A separate earth terminal connected to the supply earth is provided (Figure 7.27(b)). The low (LO) terminal may be a virtual earth or connected to earth through some impedance.

A further improvement can be achieved by the inclusion of a guard circuit. This consists of an internal screening box or an arrangement of tracks on a printed

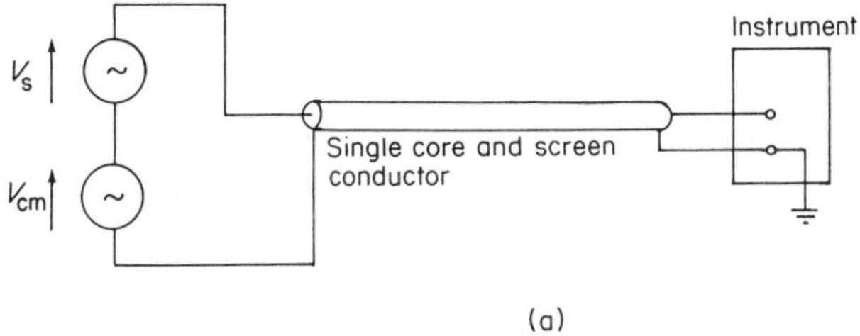

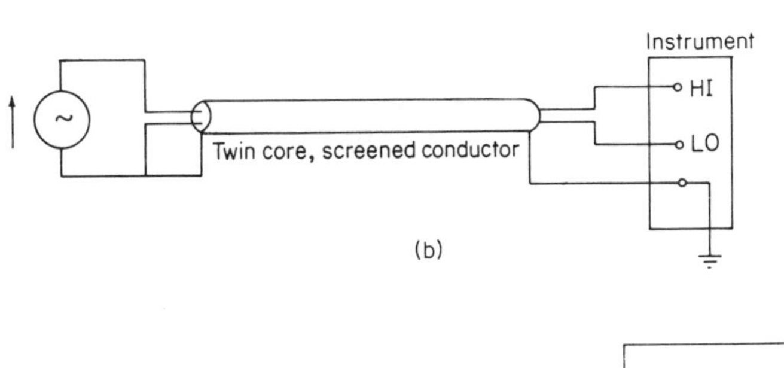

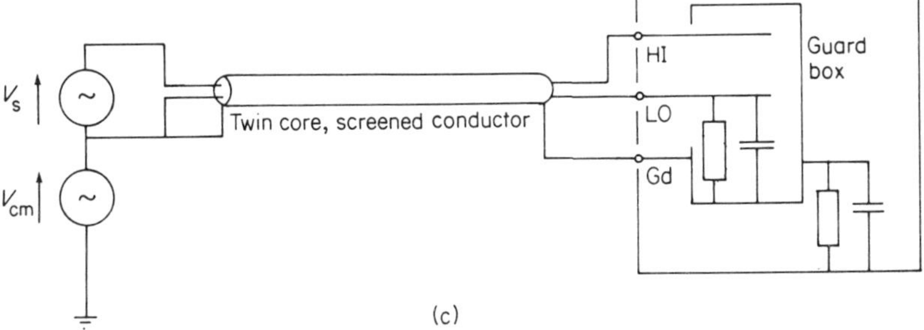

Figure 7.27. Instrument earths: (a) instrument with earthed input; (b) instrument with isolated differential input; (c) instrument with isolated differential inputs and guard

circuit board and is brought out to the guard terminal to which the screen is connected (Figure 7.27(c)). The guard is then driven at the common mode voltage, preventing the flow of any associated current between either terminal and the guard. Where a guard is provided care must be taken to ensure that it is connected, as it can otherwise present a high impedance which can pick up a.c. interference and communicate it to the measuring circuits.

Where signal conditioning circuits are used, particularly involving operational amplifiers, the use of an isolated and screened power supply is recommended. A screened isolating transformer with either a single or a double screen may be used (Figure 7.28), with the double screen offering the best results.[18] A double screen may also be used for the amplifier circuits. Figure 7.29 shows a capacitive divider with associated amplifier. The outer screen is connected to the divider earth and the lower screen to the bottom of the divider. This arrangement gives low input capacitance to earth and increases the immunity to switching surges and transients.

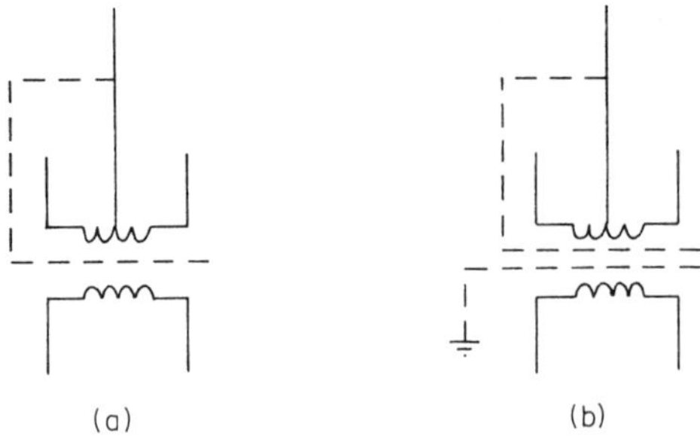

Figure 7.28. Screened transformers: (a) single screen; (b) double screen

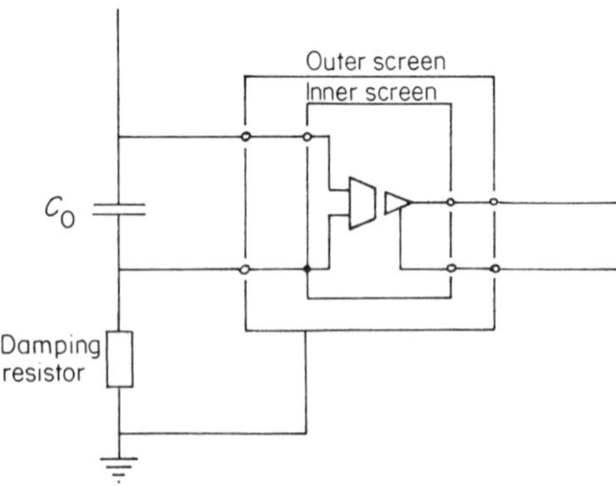

Figure 7.29. Twin screened amplifier enclosure in use with a special capacitive divider

Reduction of electromagnetic interference

If a conductor is placed in the magnetic field of an alternating current it will have an alternating voltage induced in it due to magnetic coupling. Three types of screening are used to minimize the effects of electromagnetic interference, these are as follows:[17]

(i) By positioning a conducting sheet around the conductor or component to be screened. The Eddy currents induced in this sheet by the alternating magnetic flux produce an opposing flux, which acts to prevent the penetration of the interfering flux to the screened components.

(ii) A magnetic shield, consisting of an enclosure of high permeability material, can be used to provide a low reluctance path around the screened components, thus diverting the flux away from them.

(iii) By arranging that the coupling between the interfering magnetic field is kept to a minimum and by endeavouring to cancel out such voltages as are induced. By routing connections as far as possible away from the source of the interference, particularly transformers and inductors, and by twisting leads together (twisted pair) a considerable reduction can be achieved.

RADIOFREQUENCY INTERFERENCE

If the electromagnetic interference is at radiofrequencies (r.f.) then it may be necessary to use a screened enclosure. This is usually constructed of copper sheet and care must be taken to ensure that the screen is complete with all joints electrically continuous and of low impedance, with no open circuits at doors or other openings. Power supplied to the enclosure must itself be both filtered and screened.

Signal conditioning and data transmission

Whilst it may in many situations be both possible and practical to connect the instrument directly to the transformer, in others it may be necessary and desirable to introduce some form of signal conditioning prior to the transmission of data. This may take the form of filtering to reduce the necessary dynamic range by suppressing the fundamental and low order harmonics. Instead, the signal levels may be increased by the introduction of some gain prior to transmission, or by providing an isolation amplifier. Alternatively, the voltage signal may be either converted to a current source for transmission on a current loop, or encoded for transmission on some form of modulation (analogue or digital) system.

Where the signal conditioning circuitry is located at the transformer, and requires to be provided with power from an external source, care must be taken to ensure that this source is adequately isolated, screened and protected.

ISOLATION AMPLIFIER

The isolation amplifier, as shown in Figure 7.30, achieves the required separation between its input and output stages by modulating the output of its pre-amplifier

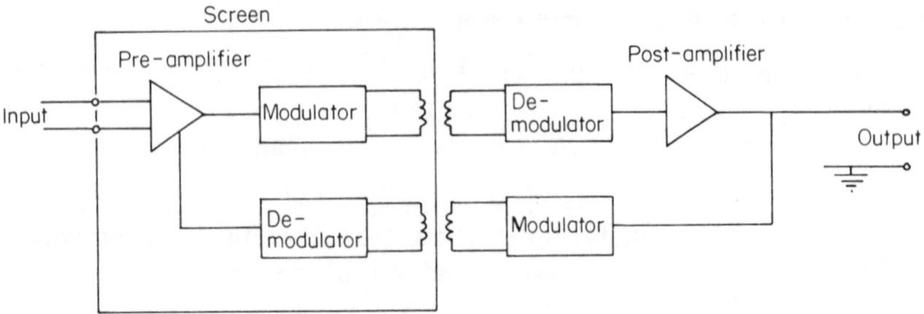

Figure 7.30. Isolation amplifier

to produce a signal which can be coupled via a transformer or optical link to the demodulator and post-amplifier. Where feedback is required, this is similarly provided by a modulator–demodulator arrangement.[19]

Such amplifiers are capable of providing isolation to the order of 2000 V, with a common mode rejection in excess of 100 dB over a frequency range from d.c. to in excess of 4 kHz.

CURRENT LOOPS

By using a signal from the measuring transformer to modulate a circulating current, the noise immunity of the system can be increased, as the majority of interference is in the form of voltage signals induced on to the conductor.

The standard current loop system uses a 4–20 mA loop current with a zero signal represented by the 4 mA level. Hence some offset must be introduced to accommodate the current range when an a.c. signal is to be transmitted.

A further problem encountered is that of dynamic range. A 72 dB dynamic range implies that signals of 4 μA are capable of being detected on a 4–20 mA loop.

MODULATION SYSTEMS

Modulation refers to the imposition of data on to a carrier signal by modification of a parameter or parameters of the carrier signal. The original data is then recovered from the carrier at the receiver by the process of demodulation. The most common forms of modulation are amplitude modulation (AM), in which the amplitude of the carrier is varied, and frequency modulation (FM), in which the instantaneous carrier frequency is varied relative to its nominal value.

Amplitude modulation produces a signal of the form shown in Figure 7.31 and

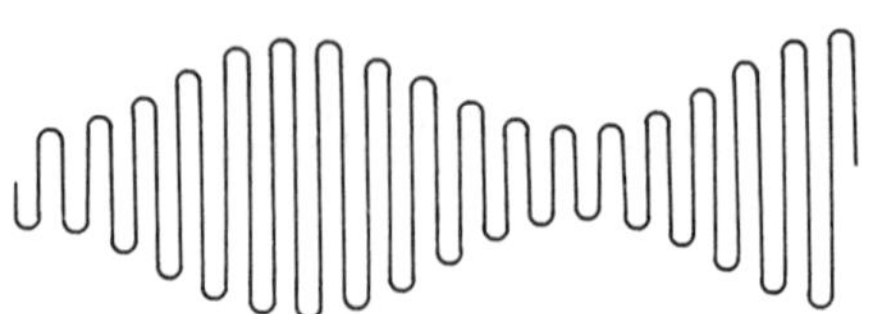

Figure 7.31. Amplitude modulation

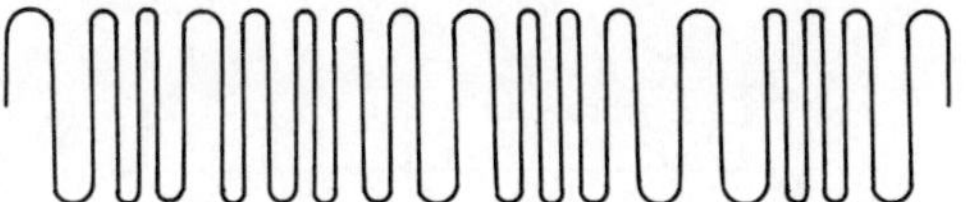

Figure 7.32. Frequency modulation

is represented by equation (7.6.1) for the modulation of a sinusoidal carrier of frequency f_c by a sine wave of frequency f_s:

$$\begin{aligned} v_c &= V_0(1 + m\cos 2\pi f_s t)\cos 2\pi f_c t \\ &= V_0(\cos 2\pi f_c t + m[\cos 2\pi(f_c + f_s)t + \cos 2\pi(f_c - f_s)t]/2), \end{aligned} \qquad (7.6.1)$$

where V_0 is the unmodulated amplitude of the carrier and m is the depth of modulation.

The modulated waveform has components at frequencies f_c, $(f_c + f_s)$ and $(f_c - f_s)$, giving on overall bandwidth of $2f_s$. Based on the Fourier series representation, the modulating waveform produces a series of sidebands about f_c. The overall bandwidth of the amplitude modulated signal is then twice that of the modulating signal.

Amplitude modulation can be achieved by using a multiplier circuit to combine the signals. The demodulator consists of a bandpass filter, usually covering one group of sidebands, a rectifier and a low pass filter.

A frequency modulated signal is shown in Figure 7.32. The frequency components of this signal extend to infinity either side of the carrier frequency and, neglecting signals outside a given bandwidth, introduces some distortion into the demodulated signal. The bandwidth required for a frequency modulated signal is given by Carson's rule, which sets the bandwidth at $2(d + f_m)$ Hz, where d is the maximum deviation of the carrier frequency from its unmodulated value and f_m is the highest frequency in the modulating signal. The required bandwidth is greater than that required for amplitude modulation.

Frequency modulation can be introduced using a variable capacitive or inductive element in the tuned circuit of an oscillator. A phase-locked loop is used for the demodulator stage.

Pulsed system[19]

A number of pulse-encoded systems that are available for data transmission are shown in Figure 7.33.

Pulse amplitude modulation (PAM) modulates the carrier with a series of pulses whose amplitudes carry the information. The spacing between the pulses is constant.

In pulse duration modulation (PDM) or pulse width modulation (PWM) pulses of constant amplitude but of varying duration are used. One edge of the pulse is fixed in time and the position of the other edge varied in relation to the amplitude of the modulating signal.

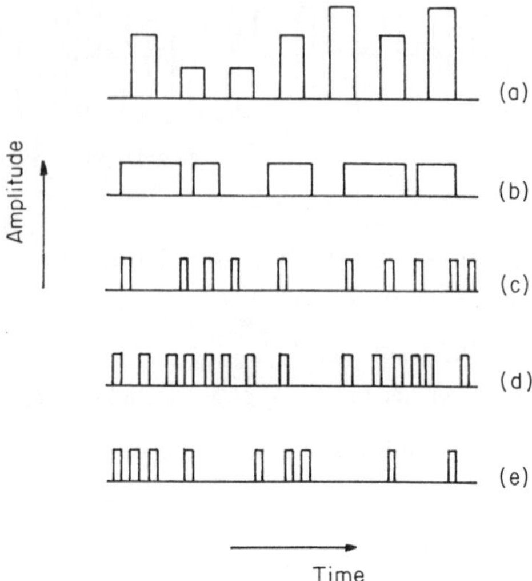

Figure 7.33. Pulse encoded forms for data transmission: (a) pulse amplitude modulation; (b) pulse width modulation; (c) pulse position modulation; (d) pulse frequency modulation; (e) pulse coded modulation

Pulse position modulation (PPM) is similar to the pulse duration and pulse width systems but instead of a continuous pulse a separate, short pulse is sent to mark the beginning and end of a time interval. A pulse position signal can therefore be formed by first differentiating and then rectifying, to ensure that all pulses are of the same sense, a pulse width modulated signal. The pulse position modulated signal absorbs less power than a pulse width modulated signal as the transmitter is active for a shorter time. The shorter pulses do, however, require a greater bandwidth for transmission.

Pulse frequency modulation (PFM) uses a train of constant amplitude, constant duration pulses with variable spacing between pulses. This spacing and hence pulse frequency is varied according to the amplitude of the modulating signal. At the receiver the pulses are detected and averaged to recover the modulating signal.

Pulse coded modulation (PCM) transmits the analogue information as a series of binary digits. The number of binary digits used then determines the dynamic range and quantizing error. The bandwidth required for pulse coded modulation depends upon the frequency content of the modulating signal and the number of binary digits used. For a modulating signal containing frequencies to f_m the minimum sampling rate must be $2f_m$ Hz. If N bits are used for each item of data then the bit generation rate is $2Nf_m$ Hz. As the receiver is only required to discriminate between 1s and 0s a bandwidth of half the bit generation rate is all that is required. This gives a bandwidth of Nf_m Hz for a pulse coded modulation system and the information carrying capacity is directly proportional to the bandwidth.

Fibre-optic Systems

The use of a fibre-optic system as a means of data transmission provides a highly noise-immune communications link with high isolation, enabling it to be used on high voltage systems.

The simplest form of data transmission on a fibre-optic link uses an intensity modulated source, equivalent to the amplitude modulation considered earlier. A limit is, however, imposed by the inherent signal to noise ratio of the fibre-optic link.

For the same attenuation between transmitter and receiver and with equivalent bandwidths the coaxial cable system will have about a 40 dB advantage in signal to noise ratio over the optical fibre.

By using pulse modulation techniques an improved signal to noise ratio can be achieved, but at the cost of increased transmission bandwidth to accommodate the pulsed system.

7.7 REFERENCES

1. Douglass, D. A. (1981). 'Current transformer accuracy with asymmetric and high frequency fault currents'. *Trans. IEEE*, **PAS-100**, 1006–1012.
2. Malewski, R., and Douville, J. (1976). 'Measuring properties of voltage and current transformers for the higher harmonic frequencies'. Paper presented at the Canadian Communication and Power Conference, Montreal.
3. Imatran Voima OY, Finland, personal communication.
4. Douglass, D. A. (1981). 'Potential transformer accuracy at 60 Hz voltages above and below rating and at frequencies above 60 Hz'. *Trans. IEEE*, **PAS-100**, 1370–1375.
5. CIGRE Working Group 36–05 (Disturbing Loads) (1981). 'Harmonics, characteristic parameters, methods of study, estimating of existing values in the network'. *Electra*, (77), 35–54.
6. Casu, A., and Simoncini, U. (1974). 'Harmonics due to rectifiers connected to the Sardinian 220 kV a.c. network'. *IEE Conf. Publ.*, **110**, 251–255.
7. Bradley, D. A., Bodger, P. S., and Hyland, P. (1984). 'Harmonic response tests on transducers for the New Zealand power system'. Paper 84SM687–0 presented at the IEEE Summer Power Meeting, Seattle, Wash.
8. Olivier, G., Bouchard, R. P., Gervais, Y., and Mukhedar, D. (1980). Frequency response of HV test transformers and the associated measurement problems'. *Trans. IEEE*, **PAS-99**, 141–146.
9. Broadhurst, T. E. (1973). 'The measurement of high voltage'. *Electrical Review*, 313–316.
10. Beg, S. (1980). 'Electrical characteristics of a 50 kV resistance potential divider'. *Indian J. Pure Appl. Phys.*, **18**, 687–691.
11. Lisser, J. (1976). 'High accuracy with amplifier type voltage transformers'. *Electrical Review*, **198** (21), 31–33.
12. Gray, F. M., Hughes, M. A., and Staewski, A. (1976). 'Monitoring of transmission line voltage for protective relaying purposes using capacitive dividers'. *IEE Conf. Publ.*, **125**, 214–221.

13. McGranaghan, *et al.* (1981). 'Measuring voltage and current harmonics on distribution systems'. Paper 81WM 126–2 presented the IEEE Winter Power Meeting, Atlanta, Ga.
14. Mouton, L., Stalewski, A., and Bullo, P. (1978). 'Non-conventional current and voltage transformers'. *Electra*, (59), 91–122.
15. Rogers, A. J. (1979). 'Optical measurement of current and voltage on power systems'. *IEE Conf. Publ.*, **174**, 22–26.
16. Böcker, H., and Sankaran, P. (1969). 'Current transformer with error compensation for transient phenomena'. *Electrotechnische Z., A*, **90**, 112–114 [in German].
17. Gregory, B. A. (1973). *An Introduction to Electrical Instrumentation*, Macmillan, London.
18. Morrison, R. (1977). *Grounding and Shielding Techniques in Instrumentation*, John Wiley, New York.
19. Jones, B. E. (1977). *Instrumentation, Measurement and Feedback*, McGraw-Hill, New York.

8

Standards for the limitation and control of power system harmonics

8.1 INTRODUCTION

Except in certain special cases such as traction substations, large rectifier/invertor installations for high voltage d.c. transmission, steelworks and the electrochemical industry, power system harmonics have until recently created relatively few problems.

This situation has changed with the growth in the number and rating of disturbing loads connected to the power system and has led to the introduction of a number of national standards.

In this chapter the term standard is used to refer to standards, guidelines, recommendations, etc., governing the generation and the permitted levels of harmonics in the power system.

The purpose of all the standards relating to the levels of power system harmonics is effectively threefold:

(i) First, the need to control the distortion of the power system current and voltage waveforms to levels that the system and its associated components can tolerate. Particularly vulnerable in this respect are power factor correction capacitors, audiofrequency power system signalling (ripple control) systems and devices using point-on-wave switching.

(ii) Second, there is the requirement that consumers connected to the power system are provided with a voltage supply waveform which is suitable for their particular needs.

(iii) Finally there is the need to ensure that the power system does not interfere with the operation of other systems such as the telephone network.

Against this common background of requirements must be set the various national priorities, for example the particular configuration of the power system and the use made of ripple control for load management. Each of these will influence to some degree the form and nature of the standards adopted.

Consider as illustration the power systems of the United Kingdom and New

Zealand. In the United Kingdom a highly interconnected power system with distributed generation serves a large number of load centres, often close to the point of generation. This contrasts with the New Zealand situation where remote centres of generation provide supply to a few, widely separated, major load centres and a large number of scattered small loads. The resulting long transmission lines and comparatively low fault levels increase the vulnerability of the system to harmonic penetration. These factors, taken together with the extensive use made of ripple control, present a very different problem than are encountered in the United Kingdom to the designers of harmonic standards.

The setting of standards has to a considerable degree been influenced by experiences gained with a range of harmonic induced problems and the measures taken to resolve them. They therefore tend to be empirical and conservative in form, as they are rarely based on a detailed study and understanding of system behaviour. This may perhaps be illustrated by consideration of the levels imposed for odd and even harmonic voltages. Historically, even harmonics have tended to be of low amplitude whilst the odd harmonics, associated with rectification, were much higher. The levels adopted reflect this condition, with those for odd harmonics being normally about double those for even harmonics. This has then led to the adoption of frequencies close to even harmonics being adopted for ripple control, with the effect of justifying those limits.

A further restriction on the development of harmonic standards has been the nature of the instrumentation available until recently. The relatively sparse data provided by analogue instrumentation acted to limit the amount of applicable data available, thus inhibited the development of standards.

From the foregoing it is apparent that any attempt to make comparisons between individual harmonic standards must be treated with care, because of the variety of factors to be taken into account in forming the standard.(1) Indeed, as each standard represents a compromise for the particular system between the requirements of maintaining an undistorted supply and the ability of a consumer to use distortion-producing equipment, no direct comparison should be expected.

Evidence available to date does not suggest that any particular approach to the setting of harmonic standards has any advantage in a particular system configuration, and that no standard is or is likely to be correct in any absolute sense. It must also be recognized that no standard relating to system harmonic content can be regarded as permanent, but rather as the current interpretation of system requirements. As understanding improves through the application of improved measurement and analysis techniques, so must the standards change.

8.2 FACTORS INFLUENCING THE DEVELOPMENT OF STANDARDS FOR THE CONTROL OF POWER SYSTEM HARMONICS

In examining the effects of harmonics on any particular power system, there are a number of specific factors which can be used to establish adequate limits.(2) In broad terms these are as follows:

(i) The definition of system harmonic by reference to the amplitudes of the individual harmonic current and/or voltages.

(ii) The total harmonic content, of either current or voltage, expressed either as an r.m.s. value or as total harmonic distortion (THD). This latter is defined by

$$\mathrm{THD} = 100\left(\sum_{n=2}^{n} U_n^2\right)^{1/2} / U_1, \qquad (8.2.1)$$

where U_1 is the fundamental component of the signal and U_2 to U_n are the harmonic components.

(iii) The system voltage at which the measurement is made and the need, if any, to distinguish between the transmission and distribution systems.

(iv) Some measure of the ability of the system to both supply and withstand harmonics. In the absence of detailed information about system harmonic impedances, this 'harmonic capacity' is often expressed by the short circuit level at the point of common coupling.

(v) The definition of the limiting harmonic levels adopted. This may be based upon the peak level or an r.m.s. level and could in turn be an instantaneous level, a sustained level over a period of time or a number of occurrences within a defined interval.

(vi) The method of measurement used and the form in which the data is obtained.

(vii) The type of disturbing load.

(viii) The possible effects on communications and associated equipment and the use of psophometric weighting factors.

(ix) The implementation of any special criteria such as the need to take account of the use of ripple control on the power system.

In practice each standard represents an amalgamation of the above factors, with the individual weightings determined by system conditions and the role the standard is intended to fulfil.

The standards themselves may be either system standards, connection standards or, more usually, some combination of the two. In a system standard, such as that introduced by New Zealand, emphasis is on the levels of harmonics that can be tolerated on the system, with little or no direct reference within the body of the standard to the sources of harmonics. Details of harmonic sources, and their likely influence on system harmonic content, are given as associated material in appendices. The Australian AS2279, and the United Kingdom's Engineering Recommendation G5/3, provide connection standards, in that they are to a large extent concerned with the definition of the equipment that can be connected to the system in a range of circumstances. Inherent to such standards must, of course, be the specification of a range of system harmonic limits.

The limits themselves may be expressed as absolute levels of current or voltage, which may not be exceeded, or as incremental limits allowing small changes to the harmonic source with limited consideration of system effects. The United Kingdom G5/3 incorporates both these approaches, as it permits the connection of certain types and ratings of convertor to the system, with no reference to existing levels of harmonic content, thus operating with incremental limits. With higher convertor ratings the existing harmonic content must be established prior to connection, and the proposed additional harmonic source considered in relation to these levels, to determine if the limits are likely to be exceeded.

The rights of consumers

A further factor in the setting of standards is the right of the consumers under the standard. The way in which the limits are defined may significantly affect the way in which they are perceived by the consumers.

The application of absolute harmonic current limits for individual consumers would seem to provide equal rights for all consumers, large or small. However, this may be considered an unfair distribution by a large consumer connected to a point of common coupling together with a number of small consumers, as it takes no account of their share of the total load. If, instead, the limit is expressed in a way which takes account of the consumers' share of total load, then this may well be seen as overly restrictive, particularly if a large consumer has little or no disturbing load.

If the limits are expressed in terms of the levels of harmonic voltage at the point of common coupling, then consumers connected to a strong point of common coupling may well benefit relative to a consumer connected to a weak point of common coupling. The adoption of harmonic voltage limits may also mean that, where existing levels are high, new consumers may be forced to install expensive additional circuitry before connection is permitted. This latter situation constitutes the 'first come, first served' approach.

The establishment of harmonic limits

The level of immunity of electrical and electronic equipment to the presence of harmonics is generally unknown, and whilst future equipment can be designed to operate with established limits this may well not be the most effective or suitable approach. Evidence available does, however, suggest that the major problem areas in the past have been the overloading of capacitors, particularly under resonance conditions, the incorrect operation of some ripple control systems and interference with communications, particularly telephones.

In considering the type and nature of the limits to be adopted, the different problems presented by the various types of load must be taken into account. There are systems containing large numbers of small disturbing loads, large consumers generating significant quantities of harmonics and a large variety of other loads between these two extremes.

The first category is characterized by domestic loads, ranging from light dimmers to washing machines. Typically the total load would be formed by the connection of large numbers of small consumers to the system, each of them providing a contribution to the whole. The harmonic generation by domestic loads is covered by various domestic appliance standards, such as CENELEC 50.006 and IEC 555, and are not considered here in detail.[3]

An industrial consumer may have a number of individual components of plant, each capable of generating significant quantities of harmonics. Such a consumer is faced with the double problem of ensuring that no harmonic induced effects occur within its own system, and maintaining the levels at the point of common coupling within the prescribed limits. Investigations have shown that in many instances where high harmonic levels exist at points within the consumer's own network, these levels do not necessarily penetrate to the external system.[4]

Between the extremes there exist a large number of systems incorporating disturbing loads which, whilst greater than the limits set by the domestic appliance standards, are not necessarily large enough for individual consideration. In these cases the overall effect produced over the range of sources cannot be assessed by the individual consumers, but only by the supply authority responsible for the power system.

The way in which the harmonics from a number of sources interact on the power system is governed by the random nature of such sources and the variation of their distribution throughout the system, both physically and with time.[5,6] The presence of resonance conditions elsewhere on the system, or the changes in resonance conditions that occur, further complicate the problem.

In certain of the standards the random nature of the sources is accounted for by introducing a diversity factor. This changes the actual ratings of the disturbing loads into an effective rating, which is then used in assessing its relationship with the standard.

System design

Ideally, through knowledge of the behaviour of the individual system components, it should be possible to define the harmonic production of any particular system. In practice such an approach is impractical because of the lack of precise definition of component behaviour at harmonic frequencies, the inadequate information on system harmonic impedances,[7,8] and the problems of dealing with the random variation and subsequent summation of harmonics from a number of sources.

8.3 NATIONAL HARMONIC STANDARDS

In order to illustrate the differences between the various national harmonic standards a number of such standards are summarized.

All the standards considered set limits on the total harmonic distortion of voltage, with the majority incorporating limits on individual harmonic voltages. These latter may take the form of a percentage limit or limits covering the full frequency range of odd and even harmonics; defined absolute or percentage limits for each individual harmonic; or some combination of these.

The permitted level of harmonic content is generally linked to the system voltage, with a further distinction sometimes made between transmission and distribution systems.

The Electricité de France approach is unusual in that it defines the total voltage distortion at a point of common coupling in terms of the individual consumer's contribution to the voltage distortion at that point. Limits are set for total harmonic distortion and for the levels of odd and even harmonics that may be attributable to any individual consumer.

Limits on harmonic current levels are set by a few of the standards. These define the levels of current harmonics that may be drawn or produced by an individual consumer. In the case of New Zealand, for the transmission system only (66 kV and above), and the United Kingdom these are absolute limits. The Finnish supply authorities,[(12)] however, relate the permitted harmonic currents to a reference current magnitude determined by the individual consumer's maximum demand.

Telephone interference is covered in the Finnish and New Zealand standards by reference to psophometrically weighted currents on the higher voltage systems. In the case of New Zealand a psophometrically weighted voltage limit is used for the transmission system instead of total harmonic distortion.

As the main source of harmonic disturbance is the static convertor, the standards usually set down some procedure for assessing the size and type of convertor that may be connected to a particular point of common coupling. The approach adopted may be that of 'first come, first served' in which consumers are allowed to connect to the point of common coupling until the limits are approached when new loads will require the incorporation of filtering or other control. On this basis it is possible for a single consumer to absorb the harmonic capacity of the point of common coupling, leaving no spare capacity for later consumers. This is the position under the Australian and United Kingdom standards. Alternatively, the harmonic capacity of the point of common coupling may be shared amongst connected consumers in some way. This is the procedure adopted in New Zealand, where each consumer is allocated a proportion of the harmonic capacity of the point of common coupling, as expressed by the system short circuit level at that point, according to their individual maximum demands.

The actual convertor size permitted for connection may be obtained by reference to tables and graphs or by calculation based on the requirements of the standard.

In assessing the total permitted distorting load, or the effect of individual distorting loads at the point of common coupling, diversity may be applied. This is usually done by adjusting the rating of the individual distorting loads by a factor which varies according to the type of distorting load.

To monitor the levels of harmonic current and voltage on the power system, measurements must be made. In general no information on measurement procedures is given in the standards, though the United Kingdom's G5/3 does specify an instrument. An exception is the New Zealand standard, which incorporates a detailed measurement schedule setting down the specification for the measurement system to be used.

France

In a document entitled 'Regulations concerning the installation of power convertors, taking into account the characteristics of the supply network',[9] Electricité de France establishes limits due to each consumer acting alone at the point of common coupling. These are 0.6% of fundamental for even harmonics, 1.0% of fundamental for odd harmonics, and 1.6% total harmonic distortion. These limits are chosen to ensure that a level of 'about 5%' total harmonic distortion at the point of common coupling is not exceeded when all consumers are connected.

The calculation of convertor sizes is based upon the use of a harmonic impedance value derived from

$$Z_h = \alpha h Z_{cc}, \tag{8.3.1}$$

where Z_h is the harmonic impedance of the network at point of connection, $Z_{cc} = U_n/S_c$ is the short circuit impedance at point of supply, U_n is the fundamental voltage amplitude, S_c is the three-phase short circuit rating in amperes, h is the harmonic number, and α is a coefficient ($=2$ for high voltage systems, $=3$ for medium voltage systems, $=1$ for low voltage systems).

The resulting values for six- and 12-pulse convertors expressed in terms of the ratio S_c/S_n (where S_n is the convertor rating) are as follows:

System voltage	Six-pulse convertor	12-pulse convertor
High voltage	240	150
Medium voltage	360	225
Low voltage	120	75

The assessment of an individual consumer's contribution is based on an analytical procedure which excludes the effect of other harmonic sources.

West Germany

The standard DIN 57160 (VDE 0160/11.81)[10] sets down a rating for harmonic generating equipment of not greater than 1% of the system fault level. A diversity factor may be applied to installations containing more than one harmonic source.

Individual harmonic levels are set at 5% of the fundamental voltage level up to the 15th harmonic; the levels decrease to 1% at the 100th harmonic according to a defining curve. Moreover, the total harmonic voltage is not to exceed 10%.

Notching of the supply voltage waveform is also considered by the standard;

the depth at the notch permitted is limited to 20% of the fundamental voltage amplitude, with oscillation at the beginning or end of the notch limited to a further 20%. The notch duration must be less than 3.89% of a cycle (14°).

Sweden

The Electricity Supply Authority in the *SEF Thyristor Committee Report*[11] limits convertor rating on systems of voltages up to 24 kV as follows:

Pulse number	Percentage of system short circuit rating (%)
< 6	0.5
6	1.0
12	2.0
> 12	3.0

Restrictions on the total harmonic distortion (THD) permitted are applied as follows:

System voltage	Percentage THD (%)
430/250 V	4.0
3.3 kV to 24 kV	3.0
Up to 84 kV	1.0

United States

The IEEE *Guide for Harmonic Control and Reactive Compensation of Static Power Convertors*[12] sets limits for the percentage total harmonic distortion only. It covers both general power systems and systems used to service only convertors and loads not affected by voltage distortion (dedicated systems). The percentage THD limits are

System voltage	General power system (%)	Dedicated power system (%)
2.4–69 kV	5.0	8.0
115 kV and above	1.5	1.5

Australia

Australian standard AS2279–1979[13] is a two-part standard, the first part of which is concerned with domestic equipment and the second with the power system.

The second part of the standard operates in three stages. Under stage 1 equipment rated up to 0.3% of the supply point short circuit level, to a maximum of 75 kVA (500 kVA for a primary distribution system above 415 V to 33 kV), can be connected directly provided the system fault level is at least 5 MVA (50 MVA for a primary distribution system). These limits also apply to a summation of a number of small loads.

Single phase equipment may also be connected, provided that it does not exceed 5 kVA at 240 V or 7.5 kVA at 415 V.

Stage 2 sets limits of THD as follows:

	System voltage	THD (%)	Odd harmonic (%)	Even harmonics (%)
Primary and secondary distribution	To 33 kV	5	4	2
Transmission	22, 33 and 66 kV	3	2	1
Sub-transmission	110 kV and above	1.5	1	0.5

Under stage 2 connection is permitted according to the chart of Figure 8.1, provided that the distortion level is within 75% of the permitted limits.

Stage 3 applies to all loads not meeting the above conditions; it requires a detailed analysis of the system and load according to a procedure set out in the standard.

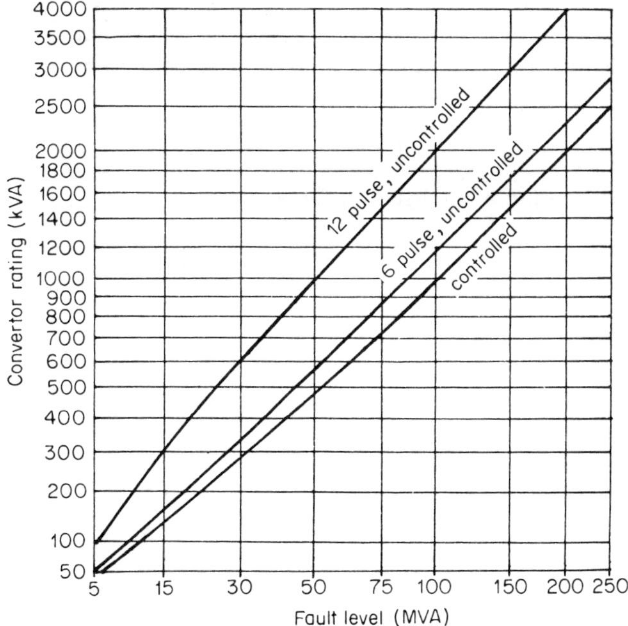

Figure 8.1. Convertor size against fault level for AS2279 (based on 1% voltage distortion for any individual odd harmonic). Reproduced by permission of the Standards Association of Australia

In assessing convertor sizes diversity factors are used for the derivation of the harmonic content.

Notching and short duration bursts of harmonics are also covered. The standard permits a notch depth of 20% of system voltage, with oscillations at the start and end of the notch restricted to a further 20%. Bursts of harmonics of less than 2 s duration, with an interval between bursts of at least 30 s, are considered as tolerable.

Finland

The Finnish power supply authorities' document *Restriction of Harmonics in Electrical Networks*[14] sets limits on the permitted total distortion and individual harmonic levels at the point of common coupling as follows:

System voltage	THD of voltage (%)	Individual harmonic level (%)
1 kV	5	4
3–20 kV	4	3
30–45 kV	3	2
110 kV	1.5	1

Limits are also imposed on the level of current harmonics that can flow in any individual consumer's connection at the point of common coupling. The limits are not expressed as absolute current levels, but as a percentage of the consumer's reference. This current is calculated from that consumer's hourly mean power (P_c) and the nominal system voltage (U_n), i.e.

$$I_{ref} = P_c/(\sqrt{3}U_n) \tag{8.3.2}$$

and the following limits are listed in the standard:

System voltage	THD current (%)	Individual harmonic current (%)
3–20 kV	10	8
30–45 kV	7	6
110 kV	5	4

Telephone interference is covered by the use of the psophometrically weighted current of equation (8.3.3), i.e.

$$I_p = 6.25 \times 10^{-5} \sqrt{\left(\sum_n (nP_nI_n)^2\right)}, \tag{8.3.3}$$

where I_p is the psophometrically weighted current, n is the order of harmonic P_n *is* the psophometric weighting factor at the nth harmonic, and I_n is the nth harmonic current.

For system voltages of 110 kV and above the maximum levels of psophometrically weighted current are

I_{ref}	I_p (amperes)
0–100	$(1 + 0.01 I_{ref})/2$
> 100	1

With permission from the operator of the supply system, the current limits, both natural and psophometrically weighted, can be exceeded if there are no other sources of harmonics present. It is then the responsibility of the consumer to reduce the harmonics to the specified levels, when other consumers require their share of the network harmonic capacity, as defined by the system short circuit level at the point of common coupling.

Based on the above the convertor sizes permitted, expressed as a percentage of the system short circuit rating, are as follows:

Pulse number	System voltage 20 kV	System voltage 30 kV
< 6	0.5	—
6	1	0.5
12	2	1
> 12	3	2

Details of the calculation procedures to estimate the effects of a particular convertor are presented as appendices.

New Zealand

The New Zealand *Limitation of Harmonic Levels Notice 1981*[15] sets individual limits of voltage and current harmonics in the transmission system for voltages of 66 kV and above. These are indicated in Tables 8.1 and 8.2 respectively.

Table 8.1. Harmonic limits for voltages of 66 kV and above

Harmonic order, n	Voltage limit (phase-to-earth harmonic voltage expressed as a percentage of the nominal phase-to-earth system voltage)
3	2.3
5	1.4
7	1.0
9	0.8
11	0.7
13	0.6
15	0.5
17–21	0.4
23–49	0.3
2	1.2
4	0.6
6	0.4
8 and 10	0.3
12–50	0.2

Table 8.2. Current harmonic limits

Harmonic order, n	Harmonic current limit (amperes at nominal system voltage)		
	220 kV	110 kV	66 kV
3	5.7	2.9	1.7
5	3.4	1.7	1.1
7	2.5	1.3	0.8
9	1.9	1.0	0.6
11	1.6	0.8	0.5
13	1.4	0.7	0.4
15	1.2	0.6	0.4
17	1.0	0.5	0.3
19 and 21	0.9	0.5	0.3
23	0.8	0.4	0.3
25–49	0.7	0.4	0.3
2	2.9	1.5	0.9
4	1.5	0.8	0.5
6	1.0	0.5	0.3
8	0.8	0.4	0.3
10	0.6	0.3	0.2
12 and 14	0.5	0.3	0.2
16 and 18	0.4	0.2	0.2
20–50	0.3	0.2	0.2

Psophometrically weighted equivalent disturbing current (EDI) and equivalent disturbing voltage (EDV) are also defined for the transmission system by equations (8.3.4) and (8.3.5).

$$\mathrm{EDI} = 6.25 \times 10^{-5} \sqrt{\left(\sum_{n=2}^{50} (nP_n I_n)^2 \right)} \tag{8.3.4}$$

and

$$\mathrm{EDV} = 6.25 \times 10^{-5} \sqrt{\left(\sum_{n=2}^{50} (nP_n U_n)^2 \right)}. \tag{8.3.5}$$

The limits for these expressions are

	66 kV	110 kV	220 kV
EDI limit (amperes)	0.8	1.3	2.6
EDV limit	1% of nominal system voltage		

The notice provides guidelines for the assessment of convertor sizes, based on the principle of sharing the harmonic capacity of the point of common coupling according to the consumers' maximum demands. Each consumer is allocated a 'harmonic allowance' defined as

$$\text{Harmonic allowance} = S_c/S,$$

Table 8.3. Multiplying factors applicable to convertors in assessing effective installation size at a three-phase point of common coupling

Type of convertor	Single-phase to neutral	Single-phase to phase	Three-phase six-pulse	Three-phase 12-pulse
Uncontrolled (no firing angle delay)	2.0	1.1	0.5	0.25
Controlled (delay firing angle)	3.7	2.2	1.0	0.65

where S_c is the consumer maximum demand in kVA and S is the supply capacity of point of common coupling in kVA.

For the distribution system, voltages below 66 kV, a limit of 5% applies to the total voltage harmonic distortion; limits are also specified for the odd and even harmonic voltage components of 4 and 2% respectively.

To determine the convertor size for a three-phase system, the 'effective installation size' for the point of common coupling is determined as the maximum rating of six-pulse-controlled convertor equipment that could be connected, without introducing distortion in excess of the limits. The individual consumer's contribution is then assessed by reducing each of their convertors to the equivalent six-pulse-controlled convertor. These are then summed and the total is compared with the consumer's proportion of the effective installation size using the harmonic allowance. A similar approach is adopted for single-phase convertor systems. The appropriate multiplying factors required to convert various convertor types to the equivalent reference convertor are displayed in Table 8.3.

The guidelines also provide information on the resolution of resonance problems and the choice and use of measuring instruments.

United Kingdom

In the United Kingdom, Engineering Recommendation G5/3[16] sets down a three-stage approach to the assessment of convertor sizes.

Stage 1 sets out the maximum sizes of different types of harmonic producing equipment that can be connected without detailed consideration of harmonic levels. These are listed in Table 8.4.

Single-phase equipments which theoretically produce no even harmonics, and not exceeding 5 kVA at 240 V or 7.5 kVA at 415/480 V, are also allowed under stage 1.

The stage 2 limits control the levels of both harmonic current and harmonic voltage distortion on the system as shown in Tables 8.5 and 8.6.

Connection of the distorting load is permitted if the current levels are not exceeded by the introduction of the addition load, provided that existing voltage

Table 8.4. Maximum sizes of individual convertor and a.c. regulator equipments under stage 1 limits

Supply system voltage (kV) at point of common coupling	Three-phase convertors			Three-phase a.c. regulators	
	Three-pulse (kVA)	Six-pulse (kVA)	12-pulse (kVA)	Six-thyristor (kVA)	Three-thyristor/three-diode (kVA)
0.415	8	12	—	14	10
6.6 and 11	85	130	250*	150	100

* This limit applies to 12-pulse devices, and to combinations of six-pulse devices always operated as 12-pulse devices, employing careful control of the firing angles and the d.c. ripple to minimize non-characteristic harmonics e.g. third, fifth and seventh.

Table 8.5. Permitted harmonic currents for any one consumer at point of common coupling under stage 2 limits*

Supply system voltage (kV) at point of common coupling	Harmonic number and current (A r.m.s.)																	
	2	3	4	5	6	7	8	9	10	11	12	13	14	15	16	17	18	19
0.415	48	34	22	56	11	40	9	8	7	19	6	16	5	5	5	6	4	6
6.6 and 11	13	8	6	10	4	8	3	3	3	7	2	6	2	2	2	2	1	1
33	11	7	5	9	4	6	3	2	2	6	2	5	2	1	1	2	1	1
132	5	4	3	4	2	3	1	1	1	3	1	3	1	1	1	1	1	1

* A tolerance of + 10% or 0.5 A (whichever is the greater) is permissible, provided that it applies to not more than two harmonics.

Table 8.6. Harmonic voltage distortion limits at any point on the system (including background levels)

Supply system voltage (kV) at point of common coupling	Total harmonic voltage distortion, V_T (%)	Individual harmonic voltage distortion (%)	
		Odd	Even
0.415	5	4	2
6.6 and 11	4	3	1.75
33 and 66	3	2	1
132	1.5	1	0.5

distortion levels prior to connection are less than 75% of those permitted. A range of typical convertor sizes under the stage 2 limits is given in Table 8.7.

Single-phase loads in excess of the stage 1 limits are discouraged but where such loads are installed they must also comply with the voltage unbalance limits for the system.

Stage 3 sets down the procedure for handling any disturbing loads not covered by stages 1 and 2 together with situations where the voltage distortion is above the 75% level. A 'first come, first served' approach is adopted using a calculation procedure set out in appendices. Diversity is included by adjusting the rating of the individual convertors by an appropriate factor.

Table 8.7. Maximum load of single-convertor installations corresponding to harmonic current limits

Supply system voltage (kV) at point of common coupling	Type of convertor	Permissible kVA capacity and corresponding effective pulse number of three-phase installations		
		Three-pulse	Six-pulse	12-pulse
0.415	Uncontrolled	—	150	300
	Half-controlled	—	65*	—
	Controlled	—	100	150
6.6 and 11	Uncontrolled	400	1000	3000
	Half-controlled	—	500*	—
	Controlled	—	800	1500
33	Uncontrolled	1200	3000	7600
	Half-controlled	—	1200*	—
	Controlled	—	2400	3800
132	Uncontrolled	1800	5200	15000
	Half-controlled	—	2200*	—
	Controlled	—	4700	7500

* These refer to three-thyristor/three-diode half-controlled bridges.

8.4 DOMESTIC STANDARDS

Control of power system disturbances by domestic appliances is covered by national and international standards such as BS5406,[17] AS2279: Part 1,[13] CENELEC 50.006[17] and IEC 555.[18]

Though control of harmonic generation by particular appliances forms a part of these standards they are intended to cover all forms of disturbance produced over a wide range of appliances. Their range of coverage includes phase-controlled and burst firing systems as well as voltage variation and flicker.

The limits on harmonic content are defined either by voltage (CENELEC 50.006) or current (IEC 555) with correspondence between the two maintained by an appropriate choice of system impedance.

Test procedures for establishing the harmonic production by individual appliances are set down in the standards.

8.5 REFERENCES

1. CIGRE (1979). 'AC harmonic filter and reactive compensation for HVDC: a general survey'. *Electra*, (63), 65–102.
2. Drouin, G. (1981). 'Present and future European harmonic standards'. Paper presented at a seminar on harmonics, UMIST, Manchester.
3. Lagostena, L., Mantini, A., and Silvestri, M. (1974). 'Disturbances produced by domestic appliances controlled by thyristors: experiments and studies conducted for the purpose of preparing adequate standards', *IEE Conf. Publ.*, **110**, 209–213.
4. Grindrod, S. J. (1981). 'A study of harmonics in electricity supply system'. PhD thesis, University of Lancaster.
5. Sherman, W. G. (1972). 'Summation of harmonics with random phase angles'. *Proc. IEE*, **119**, 1643–1648.

6. Rowe, N. B. (1974). 'The summation of randomly varying phasors or vectors with particular reference of harmonic levels'. *IEE Conf. Publ.*, **110**, 177–181.
7. Crevier, D., and Mercier, A. (1978). 'Estimation of high frequency network impedances by harmonic analysis of natural waveforms'. *IEEE Trans.*, **PAS-97**, 424–431.
8. Barnes, H., and Kearley, S. J. (1981). 'The measurement of impedance presented to harmonic currents by power distribution networks'. *IEE Conf. Publ.*, **197**, 71–75.
9. Electricité de France (1981). *Regulations Concerning the Installation of Power Convertors, Taking into Account the Characteristics of the Supply Network* [in French], EDF, Paris.
10. DIN (1981). 'Electronic Equipment to be used in Electrical Power Installation and their Assembly into Electrical Power Installations', DIN 57160, VDE 0160/11.81, DIN, Bonn.
11. SEF (1974). *SEF Thyristor Committee Report* [in Swedish], SEF, Stockholm.
12. Institute of Electrical and Electronic Engineers (1981). *Guide for Harmonic Control and Reactive Compensation of Static Power Convertors*, IEEE, New York.
13. Australian Standards Authority (1979). *Disturbances in Mains Supply Network*, Australian Standard AS2279, Parts 1 and 2, Canberra, Australian Standards Authority.
14. Finnish Association of Electricity Supply Undertakings (1978). *Restriction of Harmonics in Electrical Networks* [in Finnish], Helsinki.
15. New Zealand Ministry (1981). *Limitation of Harmonic Levels Notice*, Wellington.
16. Electricity Council (1976). *Limits for Harmonics in the United Kingdom Electricity Supply System*, Engineering Recommendation G5/3, Electricity Council, London.
17. British Standards Institution (1976). *The Limitation of Disturbances in Electricity Supply Networks caused by Domestic and Similar Appliances Equipped with Electronic Devices*, British Standard BS5406 (CENELEC 50.006), BSI, London.
18. IEC (1982). *Disturbances in Supply Systems Caused by Household Appliances and Similar Electrical Equipment*, International Electrotechnical Commission Publication 555, Parts 1, 2 and 3.

9

Harmonic penetration in a.c. systems

9.1 INTRODUCTION

Following a brief description of the experimental techniques in present use, this chapter introduces the analytical models available for the derivation of the network harmonic impedances.

This information forms the basis of harmonic penetration studies, which involve the computation of harmonic currents and voltages throughout the a.c. system in the presence of one or more current harmonic sources.

The main limitation of these types of studies is the lack of reliable information on the frequency dependence of power plant components. Subject to this limitation, this chapter also describes and compares the characteristics of single- and three-phase algorithms used in harmonic penetration studies.

9.2 HARMONIC POWER FLOW

For an intuitive appreciation of the relationships between the fundamental frequency power flow and the power flow at harmonic frequencies, the system equivalent circuit of Figure 9.1(a) has been used by Tschappu.[1] The generator G constitutes a fundamental sinusoidal voltage source and it supplies a static convertor controlled resistive load via the system impedance ($R_s + jX_s$).

With reference to the power at fundamental frequency, illustrated in Figure 9.1(b), the generator supplies power P_{g1} via the point of common coupling (PCC). The largest portion of this power (P_{l1}) flows into the load, and a small part (P_{c1}) into the convertor. This latter portion is transformed into harmonic power. In addition the losses (P_{s1}) in the system resistance (R_s) have to be supplied.

Figure 9.1(c) illustrates the equivalent circuit diagram for the harmonic power flow. Since the generator voltage is sinusoidal, it can only supply fundamental power and it is included in this diagram simply as a harmonic impedance. In this diagram the convertor appears in place of the generator as an impressed harmonic current source. The small portion of the fundamental power (P_{c1}), transformed into harmonic power, is fed as P_{sh} and P_{gh} back to the resistances of

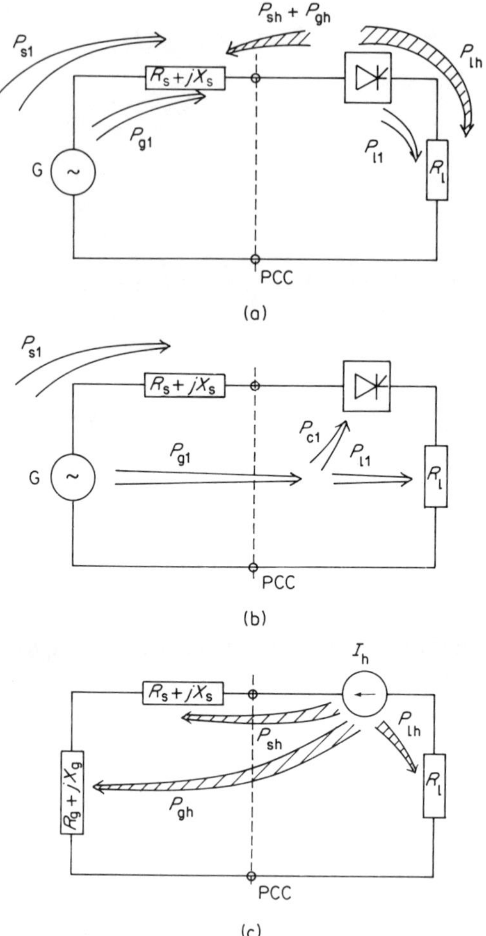

Figure 9.1. Fundamental and harmonic power flows in an a.c. system: (a) schematic diagram of power flow in the a.c. system; (b) schematic diagram of fundamental frequency power flow; (c) schematic diagram of harmonic power flow

the system and the generator respectively. The large portion (P_{lh}) goes to the load. Since R_s and R_l are connected in series, the ratio of $P_{lh}/(P_{sh} + P_{gh})$ can also be expressed as $R_l/(R_s + R_g)$. In practice, $R_l/(R_s + R_g)$ is very large, and for this reason the proportion of harmonic power propagating in the a.c. system remains small, even when the current is badly distorted.

The total loss in the system impedance consists of the fundamental power loss P_{s1}, supplied from the generator, and the harmonic power loss of the convertor, $P_{sh} + P_{gh}$. This loss in the system is larger by the ratio $(P_{sh} + P_{gh})/P_{s1}$ than the loss of a corresponding linear load.

9.3 IMPEDANCE/FREQUENCY LOCI

The impedance loci of a.c. network configurations are important tools in audiofrequency analysis. Such loci assume spiral-like shapes which illustrate that the system, although normally inductive at fundamental frequency, changes from inductive to capacitive and back as the frequency increases, producing a number of resonant points at which the system is purely resistive.

The location of those resonant points can be assessed by carrying out a cubic spline interpolation[2] in the independent axis plots, as shown in Figure 9.2, and then subdividing the intervals between the frequencies of measurement in say 20 sub-intervals, each providing an R, jX pair for the impedance/frequency locus. However, Figure 9.2 illustrates the extreme sensitivity of the system impedance near a parallel resonance. With a large transmission system being so sensitive to frequency, measurements (discussed in Section 9.4) made at discrete frequencies could completely miss an important resonance. Subsequent fluctuations in frequency and changes in system configuration could shift this resonance on to a harmonic number, with important consequences for harmonic penetration throughout the system.

With a computer solution (discussed in Section 9.5) it is possible to obtain accurate impedance information at any frequency without the need for interpolation, i.e. it is not restricted to multiples of the fundamental frequency. This effect is illustrated in Figure 9.3, which displays computed solutions at 10 and 5 Hz intervals respectively for a sparse 220 kV transmission system. This method is particularly relevant when the loci are used in non-harmonic frequency studies, such as ripple-control penetration, and in the accurate determination of resonant points for filter design.

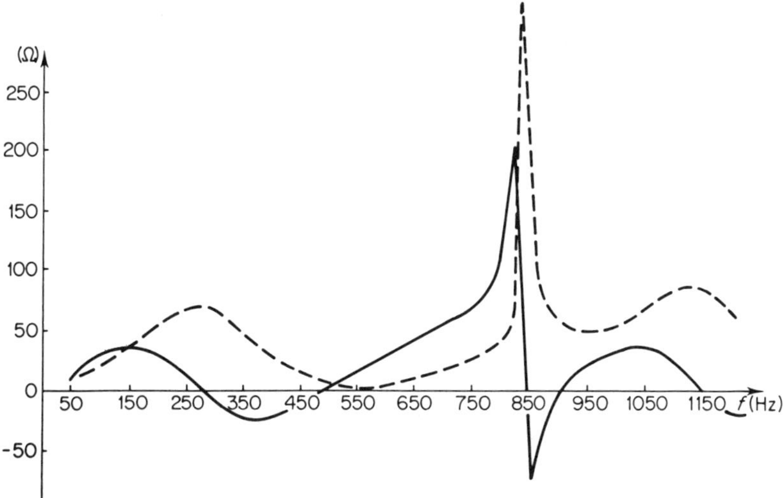

Figure 9.2. Resistance (-----) and reactance (——) versus frequency at 5 Hz intervals for a reduced 220 kV system

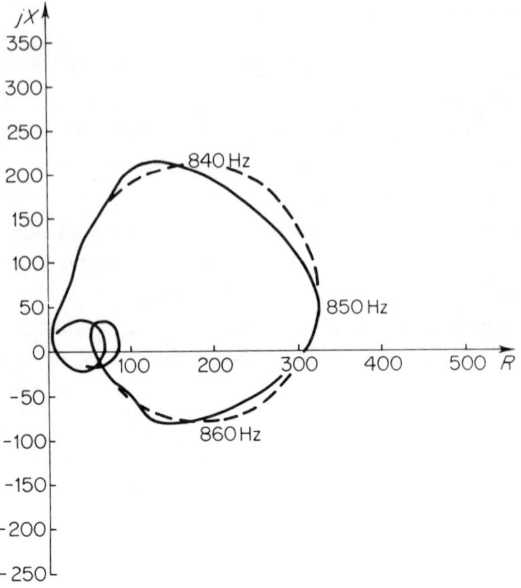

Figure 9.3. Impedance locus at 10 Hz (——) and 5 Hz (-----) intervals

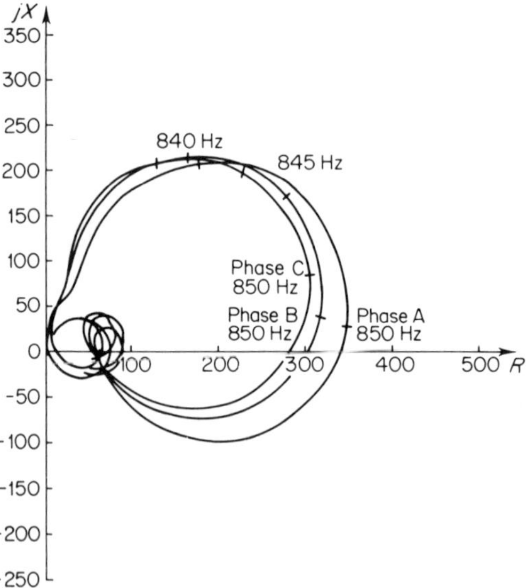

Figure 9.4. Impedance locus at 5 Hz intervals for the equivalent three-phase impedances

Another example of the capability of the digital model is illustrated in Figure 9.4, which contains the impedance loci of the three phases of the sample system. The plot clearly indicates the considerable unbalance which exists at various harmonic frequencies, and particularly at the resonant points, even though the system is reasonably balanced at fundamental frequency. It is also evident from the figure that the resonant frequencies are different in the three phases.

9.4 EXPERIMENTAL DERIVATION OF EQUIVALENT HARMONIC IMPEDANCES

In the absence of more accurate information, existing harmonic standards and recommendations often refer to harmonic impedance sources derived from the balanced short-circuit impedance of the system at fundamental frequency. The inadequacy of such an approach is becoming apparent with the help of on-line tests and computer studies.

The availability of voltage and current transducers throughout the power system provides the basis for the indirect derivation of harmonic impedances. Their assessment is therefore dependent on the performance of such transducers, which may not have been designed to respond accurately to harmonic frequencies. This is particularly the case with CVTs. The characteristic of transducers, as they apply to harmonic measurement, has already been discussed in Chapter 7.

According to the nature of the harmonic source used in the derivation of harmonic impedances, present techniques can be divided into two broad groups, which are referred to here as invasive and non-invasive.

On-line non-invasive tests

In a non-invasive test the information required is obtained purely from measurements of existing waveforms, i.e. using the harmonic content already present in the system. This is the simplest and most commonly used technique.

A recent project supported by the US Electric Power Research Institute[3] describes two non-invasive methods of estimating the harmonic impedance. One of them uses the harmonic content produced by an existing high voltage d.c. convertor station, and obtains the impedances directly from the ratio of voltage and current harmonic readings. The second method uses the harmonic sources present in the a.c. system, and estimates the harmonic impedance at a particular location by switching a shunt impedance and comparing the harmonic voltages and current before and after the switching.

Electricité de France[4] has proposed two alternative ways of deriving information of harmonic impedances, in the form of sequence components, using the principle of digital filtering rather than Fourier analysis. One method uses numerical techniques to perform the digital filtering of the physical input values, and retains only the frequencies contained in a selected bandwidth; the information is then used to identify the network with a simple impedance model valid only over a relatively narrow frequency range. The identification is carried out separately for the zero sequence and positive (negative) sequence values. The second method uses electronic filters to transform the six voltage and current values into four (two zero sequence and two positive sequence), after eliminating the 50 Hz component.

In theory non-invasive tests could provide reasonable information of harmonic impedances, provided that (i) there is a sufficiently powerful harmonic source at the point of measurement, which is necessary to overcome the unacceptably large errors that occur with low readings on instruments; (ii) all other important harmonic sources in the power system are disconnected during the test, as with two or more sources it is impossible to determine the harmonic current due just to the source of interest; (iii) the information provided by the transducers and instrumentation can be relied upon; (iv) a comprehensive range of tests is carried out under different operating conditions.

On-line invasive tests

In principle, it is possible to design a source of harmonic power which absorbs fundamental frequency power and converts it into harmonic power of the appropriate frequencies, magnitudes and phases. However, the measurement of harmonic impedances requires a distorting source of considerable capacity.

In practice, measurements are usually taken at points where there is considerable distortion due to existing harmonic sources in the network. The effect of injecting a further source, needed for the impedance measurement, is therefore superimposed with these. The result is a low signal to noise ratio which makes the measurement unreliable.

To overcome this problem the Electricity Council Research Centre (Capenhurst, UK)[5] designed a system which generates power at frequencies mid-way between the characteristic harmonic frequencies of interest, i.e. at the

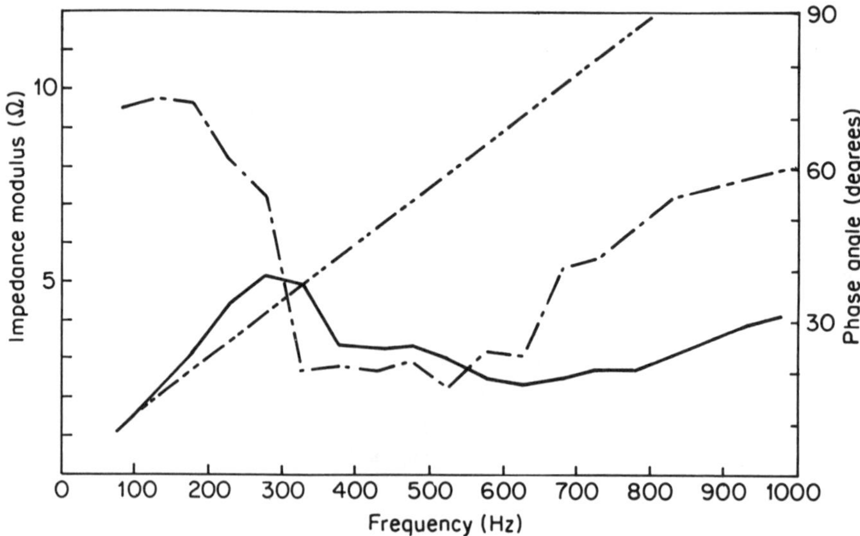

Figure 9.5. Source impedance seen from an 11 kV busbar supplying a commercial load: -·-·-·-, obtained from the short circuit impedance at 50 Hz; ———, measured impedance magnitude; — — ·—, measured impedance phase angle

odd multiples of 25 Hz, on the assumption that interpolation between these frequencies is justifiable. The power ratings of the distorting source for measurements at 11, 33 and 132 kV are 9, 36 and 180 kW respectively, and the unit consists of a switch modulator in series with a resistive load in the form of fan heaters; the 11 and 33 kV systems are portable and the 132 kV system is located in a purposely built van.

Figure 9.5 illustrates the results of typical measurements carried out with this equipment in combination with a harmonic impedance instrument,[(6)] specially designed to provide simultaneous information of harmonic voltage and current with their phase relationship.

9.5 APPLICATION OF NETWORK ANALYSIS TO HARMONIC PENETRATION

The difficulty of accurate harmonic monitoring, particularly on high voltage transmission networks, has led to the development of software models for the calculation of harmonic impedances at any specific location required.

In this section, use is made of the nodal admittance matrix and linear transformation techniques to interconnect the various power plant components of a network represented by their equivalent circuits.

Although most of the existing computer algorithms use a single-phase model, for accurate harmonic frequency analysis three-phase modelling of the power system is necessary. The harmonic currents injected into the system will in general be unbalanced and furthermore, the transmission system will include impedance imbalance and circuit coupling.

The admittance matrices for each frequency have the same non-zero element positions but their values are determined by the frequency-dependent component models.

Network subdivision

Although an element, or branch, is the basic component of a network, elements may be coupled and non-homogeneous, e.g. mutually coupled transmission lines with different tower geometries over the line length. To facilitate the inclusion of this type of element, a subsystem is defined as follows:

A subsystem is the unit into which any part of the system may be divided such that no subsystem has any mutual coupling between its constituent branches and those of the rest of the system.

The smallest unit of a subsystem is a single network element.

The subsystem unit is retained for input data organization. Data for any subsystem is input as a complete unit, the subsystem admittance matrix is formulated and then combined in the total system admittance matrix.

Subsystem admittance matrices may be derived by finding, for each section, the *ABCD* or transmission parameters; then combining these by matrix multiplication to give the resultant transmission parameters.

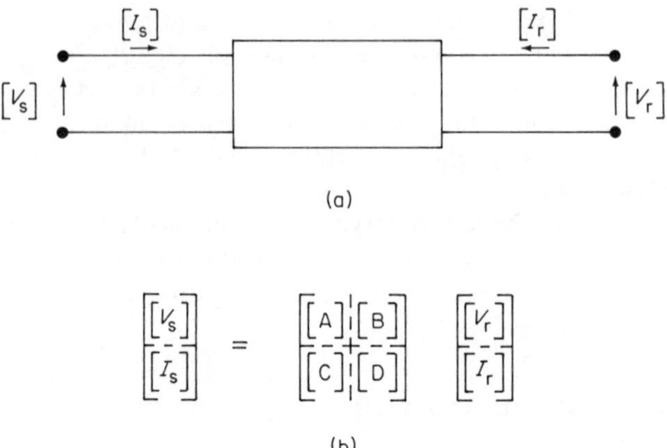

Figure 9.6. Two-port network transmission parameters: (a) multi-two-port network; (b) matrix transmission parameters. Reproduced with permission from Arrillaga, Arnold and Harker, *Computer Modelling of Electrical Power Systems*. Copyright © 1983 John Wiley & Sons Ltd.

This procedure involves an extension of the usual two-port network theory to multi-two-port networks. Current and voltages are now matrix quantities as defined in Figure 9.6. The dimensions of the parameter matrices correspond to those of the section being considered, i.e. three, six, nine or 12 for one, two, three or four mutually coupled three-phase elements respectively. All sections must contain the same number of mutually coupled three-phase elements, ensuring that all the parameter matrices are of the same order and that the matrix multiplications are executable. Uncoupled elements need to be considered as coupled ones with zero coupling to maintain correct dimensions for all matrices.

Once the resultant *ABCD* parameters have been found the equivalent nodal admittance matrix for the subsystem can be calculated from

$$[Y]=\left[\begin{array}{c|c}[D][B]^{-1} & [C]-[D][B]^{-1}[A] \\ \hline [B]^{-1} & -[B]^{-1}[A]\end{array}\right]. \tag{9.5.1.}$$

Linear transformation techniques

Linear transformation techniques are used to enable the admittance matrix of any network to be found in a systematic manner.[7,8]

Consider, for the purposes of illustration, the network drawn in Figure 9.7. Five steps are necessary to form the network admittance matrix by linear transformation.

Step 1 Label the nodes in the original network.
Step 2 Number, in any order, the branches and branch admittances.
Step 3 Form the primitive network admittance matrix by inspection.

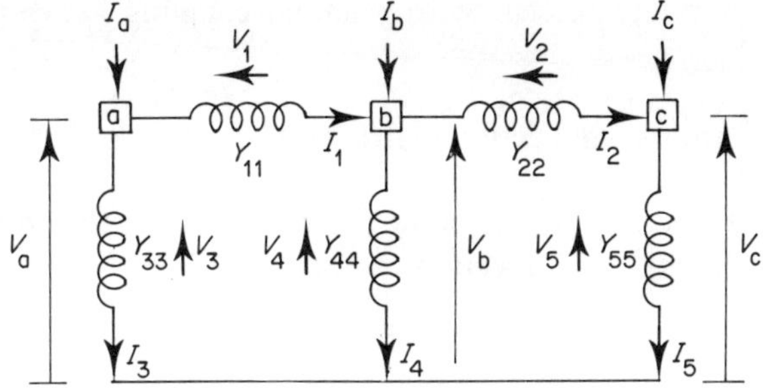

Figure 9.7. Actual connected network. Reproduced with permission from Arrillaga, Arnold and Harker, *Computer Modelling of Electrical Power Systems*. Copyright © 1983 John Wiley & Sons Ltd

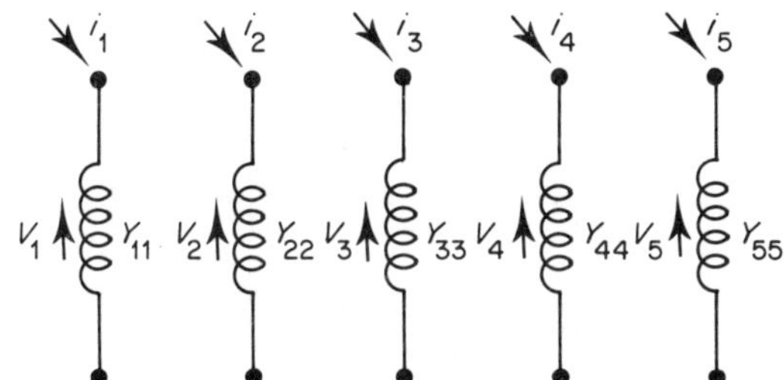

Figure 9.8. Primitive or unconnected network. Reproduced with permission from Arrillaga, Arnold and Harker, *Computer Modelling of Electrical Power Systems*. Copyright © 1984 John Wiley & Sons Ltd

This matrix relates the nodal injection currents to the node voltage of the primitive network. The primitive network is also drawn by inspection of the actual network. It consists of the unconnected branches of the original network with a current equal to the original branch current injected into the corresponding node of the primitive network. The voltages across the primitive network branches then equal those across the same branch in the actual network.

The primitive network for Figure 9.7 is shown in Figure 9.8.

The primitive admittance matrix relationship is

$$\begin{bmatrix} I_1 \\ I_2 \\ I_3 \\ I_4 \\ I_5 \end{bmatrix} = \underbrace{\begin{bmatrix} Y_{11} & & & & \\ & Y_{22} & & & \\ & & Y_{33} & & \\ & & & Y_{44} & \\ & & & & Y_{55} \end{bmatrix}}_{[Y_{\mathrm{PRIM}}]} \begin{bmatrix} V_1 \\ V_2 \\ V_3 \\ V_4 \\ V_5 \end{bmatrix}. \tag{9.5.2}$$

Off-diagonal terms are present where mutual coupling between branches is present.

Step 4 Form the connection matrix $[C]$.

This relates the nodal voltages of the actual network to the nodal voltages of the primitive network. By inspection of Fig. 9.7,

$$\begin{aligned} V_1 &= V_a - V_b, \\ V_2 &= V_b - V_c, \\ V_3 &= V_a, \\ V_4 &= V_b, \\ V_5 &= V_c, \end{aligned} \tag{9.5.3}$$

or in matrix form

$$\begin{bmatrix} V_1 \\ V_2 \\ V_3 \\ V_4 \\ V_5 \end{bmatrix} = \underbrace{\begin{bmatrix} 1 & -1 & \\ & 1 & -1 \\ 1 & & \\ & 1 & \\ & & 1 \end{bmatrix}}_{[C]} \begin{bmatrix} V_a \\ V_b \\ V_c \end{bmatrix}. \tag{9.5.4}$$

Step 5 The actual network admittance matrix which relates the nodal currents to the voltage by

$$\begin{bmatrix} I_a \\ I_b \\ I_c \end{bmatrix} = \begin{bmatrix} [Y_{abc}] \end{bmatrix} \begin{bmatrix} V_a \\ V_b \\ V_c \end{bmatrix}. \tag{9.5.5}$$

can now be derived from

$$\underset{3 \times 3}{[Y_{abc}]} = \underset{3 \times 5}{[C]^T} \; \underset{5 \times 5}{[Y_{PRIM}]} \; \underset{5 \times 3}{[C]}, \tag{9.5.6}$$

which is straightforward matrix multiplication.

Frame of reference used in three-phase system modelling

Sequence components have long been used to enable convenient examination of the balanced power system under both balanced and unbalanced loading conditions.

The symmetrical component transformation is a general mathematical

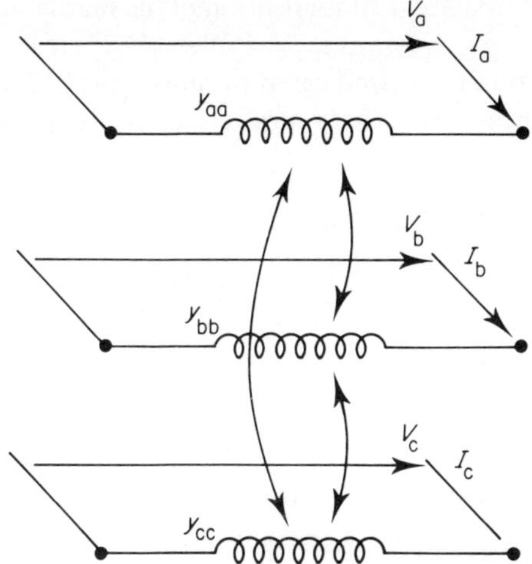

Figure 9.9. Admittance representation of a three-phase series element. Reproduced with permission from Arrillaga, Arnold and Harker, *Computer Modelling of Electrical Power Systems*. Copyright © 1983 John Wiley & Sons Ltd.

technique developed by Fortescue whereby any 'system of n vectors or quantities may be resolved when n is prime into n different symmetrical n phase systems'.[9] Any set of three-phase voltages or currents may therefore be transformed into three symmetrical systems of three vectors each. This, in itself, would not commend the method and the assumptions, which lead to the simplifying nature of symmetrical components, must be examined carefully.

Consider, as an example, the series admittance of a three-phase transmission line, shown in Figure 9.9, i.e. three mutually coupled coils. The admittance matrix relates the illustrated currents and voltages by

$$[I_{abc}] = [Y_{abc}][V_{abc}], \tag{9.5.7}$$

where

$$[I_{abc}] = [I_a I_b I_c]^T,$$
$$[V_{abc}] = [V_a V_b V_c]^T$$

and

$$[Y_{abc}] = \begin{bmatrix} y_{aa} & y_{ab} & y_{ac} \\ y_{ba} & y_{bb} & y_{bc} \\ y_{ca} & y_{cb} & y_{cc} \end{bmatrix}. \tag{9.5.8}$$

By the use of the symmetrical components transformation the three coils of Figure 9.9 can be replaced by three uncoupled coils. This enables each coil to be treated separately with a great simplification of the mathematics involved in the analysis.

The transformed quantities (indicated by subscripts 012 for the zero, positive and negative sequences respectively) are related to the phase quantities by

$$[V_{012}]=[T_s]^{-1}[V_{abc}], \tag{9.5.9}$$

$$[I_{012}]=[T_s]^{-1}[I_{abc}] \tag{9.5.10}$$

$$=[T_s]^{-1}[Y_{abc}][T_s][V_{012}], \tag{9.5.11}$$

where $[T_s]$ is the transformation matrix.

The transformed voltages and currents are thus related by the transformed admittance matrix,

$$[Y_{012}]=[T_s]^{-1}[Y_{abc}][T_s]. \tag{9.5.12}$$

Assuming that the element is balanced, we have

$$\begin{aligned} y_{aa} &= y_{bb} = y_{cc}, \\ y_{ab} &= y_{bc} = y_{ca}, \\ y_{ba} &= y_{cb} = y_{ac}, \end{aligned} \tag{9.5.13}$$

and a set of invariant matrices $[T]$ exist. Transformation (9.5.12) will then yield a diagonal matrix $[y_{012}]$.

In this case the mutually coupled three-phase system has been replaced by three uncoupled symmetrical systems. In addition, if the generation and loading may be assumed balanced, then only one system, the positive sequence system, has any current flow and the other two sequences may be ignored. This is essentially the situation with the single-phase harmonic penetration analysis.

In general, however, such an assumption is not valid. Unsymmetrical interphase coupling exists in transmission lines and to a lesser extent in transformers, and this results in coupling between the sequence networks. Furthermore, the phase shift introduced by transformer connections is difficult to represent in sequence component models.

If the original phase admittance matrix $[Y_{abc}]$ is in its natural unbalanced state then the transformed admittance matrix $[Y_{012}]$ is full. Therefore, current flow of one sequence will give rise to voltages of all sequences, i.e. the equivalent circuits for the sequence networks are mutually coupled. In this case the problem of analysis is no simpler in sequence components than in the original phase components and symmetrical components should not be used.

From the above considerations it is clear that the asymmetry inherent in all power systems cannot be studied with any simplification by using the symmetrical component frame of reference.

With the use of phase coordinates the following advantages become apparent:

(i) Any system element maintains its identity.
(ii) Features such as asymmetric impedances, mutual couplings between phases and between different system elements, and line transpositions are all readily considered.

(iii) Transformer phase shifts present no problem.

Thus phase components are retained throughout the formation and solution of the admittance matrices of harmonic penetration studies. Sequence components are only used as an aid to interpretation of results.

The use of compound admittances

When analysing three-phase networks, where the three nodes at a busbar are always associated together in their interconnections, the graphic representation of the network is greatly simplified by means of 'compound admittances', a concept which is based on the use of matrix quantities to represent the admittances of the networks.

The laws and equations of ordinary networks are all valid for compound networks by simply replacing single quantities by appropriate matrices.[7]

Consider six mutually coupled single admittances, the primitive networks of which is illustrated in Figure 9.10.

The primitive admittance matrix relates the nodal injected currents to the branch voltages as follows:

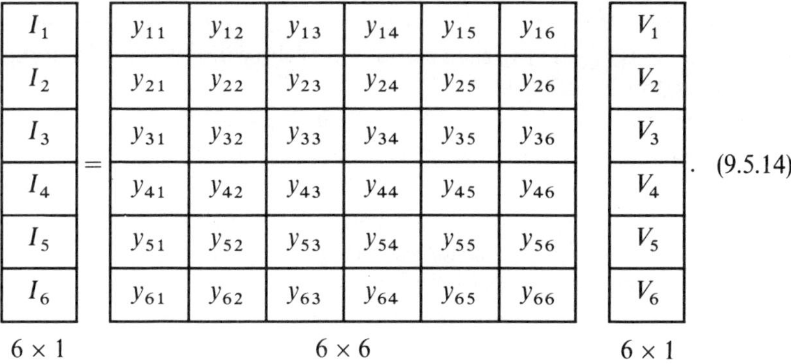

$$\underset{6\times 1}{\begin{bmatrix} I_1 \\ I_2 \\ I_3 \\ I_4 \\ I_5 \\ I_6 \end{bmatrix}} = \underset{6\times 6}{\begin{bmatrix} y_{11} & y_{12} & y_{13} & y_{14} & y_{15} & y_{16} \\ y_{21} & y_{22} & y_{23} & y_{24} & y_{25} & y_{26} \\ y_{31} & y_{32} & y_{33} & y_{34} & y_{35} & y_{36} \\ y_{41} & y_{42} & y_{43} & y_{44} & y_{45} & y_{46} \\ y_{51} & y_{52} & y_{53} & y_{54} & y_{55} & y_{56} \\ y_{61} & y_{62} & y_{63} & y_{64} & y_{65} & y_{66} \end{bmatrix}} \underset{6\times 1}{\begin{bmatrix} V_1 \\ V_2 \\ V_3 \\ V_4 \\ V_5 \\ V_6 \end{bmatrix}}. \quad (9.5.14)$$

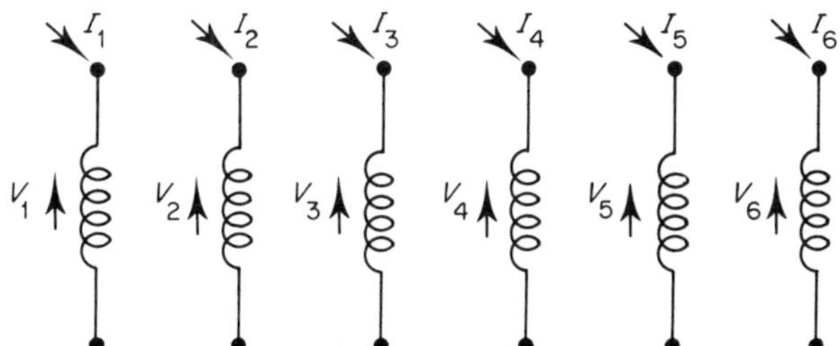

Figure 9.10. Primitive network of six coupled admittances. Reproduced with permission from Arrillaga, Arnold and Harker, *Computer Modelling of Electrical Power Systems.* Copyright © 1983 John Wiley & Sons Ltd.

Partitioning equation (9.5.14) into 3×3 matrices and 3×1 vectors, the equation becomes

$$\begin{bmatrix} [I_a] \\ [I_b] \end{bmatrix} = \begin{bmatrix} [Y_{aa}] & [Y_{ab}] \\ [Y_{ba}] & [Y_{bb}] \end{bmatrix} \begin{bmatrix} [V_a] \\ [V_b] \end{bmatrix}, \qquad (9.5.15)$$

where

$$[I_a] = [I_1\, I_2\, I_3]^T, \qquad [I_b] = [I_4\, I_5\, I_6]^T,$$

$$[Y_{aa}] = \begin{bmatrix} y_{11} & y_{12} & y_{13} \\ y_{21} & y_{22} & y_{23} \\ y_{31} & y_{32} & y_{33} \end{bmatrix}, \qquad [Y_{bb}] = \begin{bmatrix} y_{44} & y_{45} & y_{46} \\ y_{54} & y_{55} & y_{56} \\ y_{64} & y_{65} & y_{66} \end{bmatrix},$$

$$[Y_{ab}] = \begin{bmatrix} y_{14} & y_{15} & y_{16} \\ y_{24} & y_{25} & y_{26} \\ y_{34} & y_{35} & y_{36} \end{bmatrix}. \qquad [Y_{ba}] = \begin{bmatrix} y_{41} & y_{42} & y_{43} \\ y_{51} & y_{52} & y_{53} \\ y_{61} & y_{62} & y_{63} \end{bmatrix}. \qquad (9.5.16)$$

Graphically we represent this partitioning as grouping the six coils into two compound coils (a) and (b), each composed of three individual admittances. This is illustrated in Figure 9.11.

On examination of $[Y_{ab}]$ and $[Y_{ba}]$ it can be seen that

$$[Y_{ba}] = [Y_{ab}]^T$$

if, and only if $y_{ik} = y_{ki}$ for $i = 1$ to 3 and $k = 4$ to 6. That is, if and only if the couplings between the two groups of admittances are bilateral.

In this case equation (9.5.15) may be written

$$\begin{bmatrix} [I_a] \\ [I_b] \end{bmatrix} = \begin{bmatrix} [Y_{aa}] & [Y_{ab}] \\ [Y_{ab}]^T & [Y_{bb}] \end{bmatrix} \begin{bmatrix} [V_a] \\ [V_b] \end{bmatrix}. \qquad (9.5.17)$$

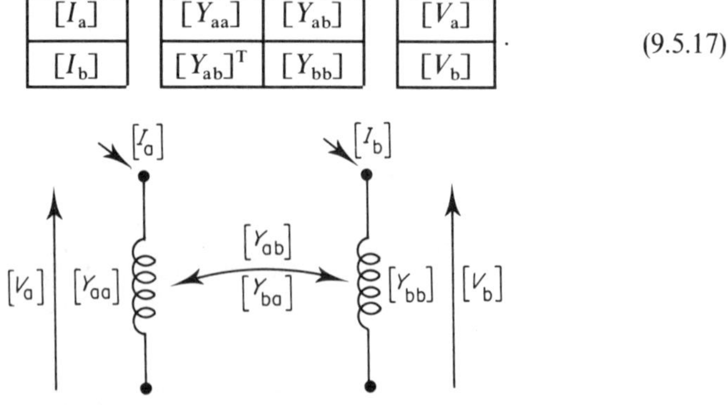

Figure 9.11. Two coupled compound admittances. Reproduced with permission from Arrillaga, Arnold and Harker, *Computer Modelling of Electrical Power Systems*. Copyright © 1983 John Wiley & Sons Ltd.

The primitive network for any number of compound admittances is formed in exactly the same manner as for single admittances, except in that all quantities are matrices of the same order as the compound admittances.

The actual admittance matrix of any networks composed of the compound admittances can be formed by the usual method of linear transformation; the elements of the connection matrix are now $n \times n$ identity matrices where n is the dimension of the compound admittances.

If the connection matrix of any network can be partitioned into identity elements of equal dimensions greater than one, the use of compound admittances is advantageous.

As an example, consider the network shown in Figures 9.12 and 9.13. This represents a simple line section. The admittance matrix will be derived using single and compound admittances to show the simple correspondence. The primitive networks and associated admittance matrices are drawn in Figure 9.14. The connection matrices for the single and compound networks are illustrated by equations (9.5.18) and (9.5.19) respectively.

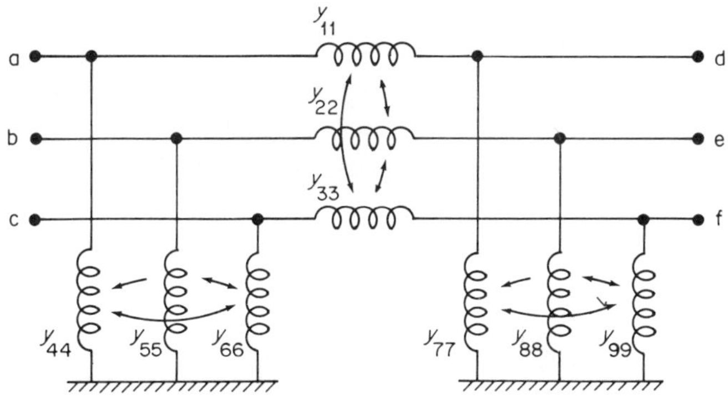

Figure 9.12. Sample network represented by single admittances. Reproduced with permission from Arrillaga, Arnold and Harker, *Computer Modelling of Electrical Power Systems*. Copyright © 1983 John Wiley & Sons Ltd.

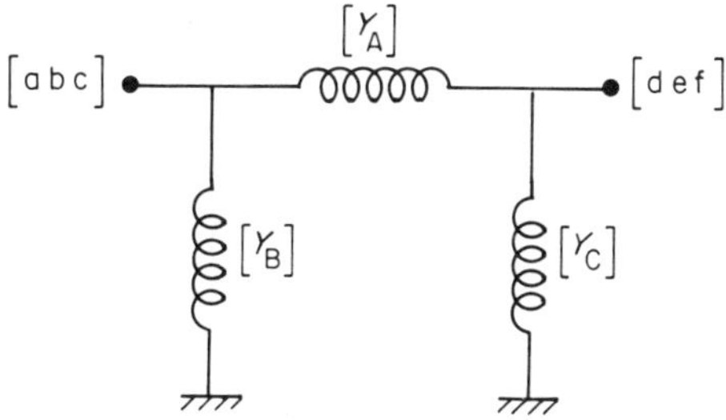

Figure 9.13. Sample network represented by compound admittances. Reproduced with permission from Arrillaga, Arnold and Harker, *Computer Modelling of Electrical Power Systems*. Copyright © 1983 John Wiley & Sons Ltd.

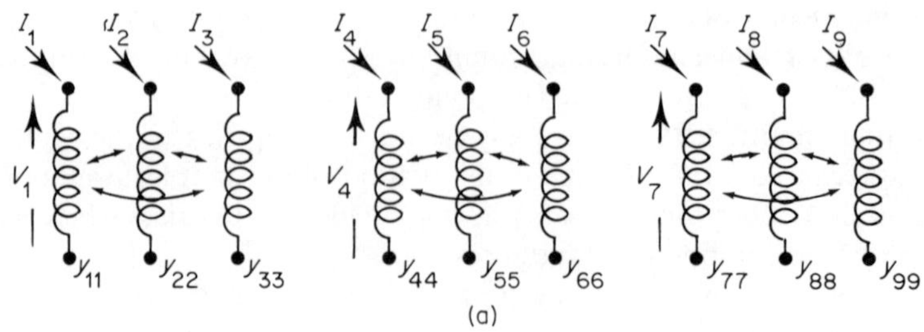

(a)

y_{11}	y_{12}	y_{13}						
y_{21}	y_{22}	y_{23}						
y_{31}	y_{32}	y_{33}						
			y_{44}	y_{45}	y_{46}			
			y_{54}	y_{55}	y_{56}			
			y_{64}	y_{65}	y_{66}			
						y_{77}	y_{78}	y_{79}
						y_{87}	y_{88}	y_{89}
						y_{97}	y_{98}	y_{99}

(b)

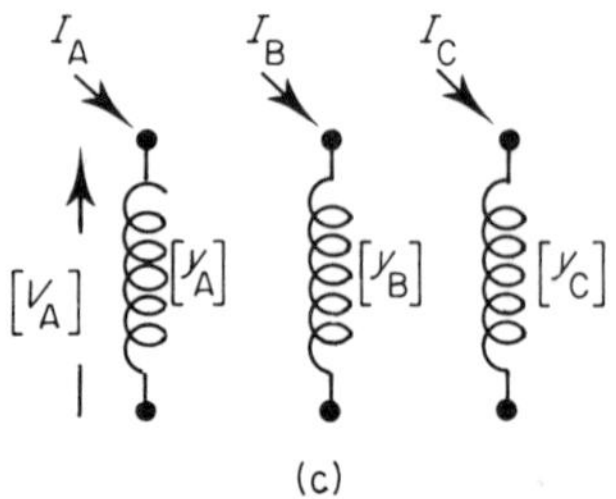

(c)

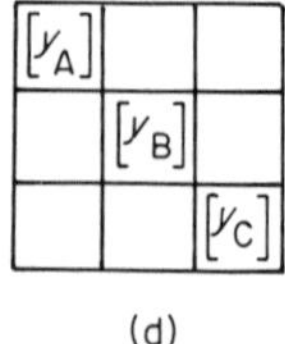

$[y_A]$		
	$[y_B]$	
		$[y_C]$

(d)

Figure 9.14. Primitive networks and corresponding admittance matrices: (a) primitive network using single admittances; (b) primitive admittance matrix; (c) primitive network using compound admittances; (d) primitive admittance matrix. Reproduced with permission from Arrillaga, Arnold and Harker, *Computer Modelling of Electrical Power Systems*. Copyright © 1983 John Wiley & Sons Ltd.

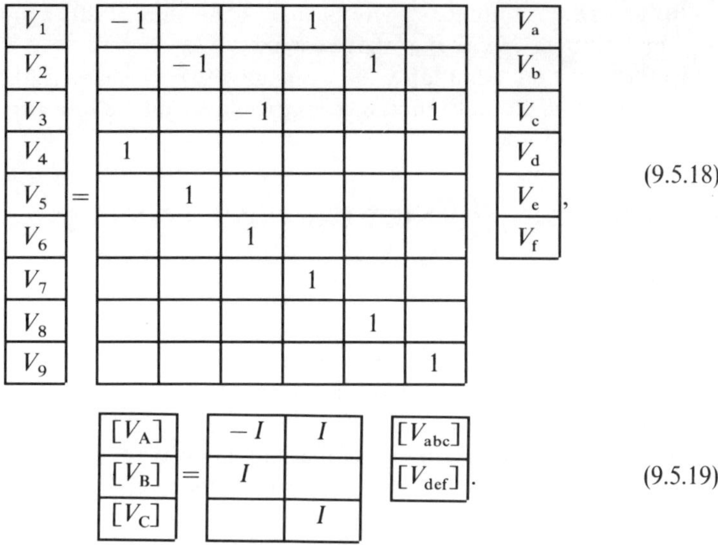

$$\begin{bmatrix} V_1 \\ V_2 \\ V_3 \\ V_4 \\ V_5 \\ V_6 \\ V_7 \\ V_8 \\ V_9 \end{bmatrix} = \begin{bmatrix} -1 & & & 1 & & \\ & -1 & & & 1 & \\ & & -1 & & & 1 \\ 1 & & & & & \\ & 1 & & & & \\ & & 1 & & & \\ & & & 1 & & \\ & & & & 1 & \\ & & & & & 1 \end{bmatrix} \begin{bmatrix} V_a \\ V_b \\ V_c \\ V_d \\ V_e \\ V_f \end{bmatrix}, \qquad (9.5.18)$$

$$\begin{bmatrix} [V_A] \\ [V_B] \\ [V_C] \end{bmatrix} = \begin{bmatrix} -I & I \\ I & \\ & I \end{bmatrix} \begin{bmatrix} [V_{abc}] \\ [V_{def}] \end{bmatrix}. \qquad (9.5.19)$$

The exact equivalence, with appropriate matrix partitioning, is clear. The network admittance matrix is given by the linear transformation equation,

$$[Y_{NODE}] = [C]^T[Y_{PRIM}][C].$$

This matrix multiplication can be executed using the full matrices or in partitioned form. The result in partitioned form is

$$[Y_{NODE}] = \begin{bmatrix} [Y_A]+[Y_B] & -[Y_A] \\ -[Y_A] & [Y_A]+[Y_C] \end{bmatrix}.$$

Rules for forming the admittance matrix of simple networks

The method of linear transformation may be used to obtain the admittance matrix of any network. For the special case of networks where there is no mutual coupling, simple rules may be used to form the admittance matrix by inspection. These rules, which apply to compound networks with no mutual coupling between the compound admittances, may be stated as follows:

(i) Any diagonal term is the sum of the individual branch admittances connected to the node corresponding to that term.
(ii) Any off-diagonal term is the negated sum of the branch admittances which are connected between the two corresponding nodes.

Where mutual coupling between elements exists, it is convenient to divide the total system into subsystems which have no interconnecting mutual couplings, and then form the total system admittance matrix using the following rules:

(i) The self-admittance of any busbar is the sum of all the individual self-admittance matrices at that busbar.
(ii) The mutual admittance between any two busbars is the sum of the individual mutual admittance matrices from all the subsystems containing those two nodes.

9.6 MODELLING OF NETWORK COMPONENTS

Transmission lines

With a three-phase, mutually coupled ground return model of a transmission line, a physical understanding of the impedances, voltages and currents is difficult and as a result, most of the reference material[9,11,12] is primarily concerned with single-phase balanced systems.

In this section the description of the standing wave effects is first approached from the single-phase transmission line model; the modelling of three-phase lines is illustrated as a progression from this.

A transmission line, illustrated in Figure 9.15, consists of distributed inductance and capacitance, which represent the magnetic and electrostatic conditions of the line and resistance and conductance which represent the line losses.

Under perfectly balanced conditions, three-phase transmission lines can be represented by their single-phase positive sequence models and nominal PI circuits. For inclusion into an admittance matrix, it is necessary to use the admittances between busbars, and from the busbars to earth as in Figure 9.16.

For long lines a number of PI models are connected in series to improve the accuracy of voltages and currents, which are affected by standing wave effects. For example, a three-section PI model provides an accuracy to1.2% for a quarter wavelength line (a quarter wavelength corresponds with 1500 and 1250 km at 50 and 60 Hz respectively).

As the frequency increases, the number of nominal PI sections to maintain a particular accuracy increases proportionally, e.g. a 300 km line requires 30

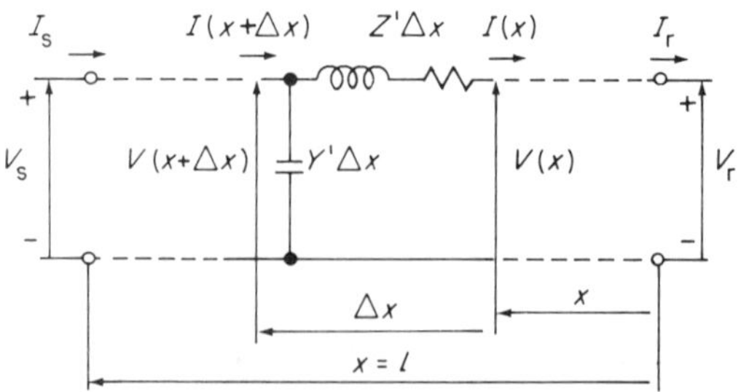

Figure 9.15. Distributed parameter transmission line. V, voltage; I, current; Z', series impedance per unit distance; Y', shunt admittance per unit distance; l, transmission line length

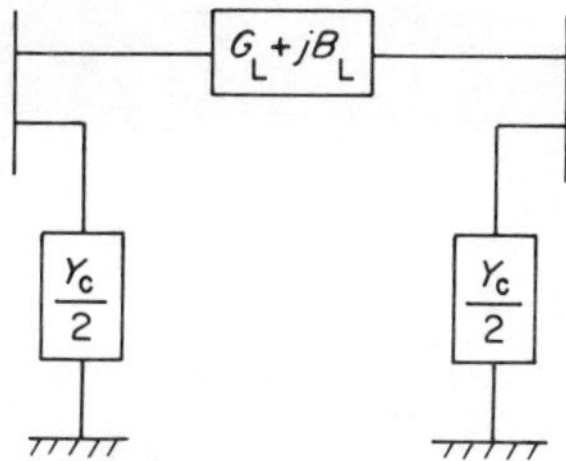

Figure 9.16. Admittance model of transmission line, where $G_L + jB_L = 1/(R_L + jX_L)$, $Y_C = 1/jX_C$, $X_L = \omega L$, and $X_C = 1/\omega C$. G, conductance; B susceptance; X, reactance; ω, frequency (in radians per second)

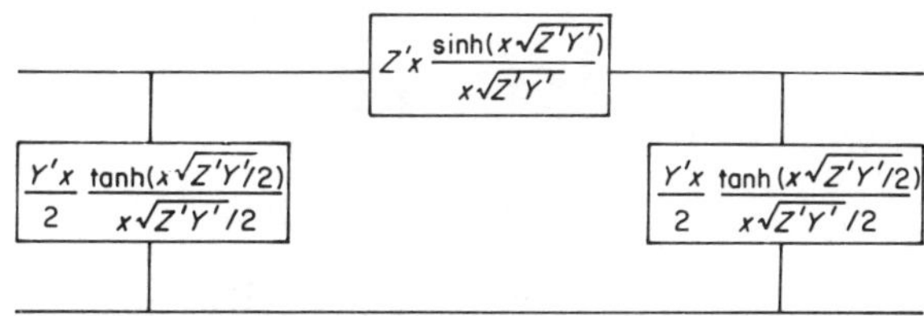

Figure 9.17. The equivalent PI model of a long transmission line

nominal PI sections to maintain the 1.2% accuracy for the 50th harmonic. However, near resonance the accuracy departs significantly from an acceptable value.

The computational effort can be greatly reduced and the accuracy improved with the use of an equivalent PI model derived from the solution of the second order linear differential equations describing wave propagation along transmission lines.[10] With reference to Figure 9.15,

$$\frac{d^2V(x)}{dx^2} = Z'Y'V(x), \qquad \frac{d^2I(x)}{dx^2} = Z'Y'I(x), \tag{9.6.1}$$

where $Z' = r + j2\pi fL$ is the series impedance per unit length and $Y' = g + j2\pi fC$ is the shunt admittance per unit length.

The equivalent PI model, Figure 9.17, is obtained from the nominal PI model by applying correction factors to the series impedance and shunt admittance, i.e.

$$\frac{\sinh(x\sqrt{Z'Y'})}{x\sqrt{Z'Y'}} \quad \text{for the series impedance,}$$

$$\frac{\tanh(x\sqrt{Z'Y'}/2)}{x\sqrt{Z'Y'}/2} \quad \text{for the shunt admittance.} \tag{9.6.2}$$

The impedances of Figure 9.18, obtained for a 220 kV, 230 km line, are plotted in Figure 9.19 against frequency. These were calculated using geometric mean

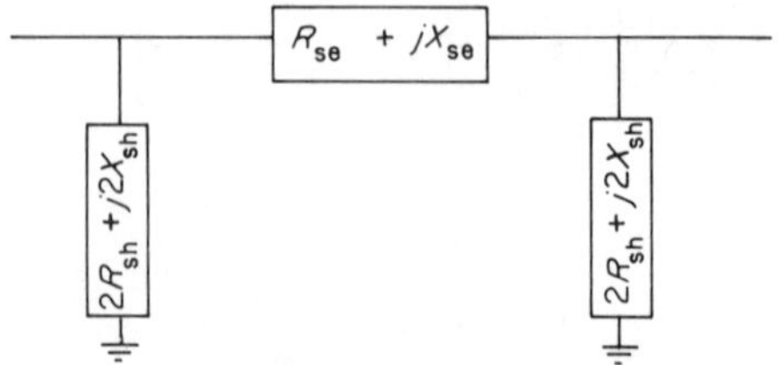

Figure 9.18. Equivalent PI impedances

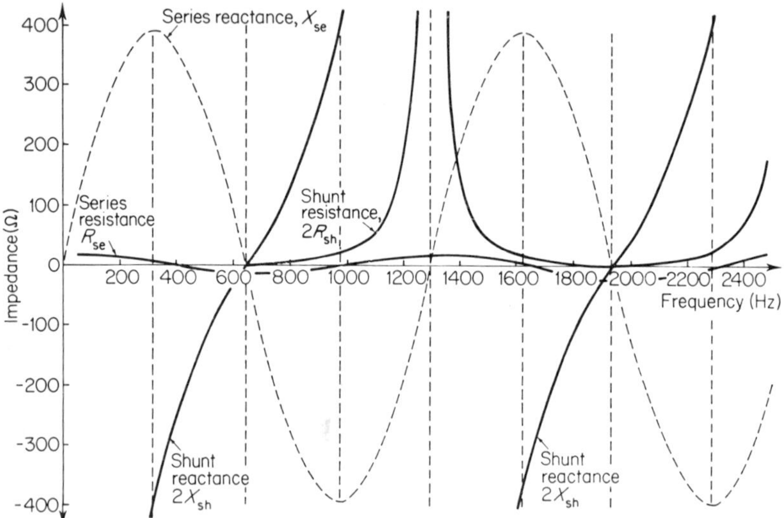

Figure 9.19. Impedance versus frequency for the equivalent PI model (skin effect included)

distances and three equal length transposition sections.[12] The shunt resistance and shunt reactance are formed by inverting the shunt admittance.

The series and shunt reactances are the predominant components of the impedances. Both have a period of 1300 Hz. The line length of 230 km corresponds with one wavelength at this frequency. The series reactance increases from its inductive 50 Hz value up to a maximum at 325 Hz (the quarter wavelength frequency) and then decreases, passing through zero at 650 Hz (the half wavelength frequency). Between the half and full wavelength frequencies the series reactance is capacitive. By contrast, the shunt reactance is capacitive and large at fundamental frequency, reducing in magnitude to zero at the half wavelength frequency. Beyond this it becomes inductive.

The series resistance is small at audiofrequencies. This is to be expected in a system designed to transmit power at fundamental frequency with minimum losses. Also, the peak magnitudes increase slowly as frequency increases. Since the series resistance does not get appreciably larger over the audiofrequency range, the attenuation does not increase significantly. Thus, currents with frequencies in this range will propagate large distances on the power system. The negative resistances are a mathematical artifice and are not physically measurable. However, they give the correct terminal conditions for a distributed parameter transmission line.

The shunt resistance, which is normally considered to be zero in a nominal PI model, has considerable effect at resonant frequencies and, as can be observed from Figure 9.19, becomes very large as the wavelength frequency is approached.

The impedance variation of a transmission line at resonance is similar to the case of a series and parallel resonating tuned circuit. In Figure 9.19, the series and shunt reactances are equal in magnitude but of opposite sign at 325 Hz, i.e. there is a series resonance (or node) with a low purely resistive impedance. This effect is better illustrated in Figure 9.20, where the impedance of the open circuited line is plotted. In this case the low impedance magnitude (series resonance) only contains the series and shunt resistances and occurs at the odd quarter wavelength frequencies.

At 650 Hz, although both the series and shunt reactances are small, the transmission line has a high impedance equivalent to a parallel resonating tuned circuit. This condition is called an antinode and can also be observed on Figure 9.20. These parallel resonances occur at the half wavelength frequencies.

Low impedance at the odd quarter wavelength frequencies and large impedance at the half wavelength frequencies, indicate the low level of attenuation of audiofrequency signals. The addition of other system components such as loads and generators must provide the harmonic damping.

The asymptotes of Figure 9.20 are calculated from a knowledge of the total series impedance, Z, and shunt admittance, Y, of the line.[10] The propagation

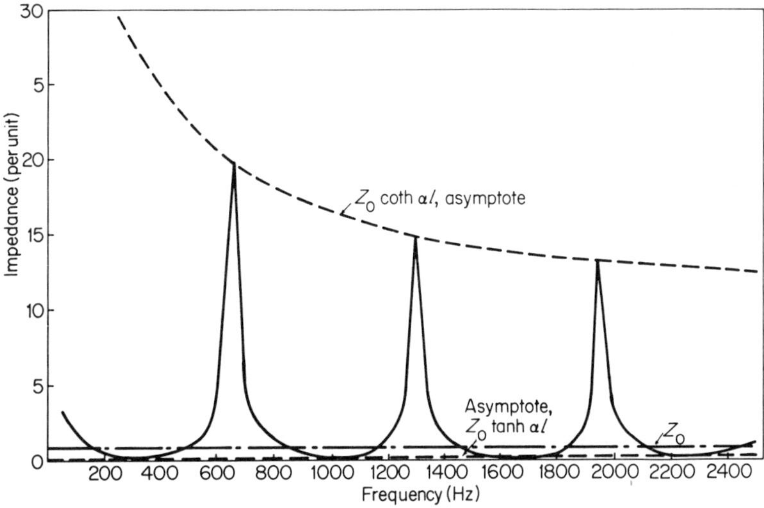

Figure 9.20. Impedance versus frequency for the open circuited Islington to Kikiwa transmission line (skin effect included)

constant γ is

$$\gamma = \sqrt{(ZY)} = \alpha + j\beta, \tag{9.6.3}$$

where α is the attenuation constant and β is the phase constant. The characteristic impedance Z_0 is

$$Z_0 = \sqrt{(Z/Y)}. \tag{9.6.4}$$

The upper asymptote or maximum impedance is

$$Z_0 \coth \alpha l \tag{9.6.5}$$

and the lower asymptote or minimum impedance is

$$Z_0 \tanh \alpha l. \tag{9.6.6}$$

The lower asymptote is small in value and slowly increases with frequency, while the upper asymptote decreases from an infinite value as frequency increases. For large frequencies these two asymptotes approach a value equal to the characteristic impedance.

Maximum values of currents and voltages along transmission lines

Due to the standing wave effect of voltages and currents on transmission lines, the maximum value of these are likely to occur at points other than at the receiving end or sending end busbars. These local maxima could result in insulation damage, overheating or electromagnetic interference. It is thus important to calculate the maximum values of currents and voltages along a line and the points at which these occur.

Knowing the receiving end current and voltage, for each harmonic frequency, the current and voltage at any point on the line can be calculated for each frequency by using the following equations:[13]

$$I(x) = \frac{I_R}{2Z_0}[(Z_R + Z_0)e^{\gamma x} + (Z_0 - Z_R)e^{-\gamma x}], \tag{9.6.7}$$

$$V(x) = \frac{I_R}{2}[(Z_R + Z_0)e^{\gamma x} + (Z_R - Z_0)e^{-\gamma x}], \tag{9.6.8}$$

where x is the distance from the receiving end, I_R is the receiving end current, V_R is the receiving end voltage, and $Z_R = V_R/I_R$. The points on the transmission line at which these are maximum, are obtained by considerations of the currents and voltages as forward (incident) and backward (reflected) travelling waves with respect to the receiving end.

For example, consider the current equation (9.6.7). The incident current at the receiving end is

$$I_R^+ = \frac{(Z_R + Z_0)}{2Z_0} I_R \tag{9.6.9}$$

and the reflected current at the receiving end is

$$I_R^- = \frac{(Z_0 - Z_R)}{2Z_0} I_R. \tag{9.6.10}$$

The angles associated with these currents, at any point along the line, are given by

$$\theta^+ = \theta_R^+ + \beta x, \qquad \theta^- = \theta_R^- - \beta x, \tag{9.6.11}$$

where θ_R^+, θ_R^- are the angles of the current at the receiving end.

The current will be a maximum for θ^+ equal to θ^-. Thus

$$\theta_R^+ + \beta x = \theta_R^- - \beta x \tag{9.6.12}$$

or

$$x = \frac{\theta_R^- - \theta_R^+}{2\beta}. \tag{9.6.13}$$

The current will also have local maxima at intervals of one half wavelength along the line.

While the total r.m.s. voltage and current (over the fundamental and all harmonics) are of greatest importance, the location of the maximum total r.m.s. voltage and current will most likely be dominated by that harmonic which is closest to a resonant frequency of the system.

Multiconductor transmission line modelling

Transmission line parameters are calculated from the line geometrical characteristics. The calculated parameters are expressed as a series impedence and shunt admittance per unit length of line. The effect of ground currents and earth wires are included in the calculation of these parameters.[14,15]

Before considering frequency dependence and long line modelling for harmonic penetration studies, the representation of a three-phase transmission line with earth wire and ground return, suitable for incorporation into a system admittance matrix, will be discussed.

SERIES IMPEDANCE

A three-phase transmission line with an overhead earth wire is illustrated in Figure 9.21(a). Each conductor has resistance and inductance and will be mutually coupled.

The total series inductance of a single-phase circuit is[16,17]

$$L = L_1 + L_2 - 2M, \tag{9.6.14}$$

where

$$L_1 = L_2 = \frac{\mu_0}{2\pi} \ln \frac{1}{\text{GMR}},$$

$$M = \frac{\mu_0}{2\pi} \ln \frac{1}{d};$$

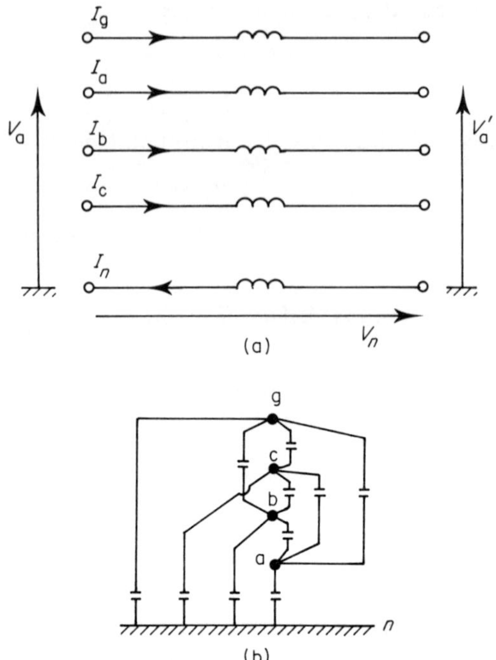

Figure 9.21. (a) Three-phase transmission series impedance equivalent and (b) three-phase transmission shunt impedance equivalent. Reproduced with permission from Arrillaga, Arnold and Harker, *Computer Modelling of Electrical Power Systems*. Copyright © 1983 John Wiley & Sons Ltd.

L_1 is the self-inductance of conductor 1, L_2 is the self-inductance of conductor 2, M is the mutual inductance between conductors 1 and 2, d is the distance between conductor centres, μ_0 is the permeability of free space, and GMR stands for the effective geometric mean distance or the radius of an infinitely thin tube that gives the same inductance as the two terms of the self-inductance.

The self-impedance per kilometre of conductor a with earth return (Z_{aa}), and the mutual impedance per kilometre between conductors a and b (Z_{ab}) are expressed as[9]

$$Z_{aa} = R_a + R_g + j(X_{aa} + X_g) \tag{9.6.15}$$

$$Z_{ab} = R_g + j(X_{ab} + X_g), \tag{9.6.16}$$

where R_a is the a.c. resistance of conductor a, X_{aa} is the self-reactance of conductor a, X_{ab} is the mutual reactance between conductors a and b, and R_g, X_g are Carson's earth return correction.[18] The effect of earth resistivity on the self-reactance X_{aa} at 50 Hz can be assessed from the approximate expression

$$X_{aa} = 0.00289 f \log\left(\frac{660\sqrt{\rho/f}}{\text{GMR}}\right) \quad \Omega/\text{km} \tag{9.6.17}$$

With respect to Figure 9.21(a), the following equation can be written for phase a:

$$V_a - V'_a = I_a(R_a + j\omega L_a) + I_b(j\omega L_{ab}) + I_c(j\omega L_{ac}) + j\omega L_{ag}\cdot I_g - j\omega L_{an}\cdot I_n + V_n, \quad (9.6.18)$$

where

$$V_n = I_n(R_n + j\omega L_n) - I_a j\omega L_{na} - I_b j\omega L_{nb} - I_c j\omega L_{nc} - I_g\cdot j\omega L_{ng}, \quad (9.6.19)$$

and substituting

$$I_n = I_a + I_b + I_c + I_g \quad (9.6.20)$$

gives

$$V_a - V'_a = I_a(R_a + j\omega L_a) + I_b j\omega L_{ab} + I_c j\omega L_{ac} + j\omega L_{ag}\cdot I_g - j\omega L_{an}(I_a + I_b + I_c + I_g) + V_n. \quad (9.6.21)$$

Regrouping and substituting for V_n, i.e.

$$\begin{aligned} \Delta V_a = V_a - V'_a \\ &= I_a(R_a + j\omega L_a - j\omega L_{an} + R_n + j\omega L_n - j\omega L_{na}) \\ &\quad + I_b(j\omega L_{ab} - j\omega L_{an} + R_n + j\omega L_n - j\omega L_{nb}) \\ &\quad + I_c(j\omega L_{ac} - j\omega L_{an} + R_n + j\omega L_n - j\omega L_{nc}) \\ &\quad + I_g(j\omega L_{ag} - j\omega L_{an} + R_n + j\omega L_n - j\omega L_{ng}) \end{aligned} \quad (9.6.22)$$

$$\begin{aligned} \Delta V_a &= I_a(R_a + j\omega L_a - 2j\omega L_{an} + R_n + j\omega L_n) \\ &\quad + I_b(j\omega L_{ab} - j\omega L_{bn} - j\omega L_{an} + R_n + j\omega L_n) \\ &\quad + I_c(j\omega L_{ac} - j\omega L_{cn} - j\omega L_{an} + R_n + j\omega L_n) \\ &\quad + I_g(j\omega L_{ag} - j\omega L_{gn} - j\omega L_{an} + R_n + j\omega L_n) \end{aligned} \quad (9.6.23)$$

or

$$\Delta V_a = Z_{aa-n} I_a + Z_{ab-n} I_b + Z_{ac-n} I_c + Z_{ag-n} I_g \quad (9.6.24)$$

and writing similar equations for the other phases and the earth wire, the following matrix equation results:

$$\begin{bmatrix} \Delta V_a \\ \Delta V_b \\ \Delta V_c \\ \hline \Delta V_g \end{bmatrix} = \left[\begin{array}{ccc|c} Z_{aa-n} & Z_{ab-n} & Z_{ac-n} & Z_{ag-n} \\ Z_{ba-n} & Z_{bb-n} & Z_{bc-n} & Z_{bg-n} \\ Z_{ca-n} & Z_{cb-n} & Z_{cc-n} & Z_{cg-n} \\ \hline Z_{ga-n} & Z_{gb-n} & Z_{gc-n} & Z_{gg-n} \end{array}\right] \begin{bmatrix} I_a \\ I_b \\ I_c \\ \hline I_g \end{bmatrix}. \quad (9.6.25)$$

Usually we are interested only in the performance of the phase conductors, and it is more convenient to use a three-conductor equivalent for the transmission line. This is achieved by writing matrix equation (9.6.25) in partitioned form as follows:

$$\begin{bmatrix} \Delta V_{abc} \\ \Delta V_g \end{bmatrix} = \begin{bmatrix} Z_A & Z_B \\ Z_C & Z_D \end{bmatrix} \begin{bmatrix} I_{abc} \\ I_g \end{bmatrix}. \quad (9.6.26)$$

From (9.6.26)

$$[\Delta V_{abc}] = [Z_A][I_{abc}] + [Z_B][I_g], \tag{9.6.27}$$

$$[\Delta V_g] = [Z_C][I_{abc}] + [Z_D][I_g]. \tag{9.6.28}$$

From equations (9.6.26) and (9.6.28), and assuming that the earth wire is at zero potential,

$$[\Delta V_{abc}] = [Z_{abc}][I_{abc}], \tag{9.6.29}$$

where

$$[Z_{abc}] = [Z_A] - [Z_B][Z_D]^{-1}[Z_C] = \begin{bmatrix} Z'_{aa-n} & Z'_{ab-n} & Z'_{ac-n} \\ Z'_{ba-n} & Z'_{bb-n} & Z'_{bc-n} \\ Z'_{ca-n} & Z'_{cb-n} & Z'_{cc-n} \end{bmatrix} \tag{9.6.30}$$

SHUNT ADMITTANCE

With reference to Figure 9.21(b) the potentials of the line conductors are related to the conductor charges by the matrix equation[14]

$$\begin{bmatrix} V_a \\ V_b \\ V_c \\ V_g \end{bmatrix} = \begin{bmatrix} P_{aa} & P_{ab} & P_{ac} & P_{ag} \\ P_{ba} & P_{bb} & P_{bc} & P_{bg} \\ P_{ca} & P_{cb} & P_{cc} & P_{cg} \\ P_{ga} & P_{gb} & P_{gc} & P_{gg} \end{bmatrix} \begin{bmatrix} Q_a \\ Q_b \\ Q_c \\ Q_g \end{bmatrix} \tag{9.6.31}$$

The following approximate relationships can be written for the potential coefficients:[16,17]

$$P_{aa} = \frac{1}{2\pi\varepsilon_0} \ln\left(\frac{2H}{R}\right) \quad \text{km/F},$$

$$P_{ab} = \frac{1}{2\pi\varepsilon_0} \ln\left(\frac{D_{ab}}{d_{ab}}\right) \quad \text{km/F}, \tag{9.6.32}$$

where P_{aa} is the potential coefficient of conductor a, P_{ab} is the potential coefficient between conductors a and b, H is the average height of the conductor above ground (in metres), R is the conductor radius, d_{ab} is the distance between conductors a and b (in metres), D_{ab} is the distance between conductor a and the image of conductor b (in metres), and ε_0 is the permitivity of free space. Similar considerations as for the series impedance matrix, lead to

$$[V_{abc}] = [P'_{abc}][Q_{abc}], \tag{9.6.33}$$

where P'_{abc} is a 3×3 matrix which includes the effects of the earth wire. The capacitance matrix of the transmission line of Figure 9.21 is given by

$$[C'_{abc}] = [P'_{abc}]^{-1} = \begin{bmatrix} C_{aa} & -C_{ab} & -C_{ac} \\ -C_{ba} & C_{bb} & -C_{bc} \\ -C_{ca} & -C_{cb} & C_{cc} \end{bmatrix}. \tag{9.6.34}$$

The series impedance and shunt admittance lumped-PI model representation of the three-phase line is shown in Figure 9.22(a) and its matrix equivalent is illustrated in Figure 9.22(b). These two matrices can be represented by compound admittances (Figure 9.22(c)) as described earlier.

Following the rules developed for the formation of the admittance matrix using

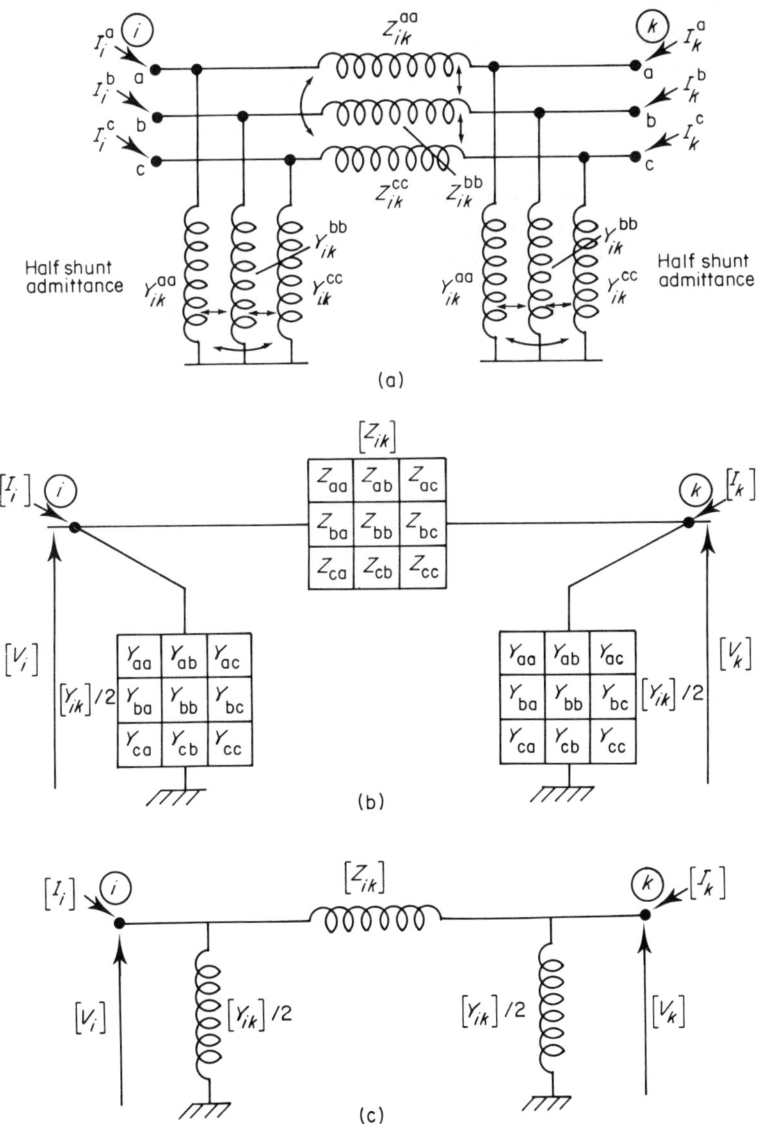

Figure 9.22. Lumped PI model of a short three-phase line series impedance: (a) full circuit representation; (b) matrix equivalent; (c) using three-phase compound admittances. Reproduced with permission from Arrillaga, Arnold and Harker, *Computer Modelling of Electrical Power Systems*. Copyright © 1983 John Wiley & Sons Ltd.

the compound concept, the nodal injected currents of Figure 9.22(c) can be related to the nodal voltages by the equation

$$\underset{6\times 1}{\begin{bmatrix} [I_i] \\ [I_k] \end{bmatrix}} = \underset{6\times 6}{\begin{bmatrix} [Z]^{-1}+[Y]/2 & -[Z]^{-1} \\ -[Z]^{-1} & [Z]^{-1}+[Y]/2 \end{bmatrix}} \underset{6\times 1}{\begin{bmatrix} [V_i] \\ [V_k] \end{bmatrix}}. \tag{9.6.35}$$

This forms the element admittance matrix representation for the short line between busbars i and k in terms of 3×3 matrix quantities.

This representation may not be accurate enough for electrically long lines. The physical length at which a line is no longer electrically short depends on the wavelength, therefore when harmonic frequencies are being considered, this physical length may be quite small. Using transmission line and wave propagation theory more exact models may be derived.

Mutually coupled three-phase lines

When two or more transmission lines occupy the same right of way for a considerable length, the electrostatic and electromagnetic coupling between those lines must be taken into account.

Consider the simplest case of two mutually coupled three-phase lines. The two coupled lines are considered to form one subsystem composed of four system busbars. The coupled lines are illustrated in Figure 9.23, where each element is a 3×3 compound admittance and all voltages and currents are 3×1 vectors.

The coupled series elements represent the electromagnetic coupling while the coupled shunt elements represent the capacitive or electrostatic coupling. These coupling parameters are lumped in a similar way to the standard line parameters.

With the admittances labelled as in Figure 9.23 and applying the rules of linear

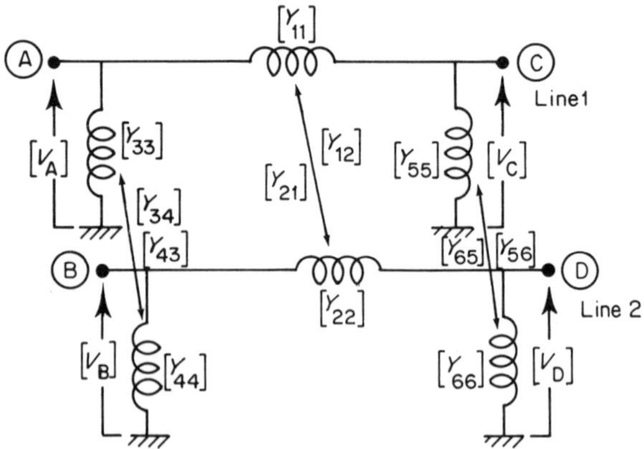

Figure 9.23. Two-coupled three-phase lines. Reproduced with permission from Arrillaga, Arnold and Harker, *Computer Modelling of Electrical Power Systems*. Copyright © 1983 John Wiley & Sons Ltd.

transformation for compound networks, the admittance matrix for the subsystem is defined as follows:

$$\begin{bmatrix} I_A \\ I_B \\ I_C \\ I_D \end{bmatrix} = \begin{bmatrix} Y_{11}+Y_{33} & Y_{12}+Y_{34} & -Y_{11} & -Y_{12} \\ Y_{12}^T+Y_{34}^T & Y_{22}+Y_{44} & -Y_{12}^T & -Y_{22} \\ -Y_{11} & -Y_{12} & Y_{11}+Y_{55} & Y_{12}+Y_{56} \\ -Y_{12}^T & -Y_{22} & Y_{12}^T+Y_{56}^T & Y_{22}+Y_{66} \end{bmatrix} \cdot \begin{bmatrix} V_A \\ V_B \\ V_C \\ V_D \end{bmatrix} .$$

$$12 \times 1 \qquad 12 \times 12 \qquad 12 \times 1 \tag{9.6.36}$$

It is assumed here that the mutual coupling is bilateral. Therefore $Y_{21} = Y_{12}^T$, etc.

The subsystem may be redrawn as in Figure 9.24. The pairs of coupled 3 × 3 compound admittances are now represented as a 6 × 6 compound admittance. The matrix representation is also shown. Following this representation and the labelling of the admittance blocks in the figure, the admittance matrix may be written in terms of the 6 × 6 compound coils as

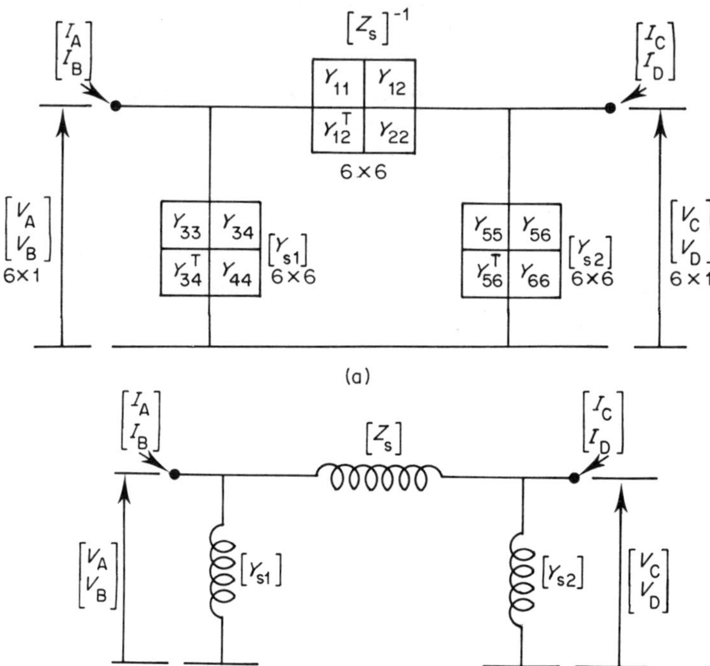

Figure 9.24. A 6 × 6 compound admittance representation of two coupled three-phase lines: (a) 6 × 6 matrix representation; (b) 6 × 6 compound admittance representation. Reproduced with permission from Arrillaga, Arnold and Harker, *Computer Modelling of Electrical Power Systems*. Copyright © 1983 John Wiley & Sons Ltd.

$$\begin{bmatrix} \begin{bmatrix} I_A \\ I_B \end{bmatrix} \\ \begin{bmatrix} I_C \\ I_D \end{bmatrix} \end{bmatrix} = \begin{bmatrix} [Z_s]^{-1} + [Y_{s1}] & -[Z_s]^{-1} \\ -[Z_s]^{-1} & [Z_s]^{-1} + [Y_{s2}] \end{bmatrix} \begin{bmatrix} \begin{bmatrix} V_A \\ V_B \end{bmatrix} \\ \begin{bmatrix} V_C \\ V_D \end{bmatrix} \end{bmatrix} . \tag{9.6.37}$$

$$12 \times 1 \qquad\qquad 12 \times 12 \qquad\qquad 12 \times 1$$

This is clearly identical to equation (9.6.36) with the appropriate matrix partitioning.

The representation of Figure 9.24 is more concise and the formation of equation (9.6.37) from this representation is straightforward, being exactly similar to that which results from the use of 3 × 3 compound admittances for the normal single three-phase line.

The data which must be available, to enable coupled lines to be treated in a similar manner to single lines, is the series impedance and shunt admittance matrices. These matrices are of order 3 × 3 for a single line, 6 × 6 for two coupled lines, 9 × 9 for three and 12 × 12 for four coupled lines.

Once the matrices $[Z_s]$ and $[Y_s]$ are available, the admittance matrix for the subsystem is formed by application of equation (9.6.37).

When all the busbars of the coupled lines are distinct, the subsystem may be combined directly into the system admittance matrix. However, if the busbars are not distinct then the admittance matrix as derived from equation (9.6.37) must be modified. This is considered in the following section.

Consideration of terminal connections

The admittance matrix as derived above must be reduced if there are different elements in the subsystem connected to the same busbar. As an example consider two parallel transmission lines as illustrated in Figure 9.25.

The admittance matrix derived previously related the currents and voltages at the four busbars A1, A2, B1 and B2. This relationship is given by

$$\begin{bmatrix} I_{A1} \\ I_{A2} \\ I_{B1} \\ I_{B2} \end{bmatrix} = \begin{bmatrix} [Y_{A1A2B1B2}] \end{bmatrix} \begin{bmatrix} V_{A1} \\ V_{A2} \\ V_{B1} \\ V_{B2} \end{bmatrix} . \tag{9.6.38}$$

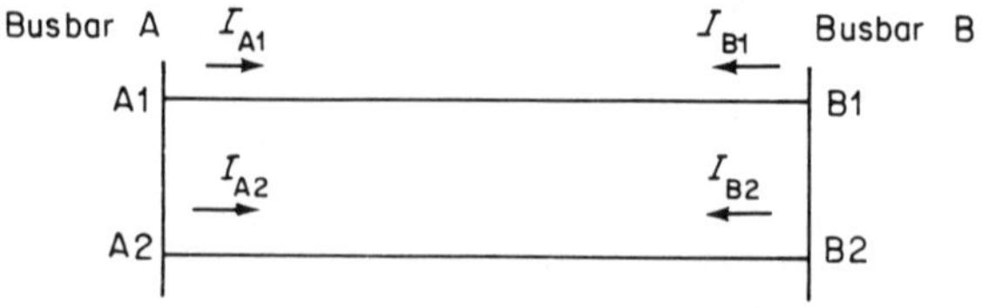

Figure 9.25. Mutually coupled parallel transmission lines. Reproduced with permission from Arrillaga, Arnold and Harker, *Computer Modelling of Electrical Power Systems*. Copyright © 1983 John Wiley & Sons Ltd.

The nodal injected current at busbar A, I_A, is given by

$$I_A = I_{A1} + I_{A2};$$

similarly

$$I_B = I_{B1} + I_{B2}.$$

Also, from inspection of Figure 9.25,

$$V_A = V_{A1} = V_{A2}, \qquad V_B = V_{B1} = V_{B2}.$$

The required matrix equation relates the nodal injected currents, I_A and I_B, to the voltages at these busbars. This is readily derived from equation (9.6.38) and the conditions specified above. This is simply a matter of adding appropriate rows and columns, and yields

$$\begin{bmatrix} I_A \\ I_B \end{bmatrix} = \begin{bmatrix} [Y_{AB}] \end{bmatrix} \begin{bmatrix} V_A \\ V_B \end{bmatrix}, \tag{9.6.39}$$

where $[Y_{AB}]$ is the required nodal admittance matrix for the subsystem.

It should be noted that the matrix in equation (9.6.38) must be retained as it is required in the calculation of the individual line currents.

Equivalent PI model of multiconductor transmission line

In the case of multiconductor transmission lines, the nominal PI series impedance and shunt admittance matrices per unit distance, $[Z']$ and $[Y']$ respectively, are square and their size is fixed by the number of mutually coupled conductors.

The derivation of the equivalent PI model for harmonic penetration studies from the nominal PI matrices, is similar to that of the single phase lines, except that it involves the evaluation of hyperbolic functions of the propagation constant which is now a matrix:

$$[\gamma] = ([Z'][Y'])^{1/2}. \tag{9.6.40}$$

There is no direct way of calculating sinh or tanh of a matrix, thus a method using eigenvalues and eigenvectors, called 'modal analysis' is employed.[19]

The basis of such a method for a multiconductor line is as follows: Consider the second order linear differential equations describing wave propagation along a single transmission line (9.6.1). These can be expanded for the multiconductor line in the form of equation (9.6.41), where the matrices are of order m, the number of phases involved:

$$\left[\frac{d^2V}{dx^2}\right] = [Z'][Y'][V], \tag{9.6.41}$$

$$\left[\frac{d^2I}{dx^2}\right] = [Y'][Z'][I]. \tag{9.6.42}$$

It should be noted that in this case the matrix products $[Z'][Y']$ and $[Y'][Z']$ are not equal, except in special cases.[20]

These equations are still difficult to solve because all phases are coupled. However, just as three-phase equations with balanced matrices can be transformed into decoupled single-phase equations using symmetrical components, it is possible to transform equations (9.6.41) and (9.6.42) into decoupled equations as well. By transforming phase voltages to 'modal' voltages,

$$[V]=[T_v][V_{mode}] \quad \text{and} \quad [V_{mode}]=[T_v]^{-1}[V] \tag{9.6.43}$$

and by choosing the proper transformation matrix $[T_v]$, equation (9.6.41) can be changed to

$$[d^2V_{mode}/dx^2]=[\Lambda][V_{mode}], \tag{9.6.44}$$

where $[\Lambda]$ is now a diagonal matrix. This diagonalization is a well-defined procedure in matrix algebra; the elements of $[\Lambda]$ are the eigenvalues of the matrix product $[Z'][Y']$, and the transformation matrix $[T_v]$ is the matrix of eigenvectors of that matrix product. Equation (9.6.42) can be diagonalized as well, with the same diagonal matrix $[\Lambda]$, i.e.

$$[d^2I_{mode}/dx^2]=[\Lambda][I_{mode}], \tag{9.6.45}$$

but the transformation matrix for currents differs from that used for voltages (in contrast to symmetrical components),

$$[I]=[T_i][I_{mode}] \quad \text{and} \quad [I_{mode}]=[T_i]^{-1}[I], \tag{9.6.46}$$

though both are related by

$$[T_i]^t=[T_v]^{-1}, \tag{9.6.47}$$

where the superscript 't' indicates a transposed matrix.

With the diagonalized equations (9.6.44) and (9.6.45), an m-phase line can now be studied as if it consisted of m single-phase lines, similar to the symmetrical component approach, except that the zero, positive and negative sequence networks now become the mode-1, mode-2 and mode-3 networks. In each mode, the single-phase long-line series impedance and shunt admittance of Figure 9.17 are used. The propagation constant of each mode is simply

$$\lambda_{mode\text{-}i}=\sqrt{\lambda_i}, \tag{9.6.48}$$

where λ_i is the ith eigenvalue or ith element in $[\Lambda]$. The modal series impedance and shunt admittance are not directly available, but must be computed from

$$[Z'_{mode}]=[T_v]^{-1}[Z'][T_i] \quad \text{and} \quad [Y'_{mode}]=[T_i]^{-1}[Y'][T_v], \tag{9.6.49}$$

with both modal matrices being diagonal. $[Y'_{mode}]$ may no longer be purely imaginary even though only shunt capacitance is modelled. This will depend on how the transformation matrices were normalized. For steady-state analysis at one particular frequency, this causes no problems. Once Z_{series} and Y_{shunt} have been calculated for each mode, the representation in phase quantities is easily obtained by transforming back, with

$$[Z_{\text{series}}] = [T_v][Z_{\text{series-mode}}][T_i]^{-1} \quad \text{and} \quad [Y_{\text{shunt}}] = [T_i][Y_{\text{shunt-mode}}][T_v]^{-1} \tag{9.6.50}$$

becoming the values of the equivalent PI model which will accurately represent the untransposed line.

In expanded form the following are expressions for the series impedance and shunt admittance of the equivalent PI model:(20)

$$[Z]_{\text{EPM}} = l[Z'][M]\left[\frac{\sinh \gamma l}{\gamma l}\right][M]^{-1}, \tag{9.6.51}$$

where l is the transmission line length, $[Z]_{\text{EPM}}$ is the equivalent PI series impedance matrix, $[M]$ is the matrix of normalized eigenvectors,

$$\left[\frac{\sinh \gamma l}{\gamma l}\right] = \begin{bmatrix} \dfrac{\sinh \gamma_1 l}{\gamma_1 l} & 0 & \cdots & 0 \\ 0 & \dfrac{\sinh \gamma_2 l}{\gamma_2 l} & \cdots & 0 \\ \vdots & \vdots & & \vdots \\ 0 & 0 & \cdots & \dfrac{\sinh \gamma_j l}{\gamma_j l} \end{bmatrix} \tag{9.6.52}$$

and γ_j is the jth eigenvalue for $j/3$ mutually coupled circuits. Similarly

$$[Y]_{\text{EPM}} = l[M]\left[\frac{\tanh(\gamma l/2)}{\gamma l/2}\right][M]^{-1}[Y'], \tag{9.6.53}$$

where $[Y]_{\text{EPM}}$ is the equivalent PI shunt admittance matrix.

An illustration of the relative accuracies achieved by the equivalent PI and nominal PI models is displayed in Figure 9.26.

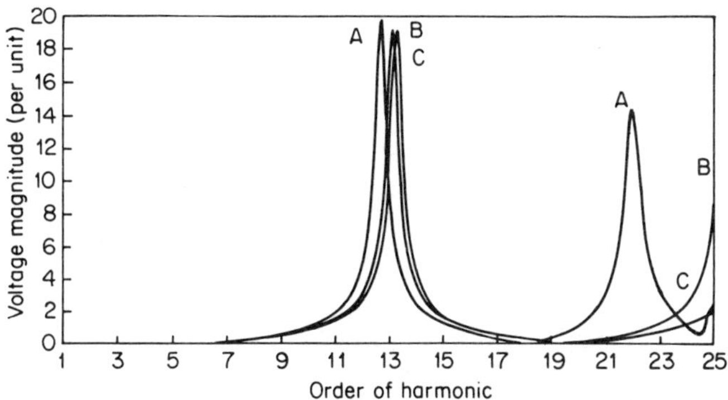

Figure 9.26. Comparison of the equivalent and nominal PI transmission line models at 5 Hz intervals: curve A, three sections; curve B, six sections; curve C, equivalent PI.

This figure shows the per unit positive sequence voltages of a 230 km, 220 kV line, the parameter information for which is shown in Figure 9.27. Standing wave effects show the different accuracies provided by the two models, with the resonant frequencies of the nominal PI model approaching those of the equivalent PI model as the number of sections increases from three to six. Accuracy of the nominal PI model decreases as the frequency increases and this can be observed at the wavelength frequency.

Cascading of nominal PI circuits requires a large number of sections for long lines at higher frequencies, to achieve acceptable accuracy. The equivalent PI model avoids the problems of determining the number of sections needed and round-off error that accumulates in this situation.

Computer derivation of the correction factors for conversion from the nominal PI to the equivalent PI model, and their incorporation into the series impedance and shunt admittance matrices, is carried out as indicated in the structure diagram of Figure 9.28. The LR2 algorithm of Wilkinson and Reinsch[21] is used with due regard for accurate calculations in the derivation of the eigenvalues and eigenvectors.

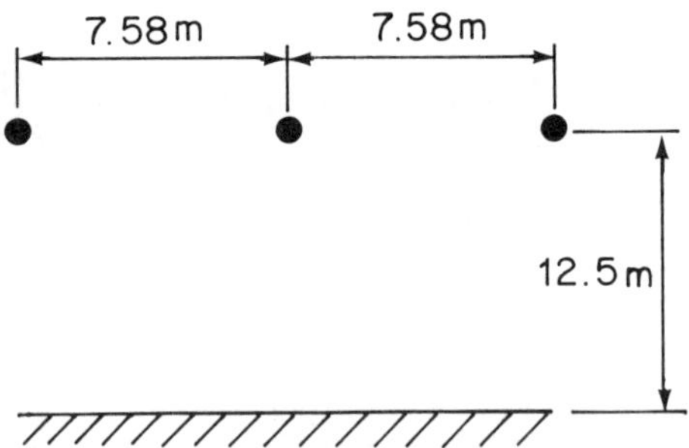

Figure 9.27. Conductor information for the Islington to Kikiwa line: conductor type, Zebra (54/3.18 + 7/3.18); length, 230 km; resistivity, 100 Ω m

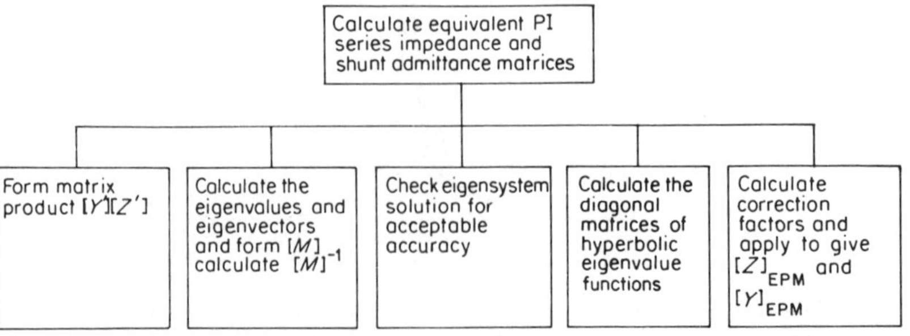

Figure 9.28. Structure diagram for calculation of the equivalent PI model

Skin effect modelling

Changes in conductor parameters due to the non-uniform current distribution inside a conductor is termed 'skin effect'. Current tends to flow on the surface or skin of a conductor and this effect increases as frequency increases. The consequence of skin effect is an increase in the resistance of the conductor and a decrease in its internal inductance.

The following expression can be derived from field theory for the a.c. resistance taking into account skin effect:[11]

$$R = \mathrm{Re}\left(\frac{j\mu\omega}{2\pi a}\cdot\frac{J_z(r)}{\partial J_z(r)/\partial r|_{r=a}}\right), \tag{9.6.54}$$

where $J_z(r)$ is the current density, and a is the outside radius of the conductor.

Changes in the internal inductance have little effect on the geometric mean radius and, therefore, on the conductor component of self-inductance and on the self-inductance with ground return. Therefore the inductive skin effect is usually ignored.[10]

When conductors are separated by small distances, the magnetic field inside the conductor is not circular. Current density is not symmetrical about the conductor axis and the net effect is an increase in series resistance and a decrease in inductive reactance. This phenomenon is called 'proximity effect'. In solid round wires, this effect is much less significant than skin effect, and it has been concluded[22] that it can be neglected for separations greater than 20 cm. Since bundled conductors are generally spaced slightly further apart than this, proximity effects are neglected for transmission lines. This is not the case for cable parameters however.

Kennelly *et al.*[22] conducted tests on solid copper and aluminium wires and simple stranded wires at various separations, up to 5 kHz. It was observed that stranded conductors, without spiralling, had similar skin effect ratios to solid conductors of the same cross-sectional area. Unfortunately, aluminium conductor steel reinforced (ACSR) cables used in power system transmission lines have steel cores and multiple layers of aluminium strands with different lays. Some current leaks from strand to strand and this leakage will vary depending upon the condition of the strand surface, the conductor tension, and length of lay, and will change over the time of service. These non-ideal parameters make calculation of the skin effect ratio over a range of frequencies difficult. A number of authors[23–26] have developed accurate models for various combinations of these effects at power frequencies, by dividing the conductors into a number of subconductors. However, the amount of data and computation required to do this for each conductor prohibits its use for a general transmission line model.

Simpler relationships developed for solid conductors, valid for various frequency ranges,[27] do not give completely accurate results for ACSR cables, as can be seen from Figure 9.29.

These approximate relationships have been applied to the resistance of the series impedance in single-phase models, i.e. the skin effect ratio is applied to the

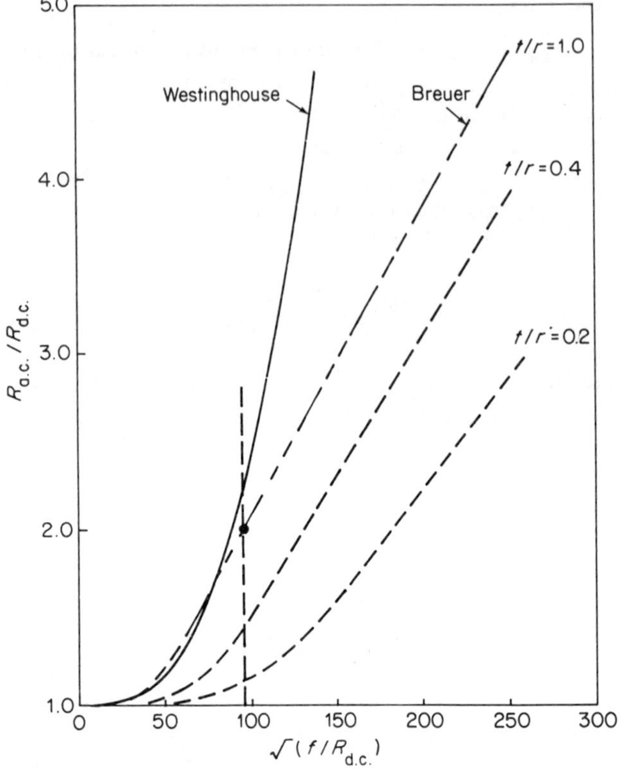

Figure 9.29. Skin effect resistance ratios for different models. The solid dot indicates the skin effect ratio for the Islington to Kikiwa line at the half-wavelength resonant frequency. Dashed curves indicate ACSR conductors with various tube ratios

resistance of an equivalent conductor and not to each conductor individually.

Lewis and Tuttle[28] presented a practical method of calculating the skin effect resistance ratios by approximating ACSR conductors to uniform tubes having the same inside and outside diameters as the aluminium conductors, see Figure 9.30. The assumption is that little current flows in the steel core.

Values are most accurate for an even number of layers of aluminium, the longitudinal magnetic field for each layer induced in the core almost cancelling. The worst case is for a single layer cable where the skin effect resistance ratio is current dependent. However, at transmission voltages, single layered cables are used infrequently.

Thickness to radius ratios for some common conductors are presented in Table 9.1 and a range of resistance ratios are plotted in Figure 9.29.

To reduce errors in the calculation of resonant peaks, more rigorous analysis using Bessel function relationships for tubular conductors must be used. There is, however, a trade-off between accuracy and computation cost; the necessary

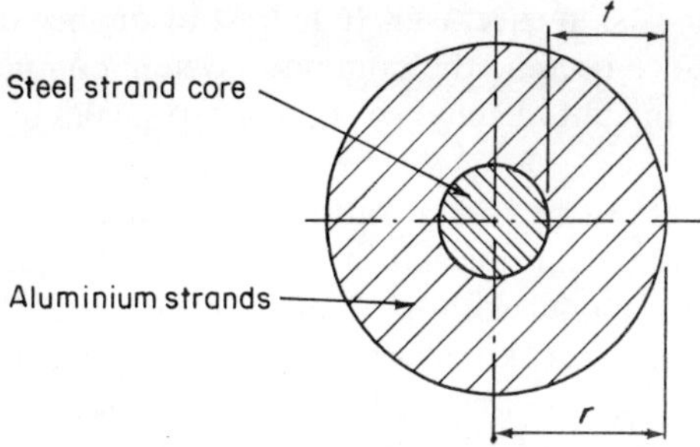

Figure 9.30. ACSR hollow tube conductor representation

Table 9.1. Tube ratios for commonly used conductors

Conductor	t/r
Chukar	0.73
Special (2)	0.67
Special (3)	0.77
Pheasant	0.66
Zebra	0.66
Goat	0.57
Coyote	0.64

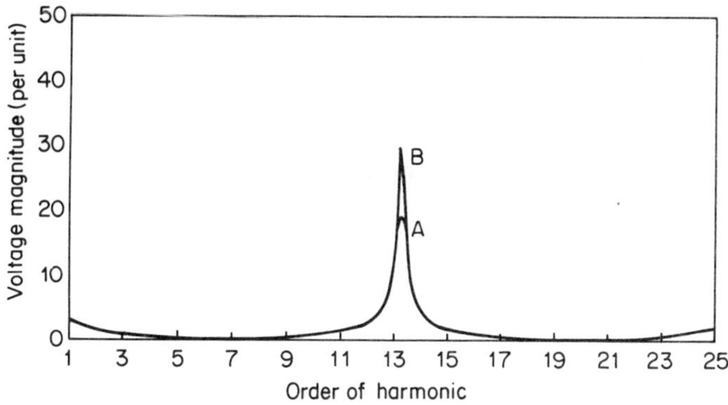

Figure 9.31. The effect of skin effect modelling: curve A, skin effect included; curve B, no skin effect

solution being an acceptance of some inaccuracy for practical computability. It is doubtful whether the use of more accurate skin effect ratios are justified in a practical power system.

For long lines, skin effect ratios and their effect on the resonant voltage magnitudes are important. Because the series resistance of a transmission line is a small component of the series impedance when the transmission line is not at resonance, the harmonic voltages, shown in Figure 9.31, do not change to any significant extent when skin effect is included. At resonance the series resistance and shunt conductance become the dominant system components. Changes in the series resistance magnitude change the voltage peaks but do not affect the resonant frequency.

Referring to Figure 9.31 the voltage ratio at resonance, between cases with and without skin effect, is 1.5 while the skin effect ratio calculated from Figure 9.29 is 2.0. In a single-phase model without ground return the ratio of voltages at resonance, with and without skin effect, is the same as the skin effect ratio. In a three-phase model the presence of shunt conductance and series resistance coupling between phases, and the different resonant frequencies of the phases, reduces the resonant peak voltages compared with single-phase modelling.

Skin effect is also taken into account in modelling the earth return as a conductor. The depth of penetration of the earth currents decreases with an increase in frequency or a decrease in earth resistivity. The series inductance decreases as a result of these changes.

To keep the change of line resistance due to skin effect in perspective, it should be noted that the combined effects of climate and loading on line temperatures also alters line resistance. If the resistance ratio of a line is defined as

$$\frac{R_2}{R_1} = \frac{T + t_2}{T + t_1}, \tag{9.6.55}$$

where t_1, t_2 are the temperatures in centigrade and T is a constant dependent on conductor type (228 for hard drawn aluminium, 241 for hard drawn copper), then, with $t_1 = 10\,^\circ\mathrm{C}$ and $t_2 = 40\,^\circ\mathrm{C}$ for aluminium,

$$\frac{R_2}{R_1} = \frac{228 + 40}{228 + 10} \simeq 1.13.$$

A 30 °C change in line temperature is not unusual and the corresponding change in d.c. resistance, and hence resonant peak magnitude voltage, is thus approximately 13%.

Since a transmission line forms part of a power system with variable conditions, it is not justifiable to include extremely accurate skin effect resistance ratios, valid in only a small number of controlled conditions.

Shunt elements

Shunt reactors and capacitors are used in a power system for reactive power control. The data for these elements are usually given in terms of their rated megavolt-amps and rated kilovolts. The equivalent phase admittance in per unit is calculated from these data.

The coupled admittances to ground at bus k are formed into a 3×3 admittance matrix as shown in Figure 9.32, and this reduces to the compound admittance representation indicated. The admittance matrix is incorporated directly into the system admittance matrix, contributing only to the self-admittance of the particular bus.

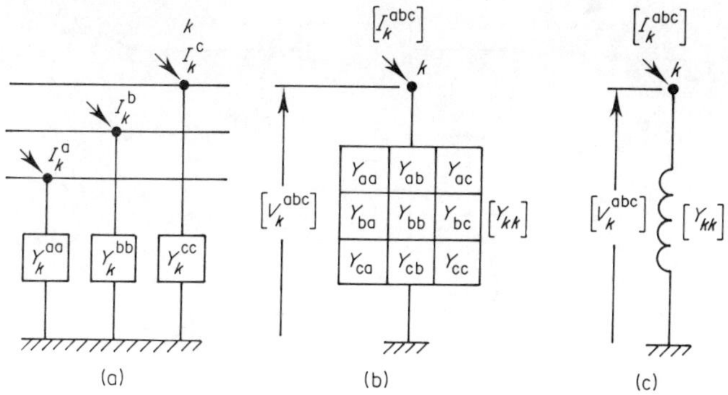

Figure 9.32. Representation of a shunt element: (a) coupled admittance, (b) admittance matrix; (c) compound admittance

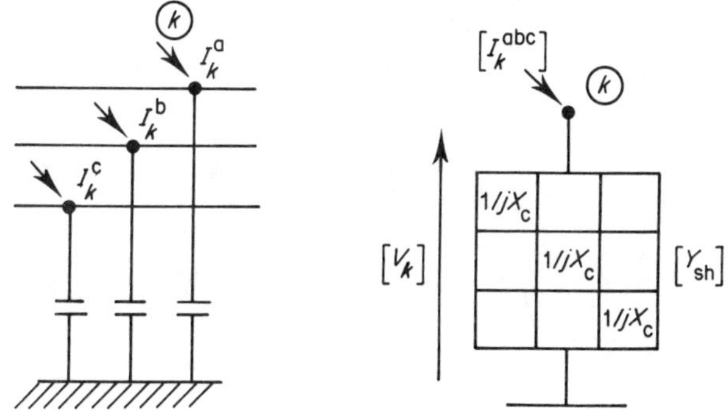

Figure 9.33. Representation of a shunt capacitor bank. Reproduced with permission from Arrillaga, Arnold and Harker, *Computer Modelling of Electrical Power Systems.* Copyright © 1983 John Wiley & Sons Ltd.

While provision for off-diagonal terms exists, the admittance matrix for shunt elements is usually diagonal, as there is normally no coupling between the components of each phase.

Consider, as an example, a three-phase capacitor bank shown in Figure 9.33. A triple representation similar to that for a line section is illustrated.

The element admittance matrix will be diagonal and proportional to frequency. The megavolt-amp rating at fundamental frequency (Q) and the nominal voltage (V) are used to calculate the capacitive reactance, i.e. $x_c = V^2/nQ$. The presence of any series inductance in the capacitor banks is ignored.

Series elements

Series elements are connected directly between two buses and for modelling purposes they constitute a subsystem in the network subdivision described earlier.

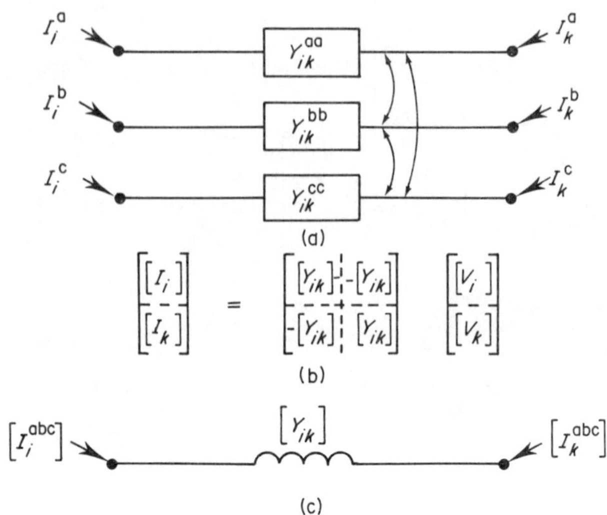

$$\begin{bmatrix} [I_i] \\ [I_k] \end{bmatrix} = \begin{bmatrix} [Y_{ik}] & -[Y_{ik}] \\ -[Y_{ik}] & [Y_{ik}] \end{bmatrix} \begin{bmatrix} [V_i] \\ [V_k] \end{bmatrix}$$

Figure 9.34. Representation of a series element: (a) coupled admittances; (b) admittance matrix; (c) compound admittance. Reproduced with permission from Arrillaga, Arnold and Harker, *Computer Modelling of Electrical Power Systems*. Copyright © 1983 John Wiley & Sons Ltd.

A three-phase coupled series admittance between two busbars i and k is shown in Figure 9.34(a), as well as its reduced nodal admittance matrix (Figure 9.34(b)) and compound admittance (Figure 9.34(c)).

The admittance matrix for the subsystem can be written by inspection as

$$[Y] = \begin{array}{|c|c|} \hline [Y_{se}] & -[Y_{se}] \\ \hline -[Y_{se}] & [Y_{se}] \\ \hline \end{array} . \tag{9.6.56}$$

The series capacitor, used for transmission line reactance compensation, is an example of an uncoupled series element; in this case the admittance matrix is diagonal.

Single-phase transformers

A transformer with turns ratio a interconnecting two nodes i, k can be represented by an ideal transformer in series with the nominal transformer leakage admittance at the required frequency as shown in Figure 9.35(a).

In harmonic penetration studies the magnetizing admittance is usually ignored since under normal operating conditions its contribution is not significant. If, however, the transformer is under severe saturation, as explained in Section 4.2 appropriate current harmonic sources must be added at the transformer terminals.

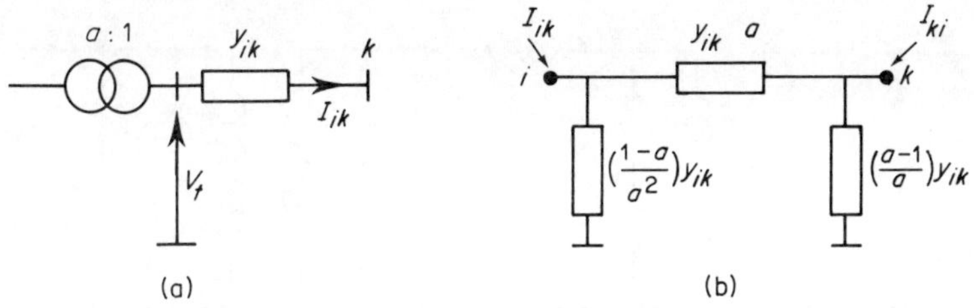

Figure 9.35. Transformer with off-nominal tap setting. Reproduced with permission from Arrillaga, Arnold and Harker, *Computer Modelling of Electrical Power Systems*. Copyright © 1983 John Wiley & Sons Ltd.

If the transformer is on nominal tap ($a = 1$), the nodal equations for the network branch in the per unit system are

$$I_{ik} = y_{ik}V_i - y_{ik}V_k, \tag{9.6.57}$$

$$I_{ki} = y_{ik}V_k - y_{ik}V_i. \tag{9.6.58}$$

In this case $I_{ik} = -I_{ki}$.

For an off-nominal tap setting and letting the voltage on the k side of the ideal transformer be V_t we can write

$$V_t = V_i/a, \tag{9.6.59}$$

$$I_{ki} = y_{ik}(V_k - V_t), \tag{9.6.60}$$

$$I_{ik} = -I_{ki}/a. \tag{9.6.61}$$

Eliminating V_t between equations (9.6.59) and (9.6.60) we obtain

$$I_{ki} = y_{ik}V_k - (y_{ik}/a)V_i, \tag{9.6.62}$$

$$I_{ik} = -(y_{ik}/a)V_k + (y_{ik}/a^2)V_i. \tag{9.6.63}$$

A simple equivalent PI circuit can be deduced from equations (9.6.62) and (9.6.63), the elements of which can be incorporated into the admittance matrix. This circuit is illustrated in Figure 9.35(b).

The equivalent circuit of Figure 9.35(b) has to be used with care in banks containing delta-connected windings. In a star–delta bank of single-phase transformer units, for example, with nominal turns ratio, a value of 1.0 per unit voltage on each leg of the star winding produces under balanced conditions 1.732 per unit voltage on each leg of the delta winding (rated line to neutral voltage as base). The structure of the bank requires in the per unit representation an effective tapping at $\sqrt{3}$ nominal turns ratio on the delta side, i.e. $a = 1.732$.

For a delta–delta or star–delta transformer with taps on the star winding, the equivalent circuit of Figure 9.35(b) would have to be modified to allow for effective taps to be represented on each side. The equivalent circuit model of the single-phase unit can be derived by considering a delta–delta transformer as comprising a delta–star transformer connected in series (back to back) via a zero-impedance link to a star–delta transformer, i.e. star windings in series. Both

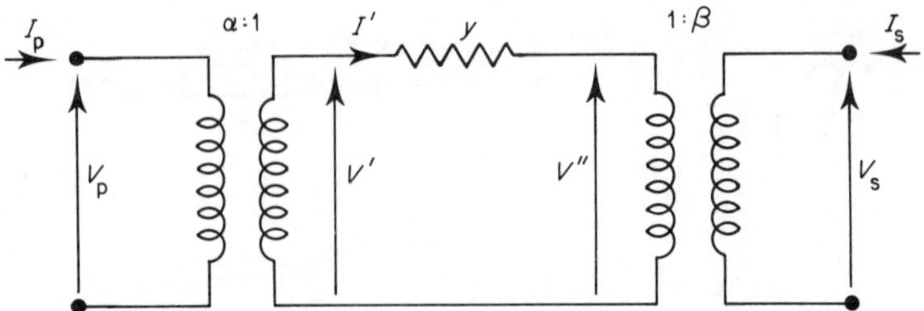

Figure 9.36. Basic equivalent circuit in per unit for coupling between primary and secondary coils with both primary and secondary off-nominal tap ratios of α and β. Reproduced with permission from Arrillaga, Arnold and Harker, *Computer Modelling of Electrical Power Systems*. Copyright © 1983 John Wiley & Sons Ltd.

neutrals are solidly earthed. The leakage impedance of each transformer would be half the impedance of the equivalent delta–delta transformer. An equivalent per unit representation of this coupling is shown in Figure 9.36. Solving this circuit for terminal currents,

$$I_p = \frac{I'}{\alpha} = \frac{(V' - V'')y}{\alpha}$$

$$= \frac{(V_p/\alpha - V_s/\beta)y}{\alpha} = \left(\frac{y}{\alpha^2}\right)V_p - \left(\frac{y}{\alpha\beta}\right)V_s, \tag{9.6.64}$$

$$-I_s = \frac{I'}{\beta} = \left(\frac{y}{\alpha\beta}\right)V_p - \left(\frac{y}{\beta^2}\right)V_s, \tag{9.6.65}$$

or in matrix form

$$\begin{bmatrix} I_p \\ I_s \end{bmatrix} = \begin{bmatrix} y/\alpha^2 & -y/\alpha\beta \\ -y/\alpha\beta & y/\beta^2 \end{bmatrix} \begin{bmatrix} V_p \\ V_s \end{bmatrix}.$$

These admittance parameters form the primitive network for the coupling between a primary and secondary coil.

To cope with phase-shifting, the transformer of Figure 9.36 has to be provided with a complex turns ratio. Moreover, the invariance of the product VI^* across the ideal transformer requires a distinction to be made between the turns ratios for current and voltage, i.e.

$$V_p I_p^* = -V'I'^*$$

or

$$V_p = (a + jb)V' = \alpha V',$$

$$I_p^* = -\frac{I'^*}{a + jb},$$

$$I_p = -\frac{I'}{a - jb} = -\frac{I'}{\alpha^*}.$$

Thus the circuit of Figure 9.36 has two different turns ratios, i.e. $\alpha_v = a + jb$ for the voltages and $\alpha_i = a - jb$ for the currents.

Solving the modified circuit for terminal currents,

$$I_p = \frac{I'}{\alpha_i} = \frac{(V' - V'')y}{\alpha_i} = \frac{(V_p/\alpha_v - V_s/\beta)y}{\alpha_i} = \left(\frac{y}{\alpha_v\alpha_i}\right)V_p - \left(\frac{y}{\alpha_i\beta}\right)V_s, \quad (9.6.66)$$

$$-I_s = \frac{I'}{\beta} = \left(\frac{y}{\alpha_v\beta}\right)V_p - \left(\frac{y}{\beta^2}\right)V_s. \quad (9.6.67)$$

Thus, the general single-phase admittance of a transformer including phase shifting is

$$[y] = \begin{bmatrix} \dfrac{y}{\alpha_i\alpha_v} & -\dfrac{y}{\alpha_i\beta} \\ -\dfrac{y}{\alpha_v\beta} & \dfrac{y}{\beta^2} \end{bmatrix}. \quad (9.6.68)$$

It should be noted that although an equivalent lattice network similar to that in Figure 9.36 could be constructed, it is no longer a bilinear network as can be seen from the asymmetry of y in equation (9.6.68). The equivalent circuit of a single-phase phase-shifting transformer is thus of limited value and the transformer is best represented analytically by its admittance matrix.

Modified transformer admittances for harmonic analysis

As the internal resonant frequencies of high voltage power transformers occur well above the range of interest for harmonic penetration studies, the interwinding capacitances and capacitances to ground of transformers, have very little effect on the accuracy of the results.

The frequency dependence of the resistance accounts for the increased transformer core losses with frequency due to skin effect. Figure 9.37 shows the change in resistance and inductance with frequency for a practical transformer.

Assuming that transformers are not operated in saturation, various representations have been suggested to replace the leakage inductance. These are shown in Figure 9.38(a)–(c).

In Figure 9.38(a) X_{50} is the leakage reactance at 50 Hz.[29]

In Figure 9.38(b)

$$R = 0.1026\, khX_{50}(J + h),$$

where J is the ratio of hysteresis to eddy current losses, taken as three for silicon steels and $k = 1/(J + 1)$. As an alternative model the values of R and X are scaled to 80% of the 50 Hz values.[30]

In Figure 9.38(c)

$$90 < V^2/SR_s < 110, \qquad 13 < SR_p/V^2 < 30,$$

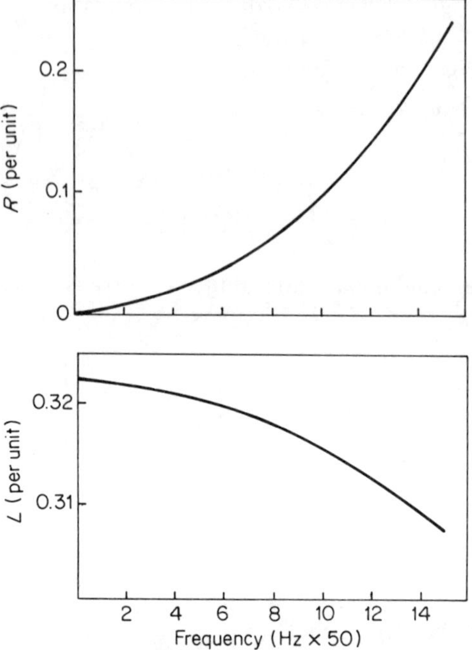

Figure 9.37. Frequency dependence of transformer model: (a) per unit resistance versus frequency; (b) per unit inductance versus frequency (note supressed zero of scale). Reproduced with permission from Arrillaga, Arnold and Harker, *Computer Modelling of Electrical Power Systems*. Copyright © 1983 John Wiley & Sons Ltd.

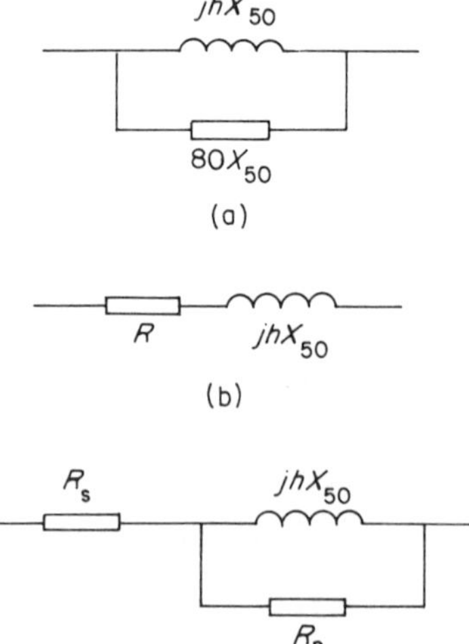

Figure 9.38. Transformer models suggested for harmonic penetration (CIGRE). Reproduced with permission from Arrillaga, Arnold and Harker, *Computer Modelling of Electrical Power Systems*. Copyright © 1983 John Wiley & Sons Ltd.

with S being the rated power of the transformer; typical values (per unit) of R_s and R_p are 0.04 and 60 for a 30 MVA rating.

Corresponding values for the case of a 100 MVA transformer are 0.01 and 20 per unit respectively.

Considering the wide range of models, further work is clearly needed in this area to provide more specific information related to particular transformer ratings and characteristics.

Whenever the effect of transformer magnetic non-linearity is considered relevant, the magnetizing current harmonics must be calculated, as described in Chapter 4, and represented as current injecting sources.

Synchronous machine models

Generally it can be assumed that synchronous generators produce no harmonic voltages and they can therefore be modelled by a shunt impedance at the generator terminals.

A linear reactance derived from either the subtransient or negative sequence inductances is often used,(31) both having similar values. The use of second order equations(32,33) is reported to give lower values of impedance at higher frequencies, although no reasons are given for this approach.

Tests conducted on a small synchronous motor(34) yielded values of measured impedances which agreed to within 15% with the subtransient reactances. However no confirmation appears to be available using power system synchronous generators.

In the absence of a generally accepted model, and until more comprehensive test results are obtained, an empirical linear model is suggested which consists of the full subtransient reactance with a power factor of 0.2.

Primitive admittance model of three-phase transformers

Many three-phase transformers are wound on a common core and all windings are therefore coupled to all other windings. Therefore, in general, a basic two-winding three-phase transformer has a primitive or unconnected network consisting of six coupled coils. If a tertiary winding is also present the primitive network consists of nine coupled coils. The basic two-winding transformer shown in Figure 9.39 is now considered, the addition of further windings being a simple but cumbersome extension of the method.

The primitive network can be represented by the primitive admittance matrix which has the following general form:

$$\begin{bmatrix} I_1 \\ I_2 \\ I_3 \\ I_4 \\ I_5 \\ I_6 \end{bmatrix} = \begin{bmatrix} y_{11} & y_{12} & y_{13} & y_{14} & y_{15} & y_{16} \\ y_{21} & y_{22} & y_{23} & y_{24} & y_{25} & y_{26} \\ y_{31} & y_{32} & y_{33} & y_{34} & y_{35} & y_{36} \\ y_{41} & y_{42} & y_{43} & y_{44} & y_{45} & y_{46} \\ y_{51} & y_{52} & y_{53} & y_{54} & y_{55} & y_{56} \\ y_{61} & y_{62} & y_{63} & y_{64} & y_{65} & y_{66} \end{bmatrix} \cdot \begin{bmatrix} V_1 \\ V_2 \\ V_3 \\ V_4 \\ V_5 \\ V_6 \end{bmatrix} \quad (9.6.69)$$

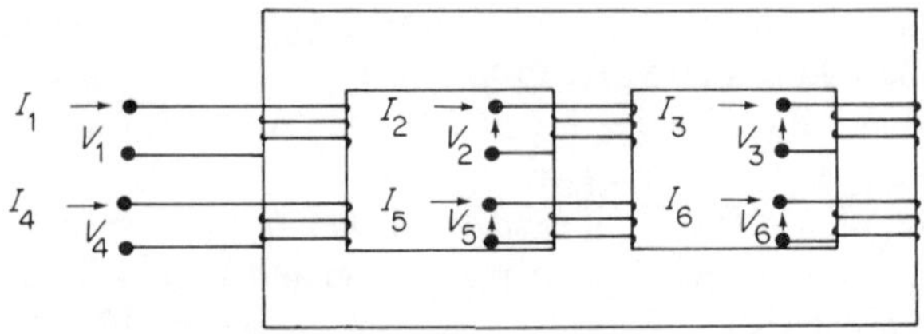

Figure 9.39. Diagrammatic representation of a two-winding transformer. Reproduced with permission from Arrillaga, Arnold and Harker, *Computer Modelling of Electrical Power Systems.* Copyright © 1983 John Wiley & Sons Ltd.

The elements of matrix $[Y]$ can be measured directly, i.e. by energizing coil i at the appropriate frequency and short circuiting all other coils, column i of $[Y]$ can be calculated from $y_{ki} = I_k/V_i$.

Considering the reciprocal nature of the mutual couplings in equation (9.6.69), 21 short circuit measurements would be necessary to complete the admittance matrix at each frequency. Such a detailed representation is seldom required.

By assuming that the flux paths are symmetrically distributed between all windings, equation (9.6.69) may be simplified to equation (9.6.70):

$$\begin{bmatrix} I_1 \\ I_2 \\ I_3 \\ I_4 \\ I_5 \\ I_6 \end{bmatrix} = \begin{bmatrix} y_p & y'_m & y'_m & -y_m & y''_m & y''_m \\ y'_m & y_p & y'_m & y''_m & -y_m & y''_m \\ y'_m & y'_m & y_p & y''_m & y''_m & -y_m \\ -y_m & y''_m & y''_m & y_s & y'''_m & y'''_m \\ y''_m & -y_m & y''_m & y'''_m & y_s & y'''_m \\ y''_m & y''_m & -y_m & y'''_m & y'''_m & y_s \end{bmatrix} \begin{bmatrix} V_1 \\ V_2 \\ V_3 \\ V_4 \\ V_5 \\ V_6 \end{bmatrix}. \quad (9.6.70)$$

where y'_m is the mutual admittance between primary coils, y''_m is the mutual admittance between primary and secondary coils on different cores, and y'''_m is the mutual admittance between secondary coils.

For three separate single-phase units all the primed values are effectively zero. In three-phase units the primed values, representing parasitic interphase coupling, do have a noticeable effect. This effect can be interpreted through the symmetrical component equivalent circuits.

If the values in equation (9.6.70) are available, then this representation of the primitive network should be used. If interphase coupling can be ignored, the coupling between a primary and a secondary coil is modelled as for the single-phase unit, giving rise to the primitive network of Figure 9.40.

The new admittance matrix equation is

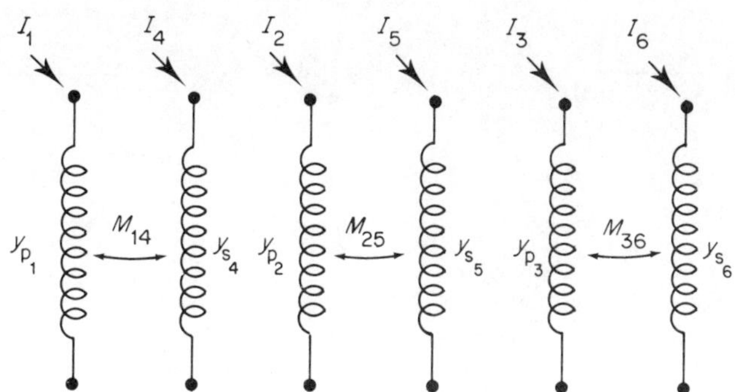

Figure 9.40. Primitive network, where $y_{p_i} = y/\alpha_i^2$, $y_{s_j} = y/\beta_j^2$ and $M_{ij} = y/\alpha_i\beta_i$ for $i = 1$, 2 or 3 and $j = 4$, 5 or 6. Reproduced with permission from Arrillaga, Arnold and Harker, *Computer Modelling of Electrical Power Systems*. Copyright © 1983 John Wiley & Sons Ltd.

$$\begin{bmatrix} I_1 \\ I_2 \\ I_3 \\ I_4 \\ I_5 \\ I_6 \end{bmatrix} = \begin{bmatrix} y_{p1} & & & M_{14} & & \\ & y_{p2} & & & M_{25} & \\ & & y_{p3} & & & M_{36} \\ M_{41} & & & y_{s4} & & \\ & M_{52} & & & y_{s5} & \\ & & M_{63} & & & y_{s6} \end{bmatrix} \begin{bmatrix} V_1 \\ V_2 \\ V_3 \\ V_4 \\ V_5 \\ V_6 \end{bmatrix}. \tag{9.6.71}$$

Models for common transformer connections

The network admittance matrix for any two-winding three-phase transformer can now be formed by the method of linear transformation.

As a simple example, consider the formation of the admittance matrix for a star–star connection with both neutrals solidly earthed in the absence of interphase mutuals. This example is chosen as it is the simplest computationally.

The connection matrix is derived from consideration of the actual connected network. For the star–star transformer illustrated in Figure 9.41, the connection matrix [C] relating the branch voltages (i.e. voltages of the primitive network) to the node voltages (i.e. voltages of the actual network) is a 6 × 6 identity matrix, i.e.

$$\begin{bmatrix} V_1 \\ V_2 \\ V_3 \\ V_4 \\ V_5 \\ V_6 \end{bmatrix} = \begin{bmatrix} 1 & & & & & \\ & 1 & & & & \\ & & 1 & & & \\ & & & 1 & & \\ & & & & 1 & \\ & & & & & 1 \end{bmatrix} \begin{bmatrix} V_p^a \\ V_p^b \\ V_p^c \\ V_s^A \\ V_s^B \\ V_s^C \end{bmatrix}.$$

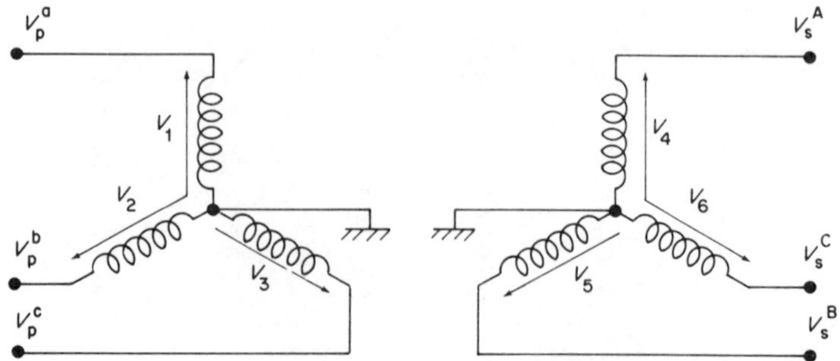

Figure 9.41. Network connection diagram for a three-phase star–star transformer. Reproduced with permission from Arrillaga, Arnold and Harker, *Computer Modelling of Electrical Power Systems.* Copyright © 1983 John Wiley & Sons Ltd.

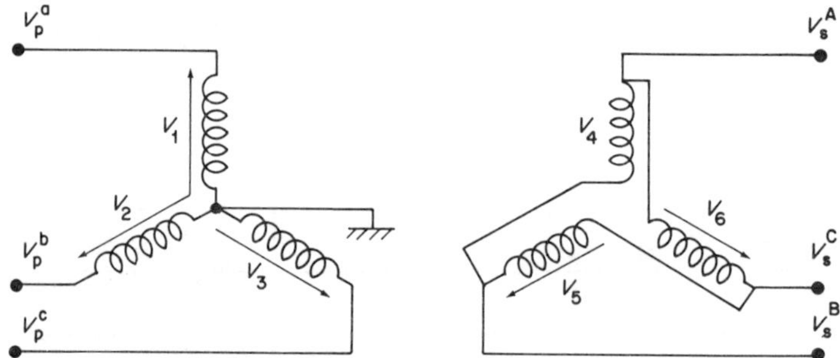

Figure 9.42. Network connection diagram for a Wye G–delta transformer. Reproduced with permission from Arrillaga, Arnold and Harker, *Computer Modelling of Electrical Power Systems.* Copyright © 1983 John Wiley & Sons Ltd.

The nodal admittance matrix $[Y]_{\text{NODE}}$ is given by

$$[Y]_{\text{NODE}} = [C]^t [Y]_{\text{PRIM}} [C]. \tag{9.6.72}$$

Substituting for $[C]$ yields

$$[Y]_{\text{NODE}} = [Y]_{\text{PRIM}}. \tag{9.6.73}$$

Let us now consider the Wye G–Delta connection illustrated in Figure 9.42.

The following connection can be written by inspection between the primitive branch voltages and the node voltages:

$$\begin{bmatrix} V_1 \\ V_2 \\ V_3 \\ V_4 \\ V_5 \\ V_6 \end{bmatrix} = \begin{bmatrix} 1 & 0 & 0 & 0 & 0 & 0 \\ 0 & 1 & 0 & 0 & 0 & 0 \\ 0 & 0 & 1 & 0 & 0 & 0 \\ 0 & 0 & 0 & 1 & -1 & 0 \\ 0 & 0 & 0 & 0 & 1 & -1 \\ 0 & 0 & 0 & -1 & 0 & 1 \end{bmatrix} \begin{bmatrix} V_p^a \\ V_p^b \\ V_p^c \\ V_s^A \\ V_s^B \\ V_s^C \end{bmatrix}. \tag{9.6.74}$$

or

$$[V]_{\text{branch}} = [C][V]_{\text{node}}. \tag{9.6.75}$$

We can also write

$$[Y]_{\text{NODE}} = [C]^{t}[Y]_{\text{PRIM}}[C] \tag{9.6.76}$$

and using $[Y]_{\text{PRIM}}$ from equation (9.6.70)

$$[Y]_{\text{NODE}} = \begin{array}{|c|c|c|c|c|c|l}
\hline
y_p & y'_m & y'_m & -(y_m + y''_m) & (y_m + y''_m) & 0 & \text{a} \\
\hline
y'_m & y_p & y'_m & 0 & -(y_m + y''_m) & (y_m + y''_m) & \text{b} \\
\hline
y'_m & y'_m & y_p & (y_m + y''_m) & 0 & -(y_m + y''_m) & \text{c} \\
\hline
-(y_m + y''_m) & 0 & (y_m + y''_m) & 2(y_s - y'''_m) & -(y_s - y'''_m) & -(y_s - y''') & \text{A} \\
\hline
(y_m + y''_m) & -(y_m + y''_m) & 0 & -(y_s - y'''_m) & 2(y_s - y'''_m) & -(y_s - y'''_m) & \text{B} \\
\hline
0 & (y_m + y''_m) & -(y_m + y''_m) & -(y_s - y'''_m) & -(y_s - y'''_m) & 2(y_s - y'''_m) & \text{C} \\
\hline
\end{array} . \tag{9.6.77}$$

Moreover, if the primitive admittances are expressed in per unit, with both the primary and secondary voltages being 1 per unit, the Wye–delta transformer model must include an effective turns ratio of $\sqrt{3}$. The upper right and lower left quadrants of matrix (9.6.77) must be divided by $\sqrt{3}$ and the lower right quadrant by 3.

In the particular case of three single-phase transformer units connected in Wye G–delta all the y' and y'' terms will disappear. Ignoring off-nominal taps (but keeping in mind the effective $\sqrt{3}$ turns ratio in per unit) the nodal admittance matrix equation relating the nodal currents to the nodal voltages is

$$\begin{array}{|c|}
\hline
I_p^a \\ \hline
I_p^b \\ \hline
I_p^c \\ \hline
I_s^A \\ \hline
I_s^B \\ \hline
I_s^C \\ \hline
\end{array}
=
\begin{array}{|c|c|c|c|c|c|}
\hline
y & & & -y/\sqrt{3} & y/\sqrt{3} & \\
\hline
 & y & & & -y/\sqrt{3} & y/\sqrt{3} \\
\hline
 & & y & y/\sqrt{3} & & -y/\sqrt{3} \\
\hline
-y/\sqrt{3} & & y/\sqrt{3} & \frac{2}{3}y & -\frac{1}{3}y & -\frac{1}{3}y \\
\hline
y/\sqrt{3} & -y/\sqrt{3} & & -\frac{1}{3}y & \frac{2}{3}y & -\frac{1}{3}y \\
\hline
 & y/\sqrt{3} & -y/\sqrt{3} & -\frac{1}{3}y & -\frac{1}{3}y & \frac{2}{3}y \\
\hline
\end{array}
\begin{array}{|c|}
\hline
V_p^a \\ \hline
V_p^b \\ \hline
V_p^c \\ \hline
V_s^A \\ \hline
V_s^B \\ \hline
V_s^C \\ \hline
\end{array} , \tag{9.6.78}$$

where y is the transformer leakage admittance in per unit, bearing in mind the modification suggested earlier to take into account the increase of resistance with frequency.

The models for the other common connections can be derived following a similar procedure.

In general, any two-winding three-phase transformer may be represented using two coupled compound coils. The network and admittance matrix for this representation is illustrated in Figure 9.43.

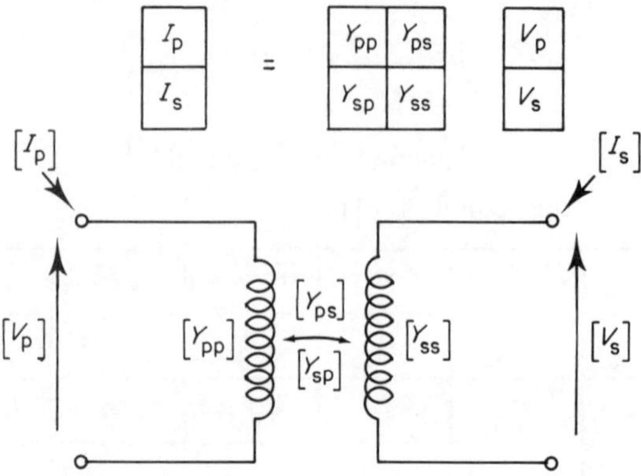

Figure 9.43. Two-winding three-phase transformer as two coupled compound coils. Reproduced with permission from Arrillaga, Arnold and Harker, *Computer Modelling of Electrical Power Systems.* Copyright © 1983 John Wiley & Sons Ltd.

Table 9.2. Characteristic submatrices used in forming the transformer admittance matrices

Transformer connection		Self-admittance		Mutual admittance
Bus *P*	Bus *S*	Y_{pp}	Y_{ss}	Y_{ps}, Y_{sp}
Wye *G*	Wye *G*	Y_I	Y_I	$-Y_I$
Wye *G*	Wye	$Y_{II/3}$	$Y_{II/3}$	$-Y_{II/3}$
Wye *G*	Delta	Y_I	Y_{II}	Y_{III}
Wye	Wye	$Y_{II/3}$	$Y_{II/3}$	$-Y_{II/3}$
Wye	Wye	$Y_{II/3}$	Y_{II}	Y_{III}
Delta	Delta	Y_{II}	Y_{II}	$-Y_{II}$

It should be noted that

$$[Y_{sp}] = [Y_{ps}]^T,$$

as the coupling between the two compound coils is bilateral.

Often, because more detailed information is not required, the parameters of all three phases are assumed balanced. In this case the common three-phase connections are found to be modelled by three basic submatrices.

The submatrices, $[Y_{pp}]$, $[Y_{ps}]$, etc., are given in Table 9.2 for the common connections in terms of the following matrices:

$$Y_I = \begin{bmatrix} y_t & & \\ & y_t & \\ & & y_t \end{bmatrix}, \quad Y_{II} = \begin{bmatrix} 2y_t & -y_t & -y_t \\ -y_t & 2y_t & -y_t \\ -y_t & -y_t & 2y_t \end{bmatrix}, \quad Y_{III} = \begin{bmatrix} -y_t & y_t & \\ & -y_t & y_t \\ y_t & & -y_t \end{bmatrix}.$$

Finally, these submatrices must be modified to account for off-nominal tap ratio as follows:

(i) Divide the self-admittance of the primary by α^2.
(ii) Divide the self-admittance of the secondary by β^2.
(iii) Divide the mutual admittance matrices by $\alpha\beta$.

It should be noted that in the per unit system a delta winding has an off-nominal tap of $\sqrt{3}$.

For transformers with ungrounded Wye connections, or with neutrals connected through an impedance, an extra coil is added to the primitive network for each unearthed neutral and the primitive admittance matrix increases in dimension. By noting that the injected current in the neutral is zero, these extra terms can be eliminated from the connected network admittance matrix.(35)

Once the admittance matrix has been formed for a particular connection, it represents a simple subsystem composed of the two busbars interconnected by the transformer.

Load modelling

When carrying out harmonic penetration studies in transmission systems, it is not usual to represent the system from generators right through to individual consumer loads. At some point down the network the elements are aggregated into an equivalent circuit. Typically, equivalent circuits are used at the points of supply (POS) to distribution authorities, who reticulate power to individual consumers within the load centres.

The methods available for determining the equivalent harmonic impedance of supply authority networks are as follows:

(i) Direct measurement, performed at a sufficient number of frequencies to enable satisfactory interpolation. This has been discussed in Section 9.4. Limitations in measurement techniques make this method very time consuming and difficult, especially for a number of points of supply.
(ii) Derivation of component characteristics, i.e. motors and industrial plant, by using statistical diversity data. This approach, while difficult, is under consideration for system stability studies(36) and could be extended to harmonic studies.
(iii) Use of the known fundamental frequency real and reactive power flow at the point of supply.

There is considerable variation in impedance with frequency and load level(5,37,38) for industrial and domestic customers. Moreover, industrial loads often have capacitors installed for power factor compensation which can cause series and parallel resonances. Various models(29,39,40) have been proposed for consumer loads, some of them relating to individual components and others as component aggregate models.

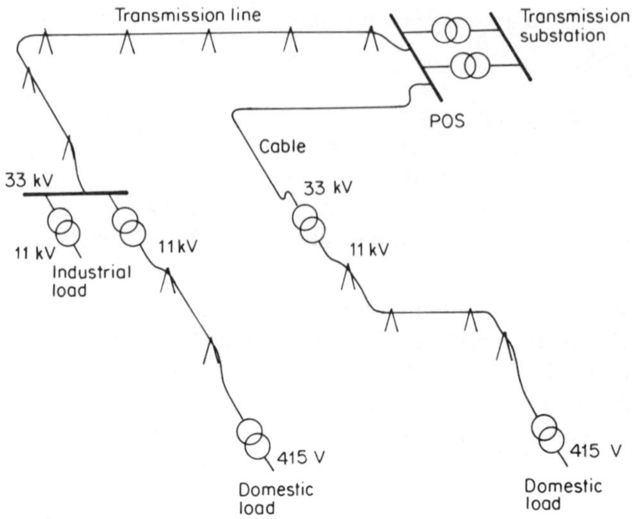

Figure 9.44. Schematic representation of a supply authority distribution network

Table 9.3. Measured impedances at two points of supply of a distribution authority

	Springton	Islington
520 Hz impedance (Ω)	57.0 $\angle 35.5^\circ$	21.3 $\angle 37.5^\circ$
50 Hz impedance (Ω)	47.0 $\angle 18.2^\circ$	19.2 $\angle 18.2^\circ$

A typical supply authority network, illustrated in Figure 9.44, shows various levels of transformation and reticulation which tend to reduce the effect of individual consumer harmonic impedance variation.

Some measured data for the Central Canterbury Electric Power Board (CCEPB) in New Zealand, confirming the relative insensitivity of harmonic impedances to customer loads, is documented in Table 9.3. It is interesting to note that the impedance at 520 Hz is inductive, and not substantially greater than that at 50 Hz at the points of supply to the Springston and Islington substations. From this information it would appear that the use of fundamental frequency data at the point of supply is reasonable, and that consumer loads need not be represented individually. The simplest model at a point of supply busbar is a shunt admittance that attracts the correct fundamental real power flow.[(29,31)] A better model includes the reactive power consumption in the admittance as well,[(41)] if this is available.

Various suggested combinations of the real and reactive power demand at fundamental frequency are shown in Figure 9.45. Converting these models into a suitable form for inclusion into the system admittance matrices is straightforward.

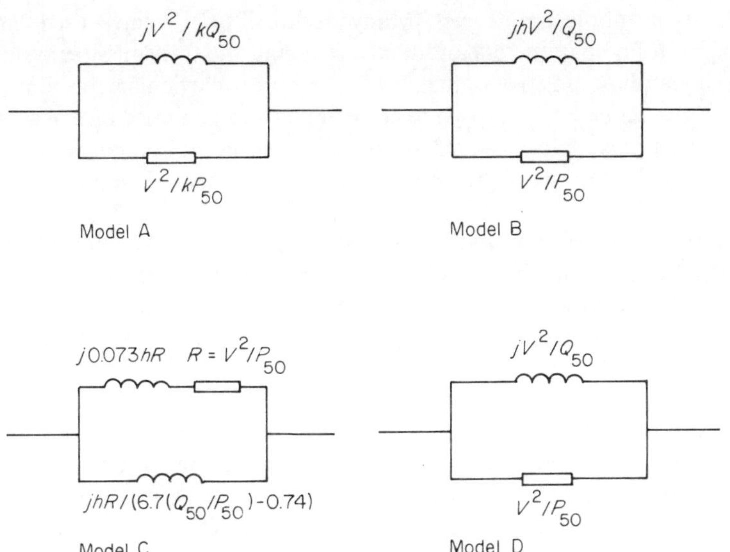

Figure 9.45. Load models for harmonic penetration studies

In model A, suggested by Pesonen *et al.*,[29] h is the harmonic order, V the nominal voltage and $k = 0.1h + 0.9$.

In model B the reactance is assumed to be frequency dependent while the parallel resistance is kept constant.

Model C was derived by measurements on medium voltage loads using audiofrequency ripple control generators.[29]

Finally in model D the load impedance, calculated at 50 Hz, remains constant for all frequencies.[37]

9.7 ALGORITHM DEVELOPMENT

Some early understanding of harmonic penetration was obtained in relation to high voltage d.c. convertors by means of analogue simulation. An impedance frequency characteristic curve was obtained by Laurent *et al.*,[42] at the French end of the cross-Channel high voltage d.c. link. However, simulation results obtained in a reduced scale model of the French 220 kV a.c. system did not match those measured.

Subsequently, Hingorani[43] used this curve to calculate an equivalent *LRC* circuit, for which filter design studies at a high voltage d.c. terminal were undertaken. Brewer *et al.*[44] also reported difficulty in matching predictive computer studies with measured results on the Kingsnorth high voltage d.c. scheme.

More recently, digital computer models have been developed with the capability of more accurate representation of the elements of the transmission system at harmonic frequencies. Many authors[41,45,46] have used balanced transmission line parameters and loads, reducing the transmission system to a linear single-phase positive sequence equivalent. In particular, the single-phase model of Breuer *et al.*[3] was used to compare harmonic impedance of a network with measured test data; however, at some frequencies the comparison was found to be poor and incapable of predicting the unbalanced impedances determined by tests.

Harmonic penetration studies have also been undertaken for distribution systems. In one case,[39] low order unbalanced characteristic and uncharacteristic harmonics were injected from a convertor load, and symmetrical component matrix analysis was used to obtain the harmonic voltages at specific network buses. Another reported study[40] applied multiphase analysis with coupling between double circuits.

This section describes the development of the harmonic penetration algorithms. Such development includes the harmonic interaction between convertors and supply systems, as well as harmonic penetration modelling, and while the results being presented only relate to the latter, the structuring of the programs has been made general to permit developments in both areas.

The basic approach to harmonic penetration investigations is very similar when using either single- or three-phase models. However, the development of three-phase software necessitates more effort on data preparation and output display.

Requirements of the algorithm for harmonic modelling

The requirements to be met for accurate harmonic modelling are:

(i) Transmission lines must be represented with provision for skin effect and standing wave phenomena.
(ii) Load, transformer, generator, shunt capacitor and filter models should be included.
(iii) Nodal admittance matrices should be formed for any range of frequencies and not restricted to harmonic multiples of the fundamental.
(iv) It should be possible to calculate system impedances at any busbar.
(v) The possibility of current injections at multiple locations in the system needs to be considered.
(vi) The network (assumed linear and passive) must be solved to obtain system voltages at all nodes for all frequencies.
(vii) Line current flows should be calculated at each frequency.
(viii) Output data needs to be plotted to facilitate interpretation.

These requirements utilize standard power system techniques involving the solution of simultaneous linear equations. However, the nature of the problem will determine which of the above features will need to be used in any particular study.

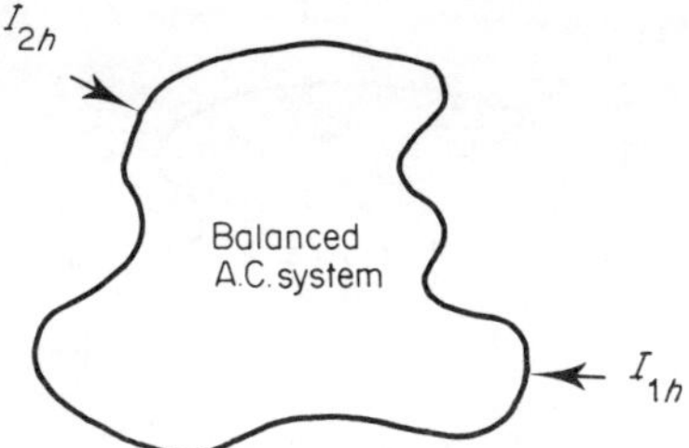

Figure 9.46. Balanced current injection into a balanced a.c. system

Balanced harmonic penetration

In Figure 9.46 two sets of balanced harmonic currents, I_{1h} and I_{2h}, of order h, are injected into any two busbars of an a.c. system; a large power system is likely to have a number of such injections. It is assumed that the a.c. system is linear and passive and therefore the principle of superposition may be applied to enable each harmonic to be considered independently.

The resultant system harmonic voltages are calculated by direct solution of the linear equation

$$[I_h] = [Y_h][V_h], \tag{9.7.1}$$

where $[Y_h]$ is the system admittance matrix.

On the assumption of a balanced a.c. system, the model will only include the positive sequence component impedances.

The above algorithm can model the steady state behaviour of a power system but unfortunately the harmonic behaviour of a physical system changes as loads, generators and line configuration alter.

Unbalanced harmonic penetration

The three-phase nature of the power system always results in some load or transmission line asymmetry as well as circuit coupling. These effects give rise to unbalanced self- and mutual admittances of the network elements.

A more accurate representation of the unbalanced conditions is illustrated in Figure 9.47. The current injections, i.e. $I_{1h} - I_{3h}$ and $I_{4h} - I_{6h}$, can be unbalanced in magnitude and phase angle. In a similar manner to the balanced system, the current injections for each frequency are presumed constant and known, and the linear equation (9.7.1) is solved directly to obtain the three-phase harmonic voltages.

For the three-phase system, the elements of the admittance matrix are themselves 3×3 matrices consisting of self- and transfer admittances. Figure 9.48 indicates the nature of the analysis where h sets of linear equations are solved.

The injected currents at most a.c. busbars will be zero, since the sources of the

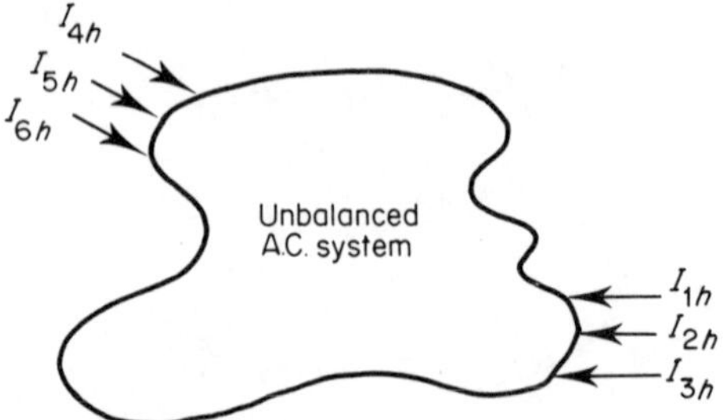

Figure 9.47. Unbalanced current injection into an unbalanced a.c. system

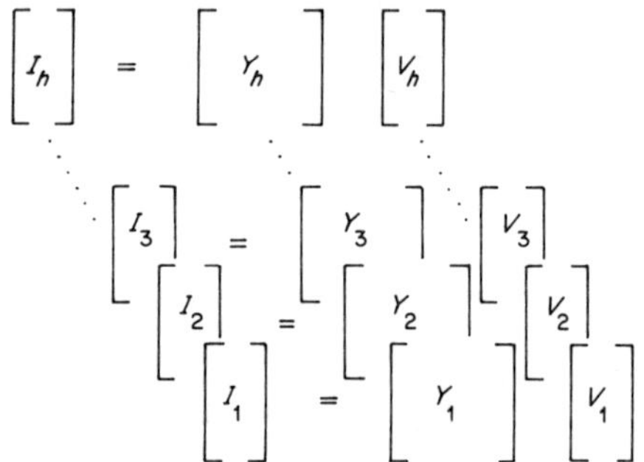

Figure 9.48. Solution of h sets of linear simultaneous equations

harmonics considered are generally from static convertors. To calculate an admittance matrix for the reduced portion of a system comprising just the injection busbars, it is necessary to form the admittance matrix with those buses, at which harmonic injection occurs, ordered last. Advantage is taken of the symmetry and sparsity of the admittance matrix,[47] using a row ordering technique to reduce the amount of off-diagonal element build-up. The matrix is triangulated using Gaussian elimination, down to but excluding the rows of the specified buses.

The resulting matrix equation for an n-node system with $n-j+1$ injection points is

$$\begin{bmatrix} 0 \\ \vdots \\ 0 \\ \hline I_j \\ \vdots \\ I_n \end{bmatrix} = \begin{bmatrix} \diagdown & \multicolumn{3}{c}{\text{(eliminated upper triangle)}} \\ \hline & Y_{jj} & \cdots & Y_{jn} \\ 0 & \vdots & & \vdots \\ & Y_{nj} & \cdots & Y_{nn} \end{bmatrix} \cdot \begin{bmatrix} V_1 \\ \vdots \\ V_{j-1} \\ \hline V_j \\ \vdots \\ V_n \end{bmatrix}. \qquad (9.7.2)$$

As a consequence, $I_j \cdots I_n$ remain unchanged since the currents above these in the current vector are zero. The reduced matrix equation is

$$\begin{matrix} I_j \\ \vdots \\ I_n \end{matrix} = \begin{bmatrix} Y_{jj} & \cdots & Y_{jn} \\ \vdots & & \vdots \\ Y_{nj} & \cdots & Y_{nn} \end{bmatrix} \cdot \begin{bmatrix} V_j \\ \vdots \\ V_n \end{bmatrix} \tag{9.7.3}$$

and the order of the admittance matrix is three times the number of injection busbars. The elements are the self- and transfer admittances of the reduced system as viewed from the injection busbars. Whenever required, the impedance matrix may be obtained for the reduced system by matrix inversion.

Reducing a system to provide an equivalent admittance matrix is an essential part of filter design where the system, as viewed from a specific bus, is required; it is also useful where a number of convertors are connected to the a.c. system at different points, as in Figure 9.49. In this example the reduced admittance matrix is of order 9.

In harmonic penetration studies the currents from the convertors are assumed to be known. In general, however,[48] any voltage distortion present at the convertor terminals affects the firing angles and hence the harmonic current injection into the system. The solution of this problem is iterative and not suited to the large matrices associated with the a.c. system. However, during each iteration only the convertor terminal voltages are required. These can be obtained in the example above by reducing the a.c. system to a three-bus equivalent system for the three convertors, as indicated in Figure 9.50, where each of the admittances represents a 3 × 3 matrix.

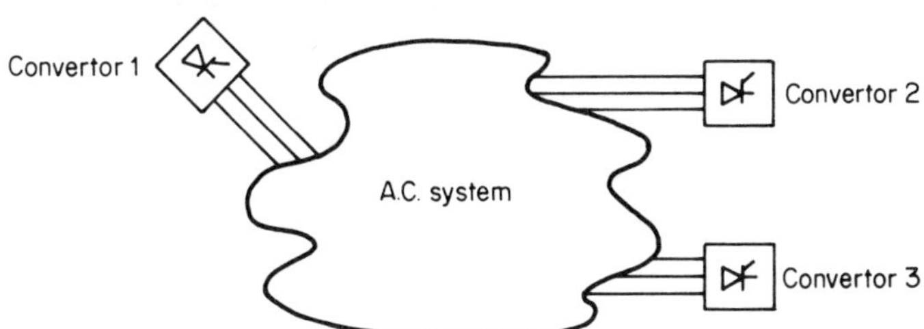

Figure 9.49. Three convertors attached to different busbars on the a.c. system

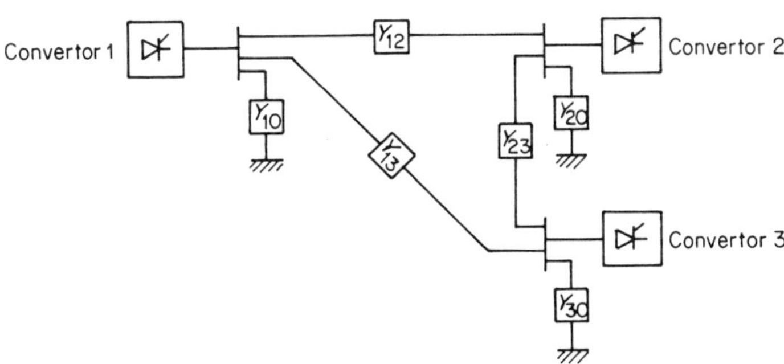

Figure 9.50. Reduced three-convertor system

Restricted measurements on the physical network limit the ability to compare a three-phase model with test results. The data obtained from live three-phase systems only includes the phase voltages and currents of the coupled phases; to compare measured and simulated impedances at a current injection busbar it is thus necessary to derive equivalent phase impedances from the 3×3 admittance matrix.

By making $I_1 = 1\underline{|0^\circ}$ per unit, $I_2 = 1\underline{|-120^\circ}$ per unit, $I_3 = 1\underline{|120^\circ}$ per unit, the matrix equation

$$\begin{bmatrix} I_1 \\ I_2 \\ I_3 \end{bmatrix} = \begin{bmatrix} Y_{11} & Y_{12} & Y_{13} \\ Y_{21} & Y_{22} & Y_{23} \\ Y_{31} & Y_{32} & Y_{33} \end{bmatrix} \cdot \begin{bmatrix} V_1 \\ V_2 \\ V_3 \end{bmatrix} \tag{9.7.4}$$

can be solved for V_1, V_2 and V_3, yielding the following equivalent phase impedances:

$$Z_1 = \frac{V_1}{I_1}, \qquad Z_2 = \frac{V_2}{I_2}, \qquad Z_3 = \frac{V_3}{I_3}. \tag{9.7.5}$$

9.8 COMPUTATIONAL REQUIREMENTS OF HARMONIC PENETRATION ALGORITHMS

Single-phase modelling

The structure of a single-phase harmonic penetration algorithm is illustrated in Figure 9.51, which involves simple and efficient software.

Only fundamental frequency system component data is used to derive information at harmonic frequencies. This information is held in the system data base and its processing requires very little extra work. Storage is only required for the non-zero elements of a single admittance matrix and this information is reformed for each frequency.

Three-phase algorithm

The structure of a three-phase algorithm, illustrated in Figure 9.52, is very similar to that used in three-phase power flow studies.[49] The harmonic penetration program is only one part of this diagram, indicating that three-phase modelling is not a direct extension of the single-phase algorithm.

Preparation of data is not trivial in three-phase harmonic modelling. This is due to the inclusion of unbalanced three-phase load data, and the frequency dependence of three-phase transmission lines.

The volume of data and the use of program blocks are of primary concern to this algorithm rather than the speed or efficiency of computation, as has been the case in the development of algorithms for power flow or transient stability simulation. The number of separate program blocks is a function of the multiple use of software, having regard for practical program debugging and maintenance.

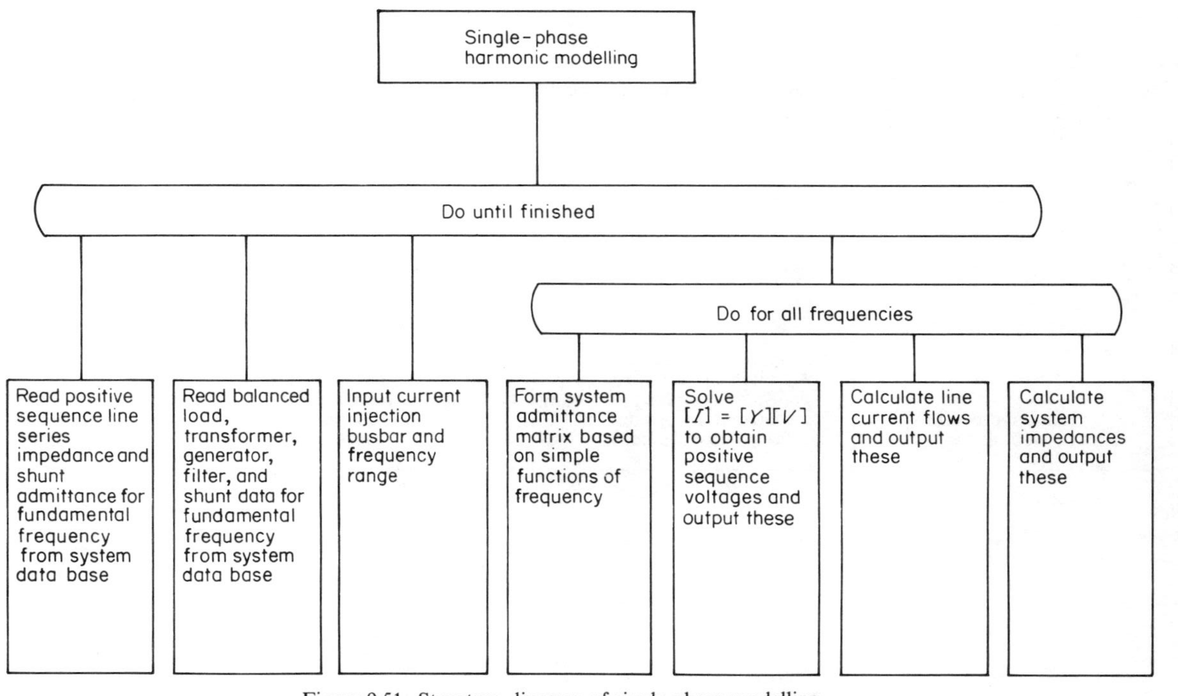

Figure 9.51. Structure diagram of single-phase modelling

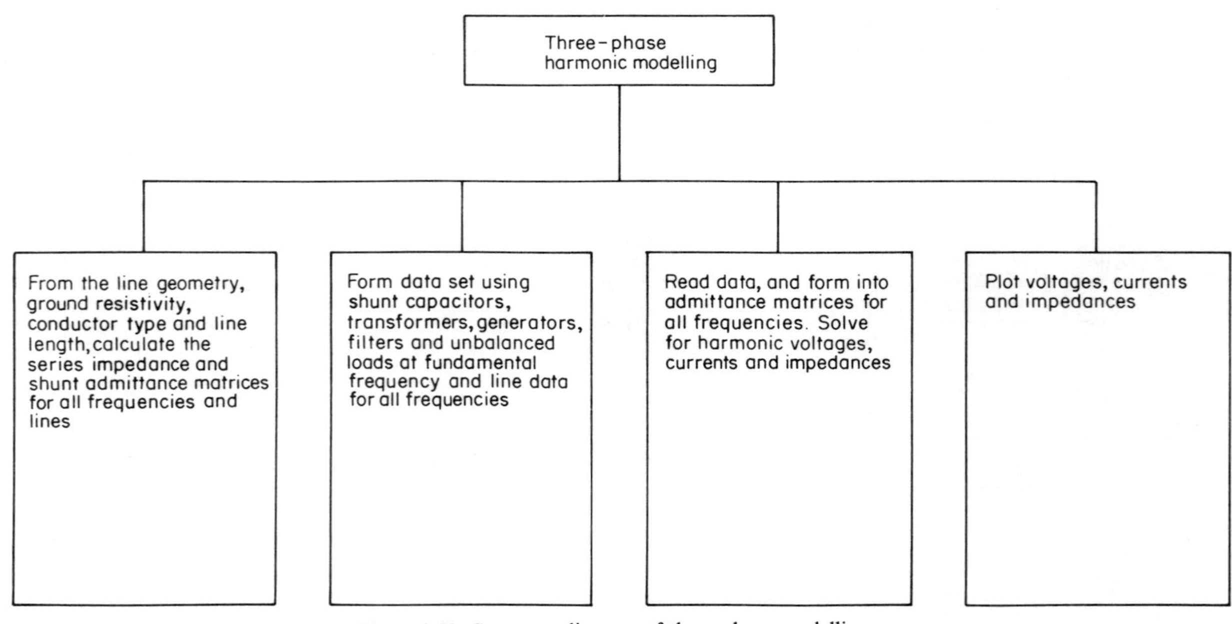

Figure 9.52. Structure diagram of three-phase modelling

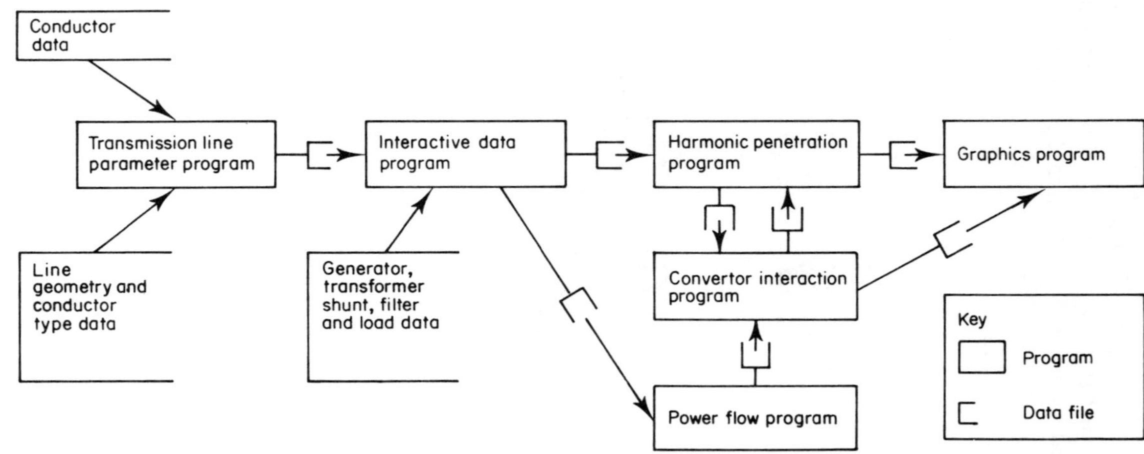

Figure 9.53. Data flow diagram for three-phase modelling

The first block of Figure 9.52 calculates the transmission line parameters for each frequency over a required range, using the equivalent PI model. The second program block completes the data base by reading line data from the first and adding it to the balanced load and other component data required. These two program blocks are unnecessary in single-phase studies due to lower data requirements and smaller software content. In this case the amount of output information generated for a typical study cannot be interpreted adequately by

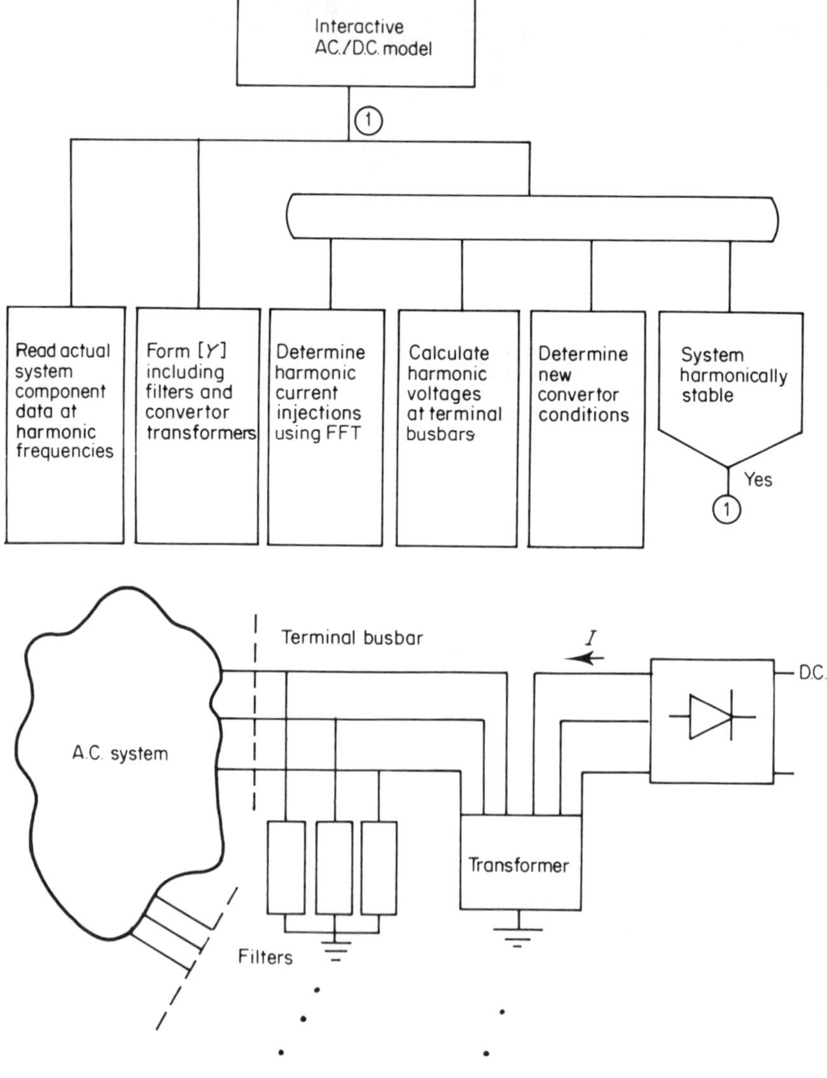

Figure 9.54. Interactive a.c.–d.c. model structure diagram

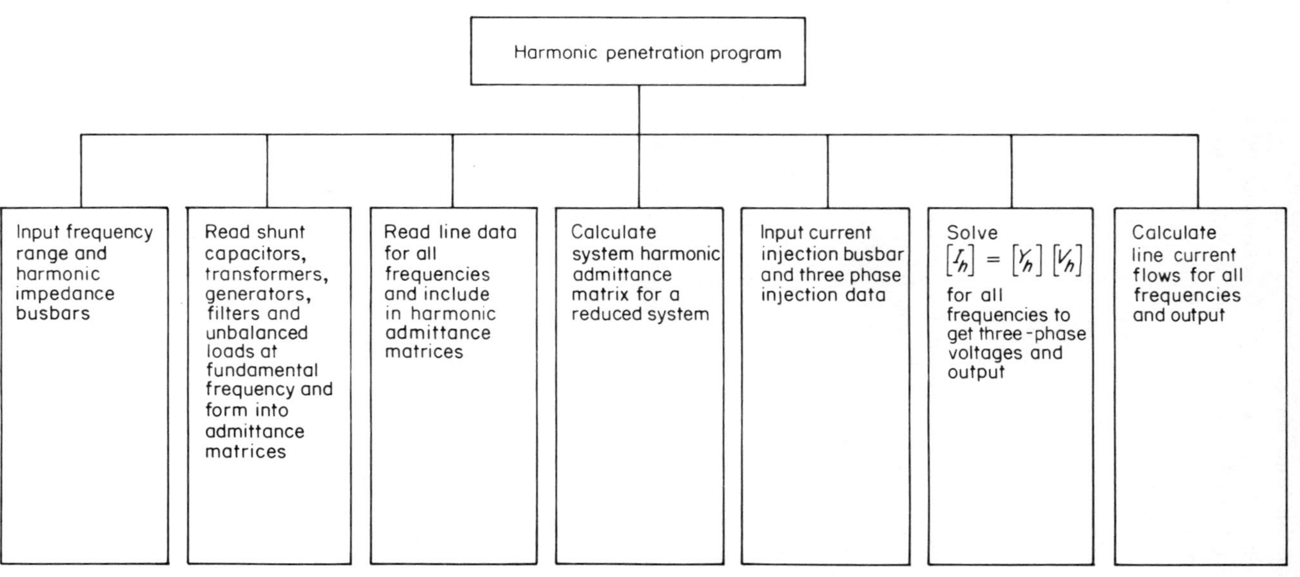

Figure 9.55. Structure diagram of three-phase harmonic penetration program

using ordinary tabular techniques. Instead, graph plotting software routines should be used for this function.

The data flow diagram of Figure 9.53 indicates the individual programs and major data files that form the basis of the three-phase algorithm. Data formation for both the harmonic penetration and three-phase power flow[50] studies is performed by the same software.

The three-phase a.c./d.c. power flow supplies fundamental frequency data and the convertor operating state[51] to the convertor interaction software. This software, illustrated in Figure 9.54, uses the three-phase system impedances, calculated by the harmonic penetration program, to solve the non-linear convertor equations obtaining harmonic voltages and currents at the convertor terminals. The current injections from a number of convertors connected at any busbars in the a.c. system are then available for the analysis of the penetration of these harmonic currents into the a.c. system.

Three-phase harmonic penetration

A three-phase harmonic penetration program, illustrated in the structure diagram of Figure 9.55, should include the following features: (i) provision for the representation of three-phase mutually coupled transmission line data; (ii) provision for the representation of three-phase unbalanced loads and other system components; (iii) formation of separate admittance matrices for each frequency; (iv) derivation of three-phase impedance matrices for a reduced portion of the network, suitable for filter design or convertor interaction studies; (v) specification of unbalanced current injections at a number of busbars on the system.

9.9 APPLICATION OF THE HARMONIC PENETRATION ALGORITHM

Harmonics generated along transmission lines

The 220 kV Islington to Kikiwa three-phase line of the New Zealand system is used to demonstrate the capability of the computer model described in previous sections.

A three-dimensional graphic representation is used to provide simultaneous information of the harmonic levels along the line. At each harmonic (up to the 25th harmonic), one per unit positive sequence current is injected at the Islington end of the line. The voltages caused by this current injection are therefore the same as the calculated impedances, i.e. V_+ gives Z_{++}, V_- gives Z_{+-} and V_0 gives Z_{+0} (the subscripts +, −, 0 refer to the positive, negative and zero sequences respectively).

Figures 9.56–9.58 illustrate the effect of two extreme cases of line termination (at Kikiwa), i.e. the line open circuited and short circuited respectively. The differences in harmonic magnitudes along the line are due to standing wave effects and shifting of the resonant frequencies caused by line terminations.

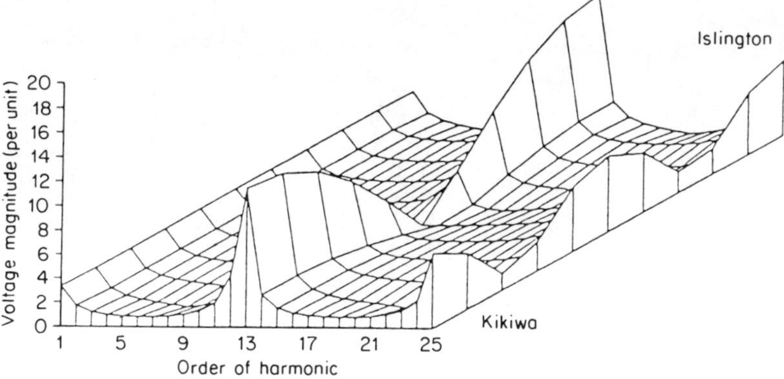

Figure 9.56. Positive sequence voltage versus frequency along the open ended Islington to Kikiwa line

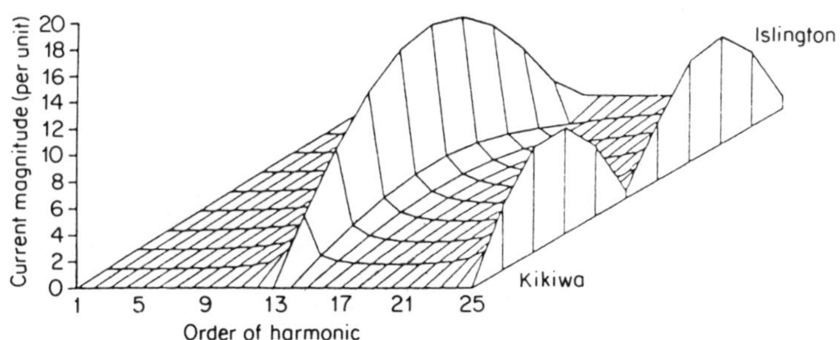

Figure 9.57. Positive sequence current along the open ended line for a 1 per unit positive sequence current injection at Islington

Figure 9.56 indicates the existence of high voltage levels at both ends of the open circuited line at the half wavelength frequencies. The 25th harmonic clearly illustrates the standing wave effect, with voltage maxima and minima alternating at a quarter of the wavelength intervals.

At any particular frequency, a peak voltage at a point in the line will indicate the presence of a peak current of the same frequency at a point about a quarter wavelength away. This is clearly seen in Figure 9.57.

When the line is short circuited at the extreme end, the harmonic current penetration is completely different, as shown in Figure 9.58(a). The high current levels at the receiving end of the line are due to the short circuit condition. Figure 9.20, earlier in this chapter, indicated that the resonant maxima decrease as frequency increases. However, this does not appear to be the case in Figure 9.58(a). The reason is that the points plotted correspond only to harmonic frequencies and resonances do not fall exactly on these frequencies; i.e. the peak magnitudes at non-harmonic frequencies can be greater than the values plotted in the figure.

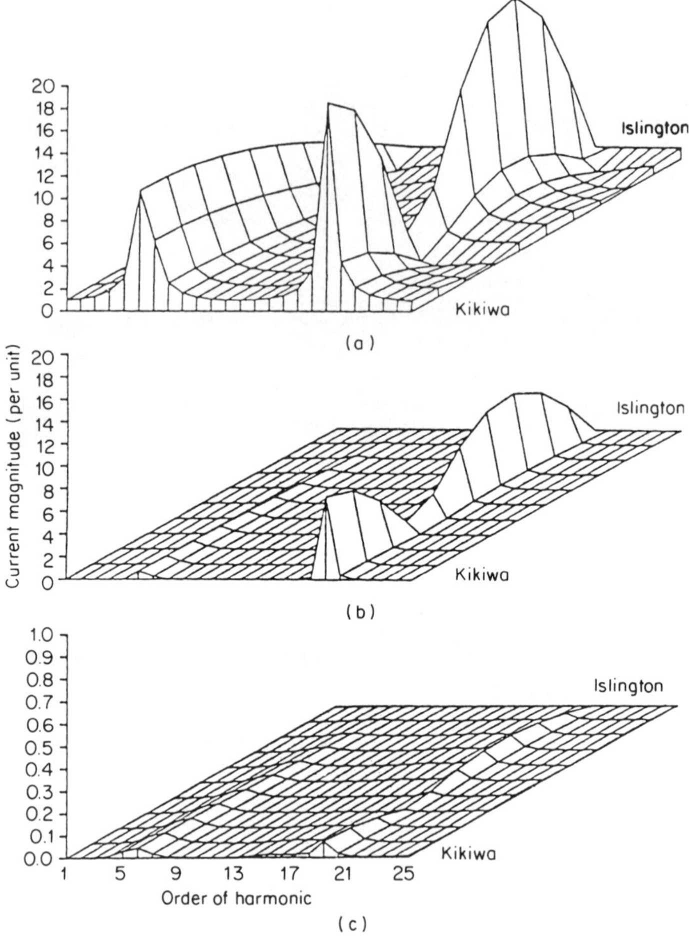

Figure 9.58. Sequence currents along the short circuited line for a 1 per unit positive sequence current injection at Islington: (a) positive sequence current; (b) negative sequence current; (c) zero sequence current

Zero sequence harmonics in transmission lines connected to static convertors

It is the zero sequence penetration, rather than the positive sequence, that provides relevant information for the assessment of possible harmonic interference in neighbouring telephone systems. The presence of zero sequence in a transmission line connected to a convertor bridge is entirely due to asymmetries in either the convertor a.c. plant components or the transmission line itself.

In Figure 9.58 the locations of maximum zero sequence current (Figure 9.58(c)) coincide with those of the positive sequence (Figure 9.58(a)), and the highest level produced in the test line, about 10% of the injected positive sequence current, occurs at the 19th harmonic, at the Kikiwa end of the short circuited line.

However, the levels of zero sequence current are low (notice the scale change between positive and zero sequence plots).

Differences in phase voltages

In conventional harmonic analysis using single-phase positive sequence models,[3] a transmission line is assumed to have one resonant frequency. However, the use of the three-phase algorithm to model the Islington–Kikiwa unbalanced transmission line shows that the resonant frequencies are different for each phase. In this case the spread of frequencies can be seen from Figure 9.59 to be approximately 6 Hz.

The different magnitudes of the resonant frequencies (up to 30%) of the three phases, partly explains the problems encountered with correlating single-phase modelling and measurement on the physical network. The results clearly indicate that harmonics in the transmission system are unbalanced and three-phase in nature.

Normal transposition of a transmission line into three equal length sections, to balance the line at fundamental frequency, can have a detrimental effect at harmonic frequencies. For instance the modelling of transpositions in the Islington to Kikiwa line produces the results illustrated in Figure 9.60, which

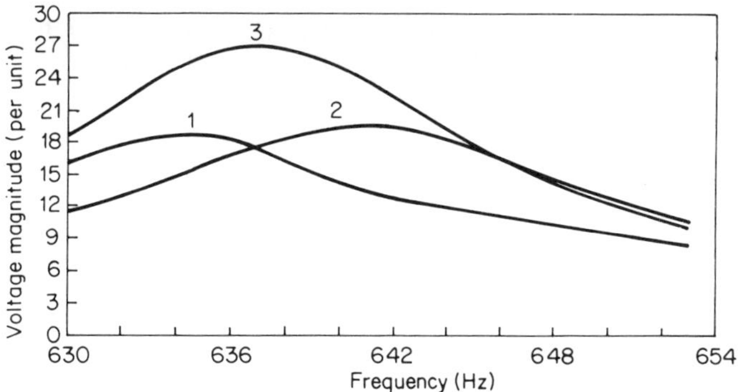

Figure 9.59. Three-phase resonant frequencies of the Islington to Kikiwa line with a 1 per unit positive sequence current injection (skin effect included)

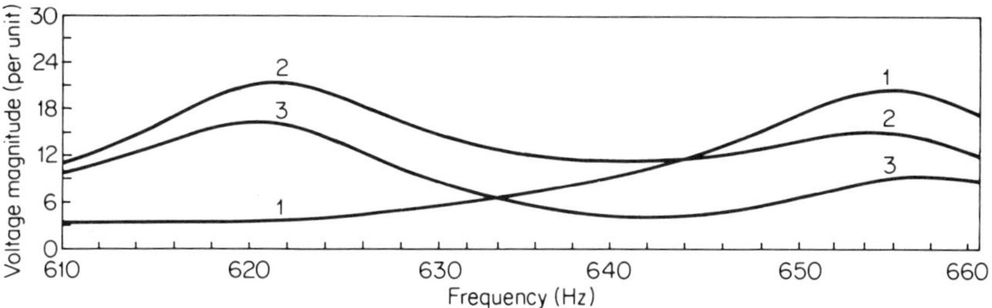

Figure 9.60. Three-phase resonant frequencies for the transposed line

shows the existence of two resonant peaks separated by almost 40 Hz for the half wavelength (i.e. at 620 and 656 Hz for phases 3 and 1 respectively).

Effect of mutual coupling in double circuits

The unbalanced behaviour of double circuit lines is well documented at fundamental frequency.[52,53] The three-phase harmonic penetration algorithm is used in this section to determine the importance of modelling mutual coupling at harmonic frequencies. The line used is shown in Figure 9.61.

Figure 9.62 displays the harmonic impedance (Z_{++}, Z_{00}) seen from the point of harmonic injection, both for a 1 per unit positive sequence current and 1 per unit zero sequence injections respectively. The figure also displays the coupling between the positive sequence and the other sequence networks, i.e. Z_{+-} and Z_{+0}.

Results for the case of a coupled line are illustrated in Figure 9.62(a) and those of two single circuit lines in Figure 9.62(b).

The magnitudes and resonant frequencies of the Z_{++} and Z_{+0} impedances are not affected by the modelling of mutual coupling. However, the level of Z_{+-} has changed substantially at resonance showing appreciable imbalance. Moreover, the magnitude and resonant frequency of Z_{00} is very different in the two cases.

Robinson[54] reported that telephone interference caused by zero sequence currents did not coincide with high levels of power system harmonics. This can partly be explained by the different resonant frequencies of the Z_{++} and Z_{00} observed.

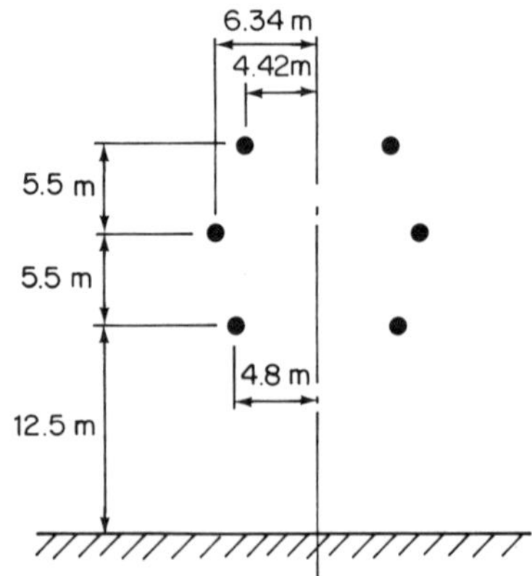

Figure 9.61. Line geometry of a double circuit line. Length, 167 km; earth resistivity, 100 Ωm; two conductors per bundle; bundle spacing, 0.45 m; conductor, 30/3.71 + 7/3.71

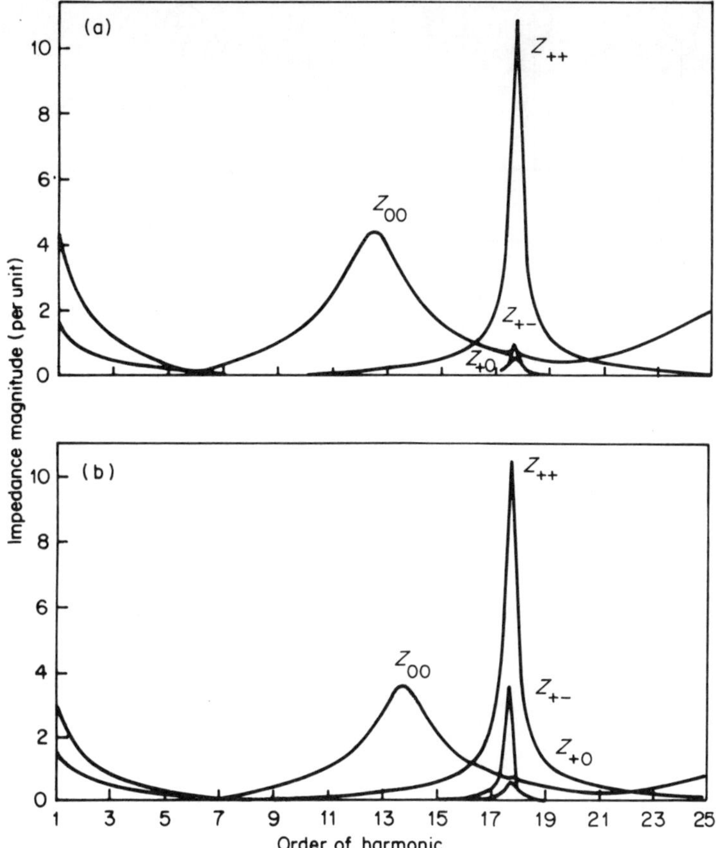

Figure 9.62. Sequence impedance magnitude versus frequency: (a) double circuit coupled line; (b) two single circuit lines

Harmonic impedances of an interconnected system

This section considers the progressive formation of the harmonic impedances of an interconnected system from the individual component characteristics. It is hoped that this will provide some understanding of the network modelling requirements at harmonic frequencies, in a situation where intuitive reasoning is not possible.

The test system, shown in Figure 9.63, is a nine-bus network comprising the 220 kV transmission system below Roxburgh in the South Island of New Zealand. The current harmonic source is an aluminium smelter at the Tiwai bus.

The double circuit lines are symmetrical about the tower axis and the transformers have star or delta connections depending on their location in the system, as indicated in Figure 9.64.

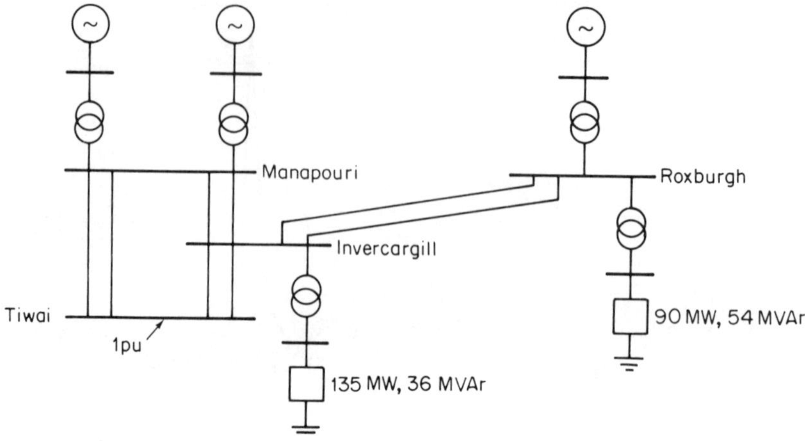

Figure 9.63. Test system including load and generation

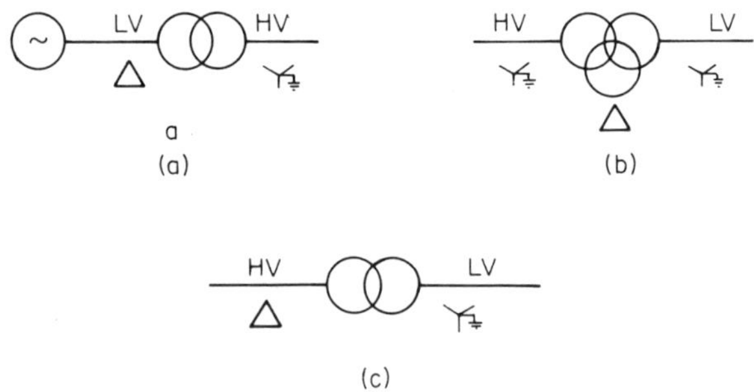

Figure 9.64. Transformer connections: (a) generating station; (b) transmission substation; (c) distribution substation

Generator transformers have deltas on the generator, or low voltage side, and grounded star connection on the high voltage side. Transmission substation transformers have grounded star on the high voltage and low voltage windings with delta connected tertiaries. Distribution transformers supplying the electrical supply authorities are delta connected on the high voltage and grounded star on the low voltage side.

The connection is important in considering the flow of zero sequence harmonic currents. A delta connected winding will act as an open circuit and a star connected winding, with neutral point grounded, as a short circuit to the zero sequence harmonic currents. The zero sequence impedance of the system will thus be considerably different to that presented to positive or negative sequence currents.

GENERATOR, TRANSFORMER AND LOAD IMPEDANCES AT ROXBURGH

With reference to Figure 9.63 a step by step formation of the system impedances is initiated by examining the effect of the various components at Roxburgh. The

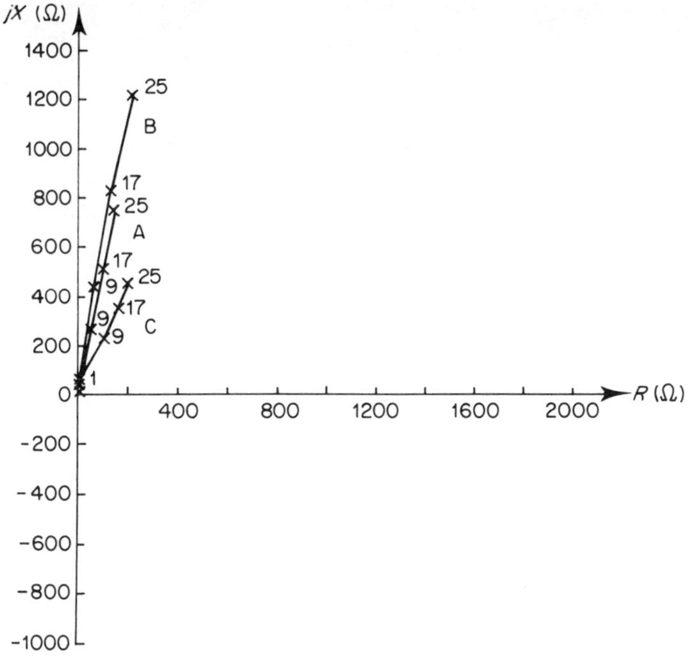

Figure 9.65. Polar plot of the generator, transformer and load impedances at Roxburgh: curve A, generator only; curve B, generator and generator transformer; curve C, generator, generator transformer, load (100%) and load transformer

harmonic impedance locus of the generator, considered in isolation, is shown in Figure 9.65 (curve A). The addition of the generator transformer produces the impedance locus of Figure 9.65 (curve B).

Finally, Figure 9.65 (curve C) illustrates the damping effect of a 90 MW and 54 MVAr load connected through a transformer to the Roxburgh bus.

INTERCONNECTION BETWEEN INVERCARGILL AND ROXBURGH

The double 220 kV transmission line between Invercargill and Roxburgh in isolation (i.e. open circuited at Roxburgh) has the impedance locus of Figure 9.66. At fundamental frequency the line is capacitive, although this is difficult to observe. As the frequency increases the line approaches a series resonance, at which point the impedance is very small and purely resistive, the phase angle becoming inductive. From this point the impedance increases in magnitude in a clockwise direction with increasing frequency. Somewhere between the 11th and 12th harmonics a parallel resonance occurs, manifested by a large and purely resistive impedance. As frequency increases further the line again becomes capacitive.

The effect of line termination is shown in Figure 9.67 for a 1 per unit harmonic injection at Invercargill. When the line is isolated (i.e. corresponding to the locus

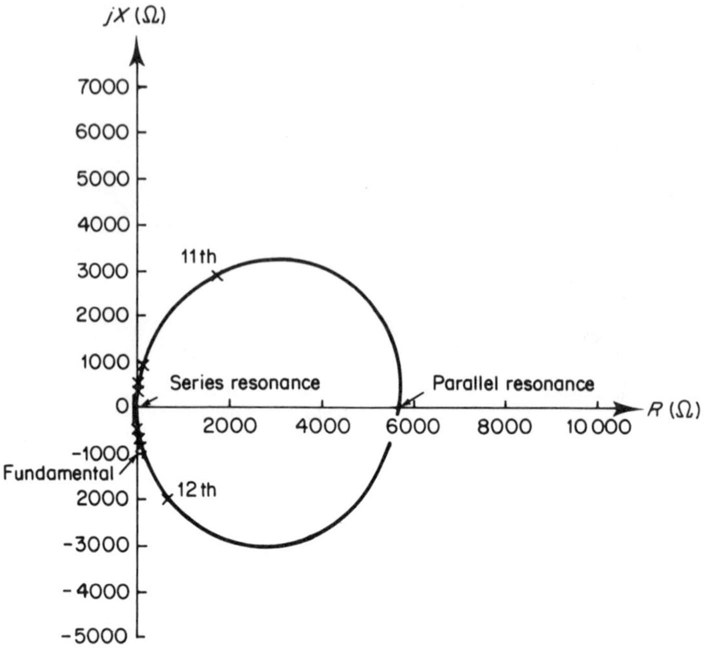

Figure 9.66. Polar plot of the impedance of the open circuited Invercargill to Roxburgh lines with 50 Hz intervals marked

of Figure 9.66) the per unit voltages of the various harmonics are illustrated in curves A in Figure 9.67. Figure 9.67(a) gives the voltage magnitudes and Figure 9.67(b) their phase angles.

The same graphs show corresponding voltage magnitudes and phases when the line is short circuited (Figure 9.67, curves B), loaded by the generator–transformer group (Figure 9.67, curves C) and by the complete system at Roxburgh (Figure 9.67, curves D).

LEFT-HAND SIDE OF THE SYSTEM

Referring now to the left-hand side part of the system, with the lines between Invercargill and Roxburgh open, Figure 9.68 illustrates the effect of 1 per unit harmonic current injections at Tiwai. Curves A and B of Figure 9.68 show the voltage spectra at Tiwai with the rest of the system open and short circuited respectively.

When, first, generation (Figure 9.68, curve C) and then load (Figure 9.68, curve D) are added with the associated transformers, similar effects to the previous section are observed. The resonant points lie between those of the open and short circuit cases, with reduced magnitudes as compared with the extreme cases of termination.

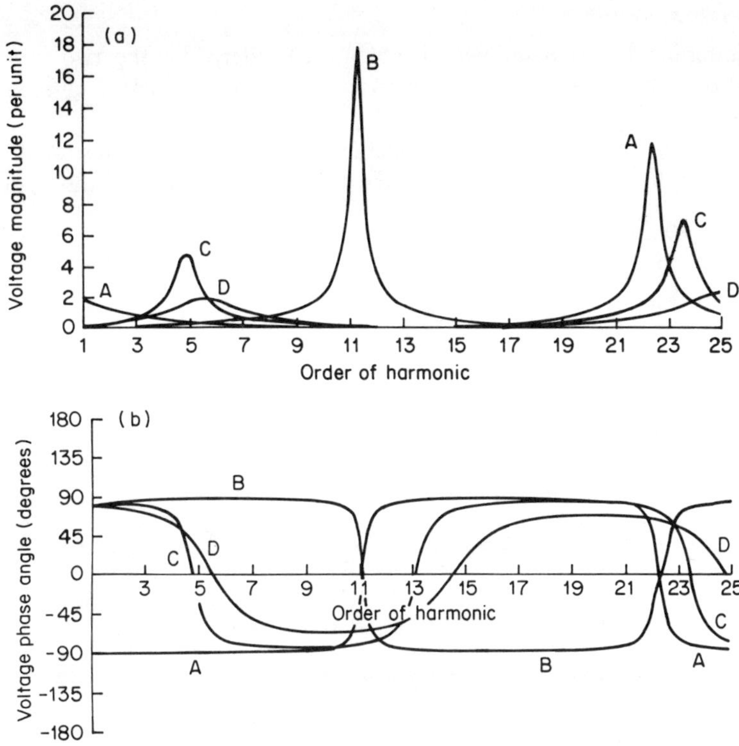

Figure 9.67. Positive sequence voltages at Invercargill versus frequency for different terminations of the Roxburgh to Invercargill lines: (a) voltage magnitudes; (b) voltage phase angles. Curves A, open circuit; curves B, short circuit; curves C, generator and generator transformer; curves D, generator, generator transformer, load and load transformer

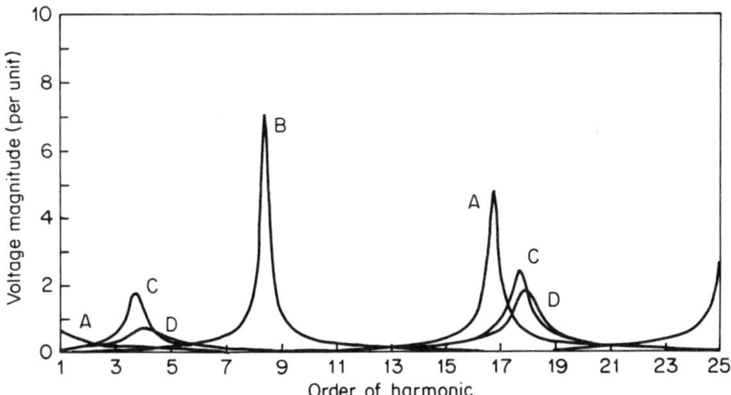

Figure 9.68. Positive sequence voltage magnitude at Tiwai versus frequency for different terminations: curve A, open circuit; curve B, short circuit; curve C, generator and generator transformer; curve D, generator, generator transformer, load and load transformer

COMPLETE TEST SYSTEM

By combining the two individual systems considered in the two preceding subsections the progressive formation of the test network (Figure 9.63) is completed.

Figure 9.69 compares the voltage magnitudes at the Tiwai bus for different loading conditions. Curve A shows the effect of the transmission system in

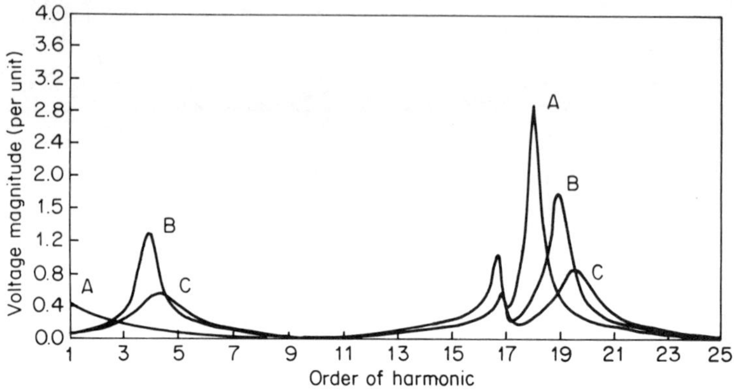

Figure 9.69. Positive sequence voltages at Tiwai versus frequency for different terminations: curve A, open circuit; curve B, generators and generator transformers; curve C, generators, generator transformers, loads and load transformers

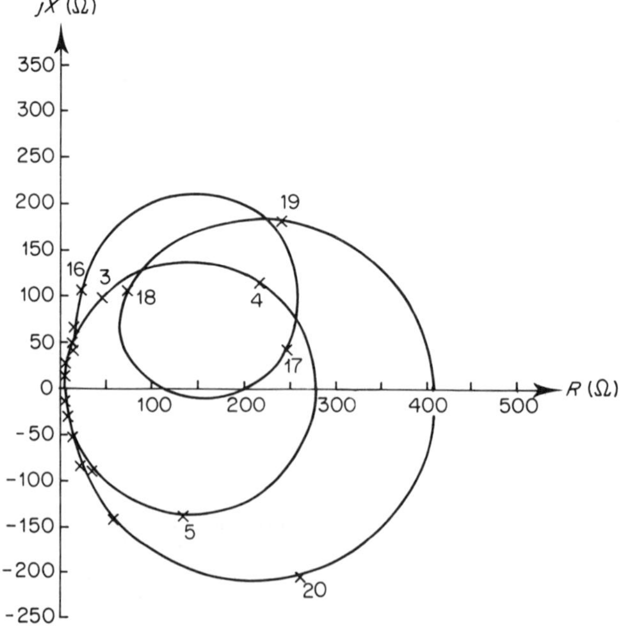

Figure 9.70. Positive sequence impedances of the test system from Tiwai with harmonic intervals indicated

isolation (i.e. with all the generators and loads disconnected); the resulting resonance frequencies of the interconnected system do not correspond to those of the two individual parts of the system described in the two preceding subsections. There are now two resonances around the 18th harmonic, one smaller in magnitude. The effect of this latter resonance is to create an extra loop in the impedance locus (as shown in Figure 9.70).

Figure 9.70 illustrates the progressive complexity of the impedance locus as the a.c. system increases.

An appreciation of the voltages over the whole system is obtained using the three-dimensional diagram of Figure 9.71. Lower voltages are observed on the transformer secondaries, namely at the generator and load busbars. Two major resonances occur at all the busbars. The first is between the fourth and fifth harmonics and the second between the 19th and 20th harmonics; these correspond with those observed in Figure 9.69 (curve A) at Tiwai.

VOLTAGE SENSITIVITY TO LINE PARAMETER VARIATION

By selectively reducing the line lengths and observing the voltages at Tiwai, the lines which most affect resonant conditions are determined. The results are illustrated in Figure 9.72.

A 5% reduction in length in the Tiwai to Manapouri lines (Figure 9.72, curve B) causes the two resonant frequencies around the 18th harmonic to shift by approximately 20 Hz. A 5% decrease in the lengths of the lines from Roxburgh to Invercargill has practically no effect on the smaller resonant frequency, but changes that between the 19th and 20th harmonics by approximately 10 Hz.

The lines close to the point of the current injection appear to have the largest effect on the system resonance frequencies, and thus their parameters must be represented with a greater level of accuracy than the lines more distant in the network.

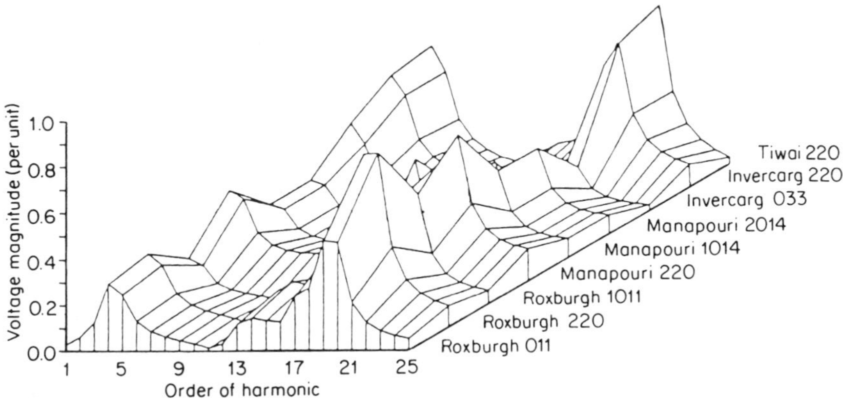

Figure 9.71. Positive sequence voltage magnitude versus frequency at all test system busbars

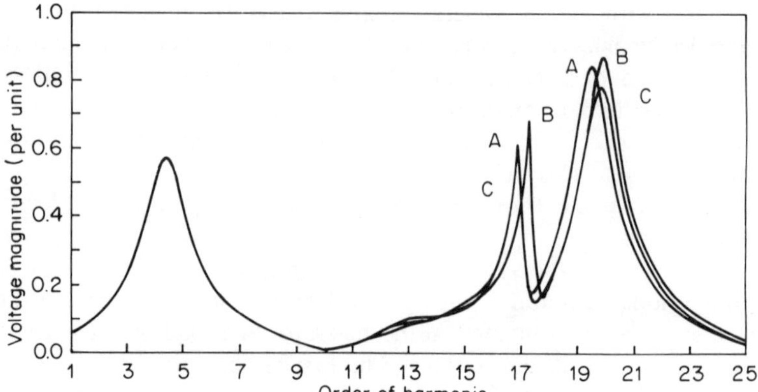

Figure 9.72. Positive sequence voltage magnitude at Tiwai versus frequency for line length variations: curve A, test system including generation and loads; curve B, Tiwai to Manapouri lines 5% shorter; curve C, Invercargill to Roxburgh lines 5% shorter

THREE-PHASE IMPEDANCES OF THE TEST SYSTEM AT TIWAI

The unbalanced nature of the transmission network can be illustrated by plotting the three individual equivalent phase impedances. Figure 9.73 shows that the imbalance is low at fundamental frequency, but increases towards the first parallel resonance which occurs between the fourth and fifth harmonics, where the magnitude differences are of the order of 30%. This effect is mainly caused by differences in the mutual impedances between phases, resulting from the asymmetry in transmission line conductor geometries.

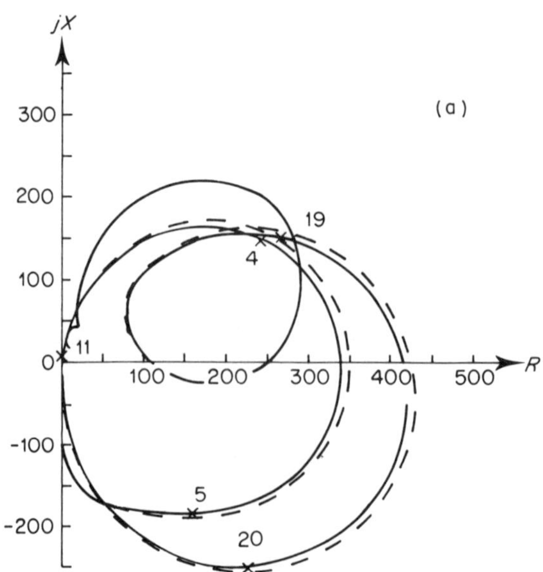

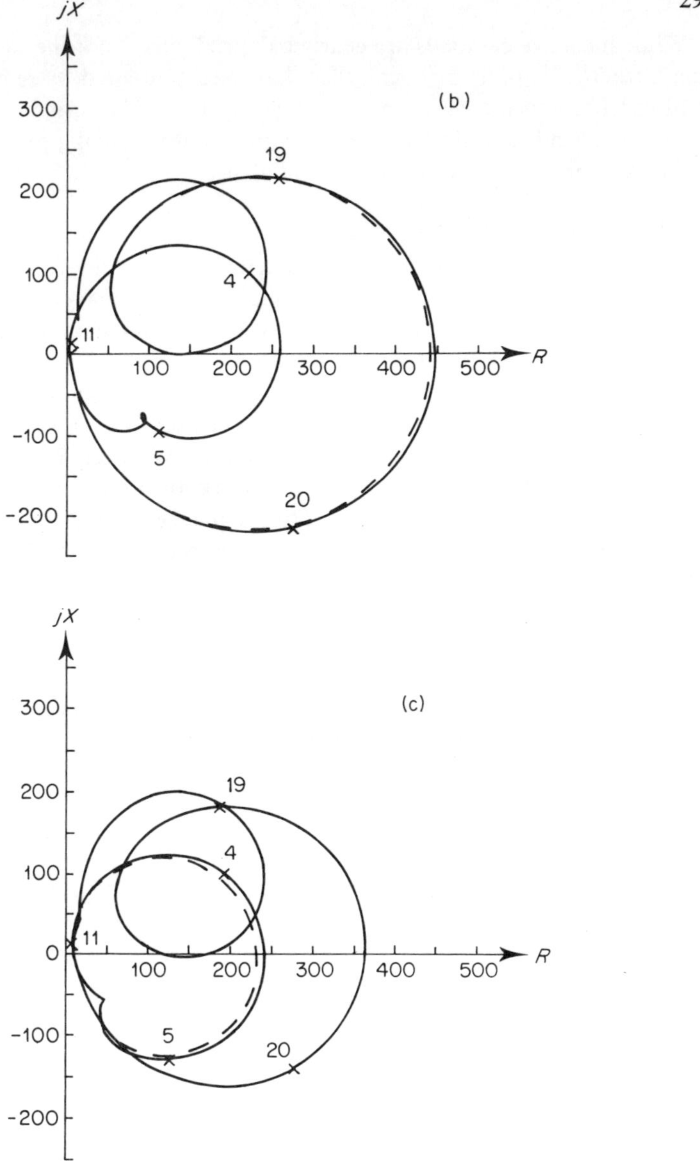

Figure 9.73. Equivalent phase impedances for the test system: (a) red phase; (b) yellow phase; (c) blue phase. ——, balanced load; -----, unbalanced load

The series resonance at the 11th harmonic exhibits low levels of imbalance, and the second parallel resonance between the 19th and the 20th harmonics, again shows considerable differences in the impedances between phases. High levels of imbalance at parallel resonant frequencies assist in explaining the difficulties being experienced with correlating single-phase simulation results with measured tests.[3,40]

While most system loads are nearly balanced, this is not the case with single-phase traction supplies.[55] This effect has been simulated by reducing phase 1 load by 10% and increasing phase 3 load by 10%. The results, also plotted in Figure 9.73, indicate that the level of impedance imbalance at the parallel resonant points increases with load imbalance.

HARMONIC IMBALANCE FACTOR

Any harmonic imbalance in the a.c. system will affect the three-phase voltage waveforms, and hence the zero crossings used in convertor control. As a consequence, harmonic imbalance will affect the non-characteristic harmonics produced by convertor plant.

At fundamental frequency an 'imbalance factor' has been used to indicate the extent of unbalanced system conditions.[56] The use of such an index at harmonic frequencies gives a simple measure of imbalance. With reference to the reduced system of Figure 9.63, Table 9.4 lists the values of the imbalance factor at parallel resonances, defined as the ratio of negative sequence voltage to positive sequence voltage, for a balanced 1 per unit harmonic current injection. The values of the imbalance factor are largest at parallel resonant frequencies at hence Table 9.4 represents maximum levels.

The imbalance factor is greatest with a network comprising only the transmission lines. The connection of both generation and load, the models of which are balanced, reduces the imbalance factor. At periods of light load, therefore, the a.c. system tends to be more unbalanced than at times of high load, and the generation of convertor uncharacteristic harmonics will increase

Table 9.4. Values of the imbalance factor at parallel resonances for various terminations of the reduced system

	Harmonic	Imbalance factor (%)
Reduced system, no generation or load	17th	24.8
Reduced system, generation only	19th	10.1
Reduced system, generation and 100% load	20th	6.2
with phasing the same on each circuit	20th	17.2

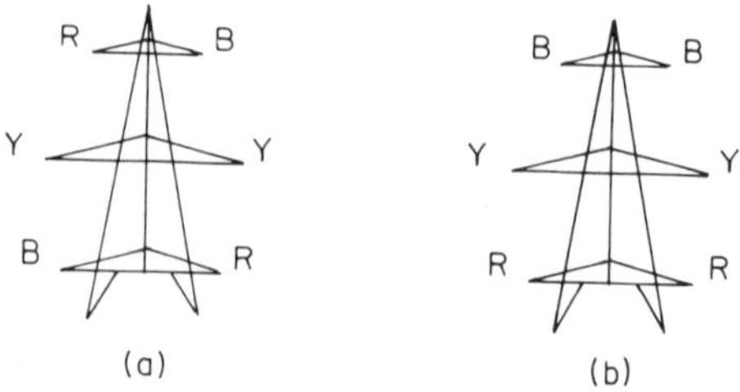

Figure 9.74. Examples of different tower conductor arrangements: (a) common arrangement; (b) phase positions the same in each circuit

accordingly. At light load, however, the level of harmonics will be higher and so imbalance is only one factor in the production of these uncharacteristic harmonics.

The last two entries in Table 9.4 are for the system with both load and generation, for line phasings as indicated in Figures 9.74(a) and (b) respectively. The position of phases in double circuit lines is normally varied in each circuit as in Figure 9.74(a) to minimize the fundamental frequency imbalance. When the phase positions are altered to be the same in each circuit, as illustrated in Figure 9.74(b), the imbalance factor rises substantially to 17.2%. Conductor geometries are thus important in unbalanced harmonic penetration analysis.

9.10 REFERENCES

1. Tschappu, F. (1981). 'Problems of the exact measurement of electrical energy in networks having harmonic content in the current'. *Landis and Gur Review*, **28** (2), 8–15.
2. 'Cubic spline program', Department of Mathematics, University of Canterbury, New Zealand, 1980.
3. Breuer, G. D., Chow, J. H., Gentile, T. J., *et al.* (1982). 'HVDC–AC harmonic interaction. I. Development of a harmonic measurement system hardware and software'. *IEEE Trans.*, **PAS-101**, 701–708.
4. Lemoine, M. (1977). 'Methods of measuring harmonic impedances', IEE Conference Publication 151, CIRED, London, 5–7.
5. Baker, W. P. (1981). 'The measurement of the system impedance at harmonic frequencies'. Paper presented at an international conference on Harmonics in Power Systems, UMIST, Manchester.
6. Barnes, H. (1974). 'An automatic harmonic analyser for the supply industry'. Paper presented at an IEE conference on Sources and Effects of Power System Disturbances, London.
7. Kron, G. (1965). *Tensor Analysis of Networks*, MacDonald, London.
8. Brameller, A., John, M. N., and Scott, M. R. (1969). *Practical Diakoptics for Electrical Networks*, Chapman and Hall, London.
9. Clarke, E. (1943). *Circuit Analysis of A.C. Power Systems*, Vol. I, John Wiley, New York.
10. Kimbark, E. W. (1950). *Electrical Transmission of Power and Signals*, John Wiley, New York.
11. Chipman, R. A. (1968). *Theory and Problems of Transmission Lines*, McGraw-Hill, New York.
12. Elgerd, O. (1971). *Electric Energy Systems Theory: An Introduction*, McGraw-Hill, New York.
13. Shultz, R. D., Smith, R. A., and Hickey, G. L. (1983). 'Calculation of maximum harmonic currents and voltages on transmission lines'. *IEEE Trans.*, **PAS-102**, 817–821.
14. Chen, M. S., and Dillon, W. E. (1974). 'Power system modelling'. *Proc. IEEE*, **62**, 901.
15. Laughton, M. A. (1968). 'Analysis of unbalanced polyphase networks by the method of phase co-ordinates. I. System representation in phase frame of reference'. *Proc. IEE*, **115**, 1163–1172.
16. Hesse, M. H. (1963). 'electromagnetic and electrostatic transmission line parameters by digital computer'. *IEEE Trans.*, **PAS-82**, 282–291.
17. Coleman, D., Watts, F., and Shipley, R. B. (1958). 'Digital calculation of overhead transmission line constants'. *Trans. AIEE*, **77**, 1266–1268.

18. Carson, J. R. (1926). 'Wave propagation in overhead wires with ground return'. *Bell Systems Tech J.*, **5**, 539–554.
19. Wedepohl, L. M. (1963). 'Application of matrix methods to the solution of travelling-wave phenomena in polyphase systems'. *Proc. IEE*, **110**, 2200–2212.
20. Bowman, W. I., and McNamee, J. M. (1964). 'Development of equivalent PI and T matrix circuits for long untransposed transmission lines'. *IEEE Trans.*, **PAS-84**, 625–632.
21. Wilkinson, J. H., and Reinsch, C. (1971). *Handbook for Automatic Computations*, Vol. II, *Linear Algebra*, Springer-Verlag, Berlin.
22. Kennelly, A. E., Laws, F. A., and Pierce, P. H. (1915). 'Experimental researches on skin effects in conductors'. *Trans. AIEE*, **34**, 1953–2018.
23. Zaborszky, J. (1953). 'Skin and spiralling effect in stranded conductors'. *Trans. AIEE*, **72**, 599–603.
24. Silvester, P. (1969). 'Impedance and nonmagnetic overhead power transmission conductors'. *IEEE Trans.*, **PAS-88**, 731–737.
25. King, S. Y. (1970). 'Electromagnetic characteristics of isolated bundle conductors'. *Proc. IEE*, **117**, 753–760.
26. Comellini, E., Invernizzi, A., and Manzoni, G. (1972). 'A computer program for determining electrical resistance and reactance of any transmission line.' *IEEE Trans.*, **PAS-92**, 308–314.
27. Westinghouse Electric Corporation (1950). *Electrical Transmission and Distribution Reference Book*, Westinghouse, New York.
28. Lewis, V. A., and Tuttle, P. D. (1958). 'The resistance and reactance of aluminium conductors steel reinforced'. *Trans. AIEE*, **PAS-77**, 1189–1215.
29. CIGRE Working Group 36–05 (Disturbing Loads) (1981). 'Harmonics, characteristic parameters, methods of study, estimating of existing values in the network'. *Electra*, (77), 35–54.
30. Baird, J. (1981). 'ELAFANT – audio frequency analysis of a power system', New Zealand Electricity, Wellington.
31. Ross, T. W., and Smith, R. M. A. (1948). 'Centralised ripple control on high voltage networks'. *Proc. IEE*, **95**, 470–479.
32. Northcote-Green, J. E. D. Baron, J. A., and Vilks, P. (1973). 'The computation of impedance frequency loci and harmonic current penetration for AC systems adjacent to HVDC conversion equipment', *IEE Conf. Publ.*, **107**, 97–102.
33. Clarke, C. D. (1973). Discussion on reference 32. *IEE Conf. Publ.*, **107**, 100.
34. Arseneau, R., Hill, E. F., Szabados, B., and Ferry, D. (1979). 'Synchronous motor impedances measured at harmonic frequencies'. Paper A79 514–1 presented at the IEEE PES Summer Meeting, Vancouver.
35. Dillon, W. E., and Chen, M. S. (1972). 'Transformer modelling in unbalanced three phase networks'. Paper C72 460–4 presented at the IEEE PAS Summer Power Meeting, Vancouver.
36. Concordia, C., and Ihara, S. (1982). 'Load representation in power system stability studies'. *IEEE Trans.*, **PAS-101**, 969–975.
37. Ross, N. W. (1972). 'Modelling a ripple control system'. *The New Zealand Electrical Journal*, **1972**, 116–123.
38. Barnes, H., and Kearley, (1981). 'The measurement of the impedance presented to harmonic current by power distribution networks'. *IEE Conf.*, **197**, 71–80.
39. Pileggi, D. J., Harish Chandra, N., and Emanuel, A. E. (1981). 'Prediction of harmonic voltage in distribution systems'. *IEEE Trans.*, **PAS-100**, 1307–1315.
40. McGranaghan, M. F., Dugan, R. C. and Sponsler, W. L. (1981). 'Digital simulation of distribution system frequency-response characteristics.'. *IEEE Trans.*, **PAS-100**, 1362–1369.

41. Campbell, L. C., and Murray, N. S. (1970). 'Harmonic penetration into power systems'. Paper presented at the 5th Universities Power Engineering Conference, Swansea, Wales.
42. Laurent, P. G., Gary, C., and Clade, J. (1962). 'D. C. interconnection between France and Great Britain by submarine cables', CIGRE, V, Part III, Paper 331.
43. Hingorani, N. G., and Burberry, M. F. (1970). 'Simulation of a.c. system impedance in HVDC system studies', *IEEE Trans.*, **PAS-89**, 820–828.
44. Brewer, G. L., Coates, R., and Clarke, C. D. (1974). 'Initial operating experience with the filters for the Kingsnorth d.c. scheme'. *IEE Conf. Publ.*, **110**, 156–161.
45. Kaban, I. H., and Parten, J. E. (1970). 'Computation of the flow of harmonic current in networks'. Paper presented at the 5th Universities Power Engineering Conference, Swansea, Wales.
46. Mohmoud, A. A., and Shultz, R. D. (1982). 'A method for analyzing harmonic distribution in a.c. power systems'. *IEEE Trans.*, **PAS-101**, 1815–1824.
47. Zollenkopf, K. (1960). 'Bi-factorization – basic computational algorithm and programming techniques'. Paper presented at a conference on Large Sets of Sparse Linear Equations, Oxford.
48. Yacamini, R., and de Oliveira, J. C. (1980). 'Harmonics in multiple convertor systems: a generalised approach'. *Proc. IEE, B*, **127**, 96–106.
49. Arrillaga, J., Arnold, C. P., and Harker, B. J. (1983). *Computer Modelling of Electrical Power Systems*, John Wiley, Chichester.
50. Harker, B. J. and Arrillaga, J. (1979)' 'Three phase a.c./d.c. loadflow'. *Proc. IEE*, **126**, 1275–1281.
51. Harker, B. J. (1980) 'Steady state analysis of integrated a.c. and d.c. systems'. PhD thesis, University of Canterbury, New Zealand.
52. Hesse, M. H. (1966). 'Circulating currents in parallel untransposed multicircuit lines. I. Numerical evaluations'. *IEEE Trans.*, **PAS-85**, 802–811.
53. Edison Electric Institute (1968). *EHV Transmission Line Reference Book*, Edison Electric Institute, New York.
54. Robinson, G. H. (1966). 'Harmonic phenomena associated with the Benmore–Haywards h.v.d.c. transmission scheme'. *New Zealand Engineer*, **21**, 16–29.
55. Winthrop, D. A. (1983). 'Planning for a railway traction load on the New Zealand power system'. Paper 62 presented at the IPENZ Conference.
56. Rober, R. D., and Leedham, P.J. (1974). 'A review of the cause and effects of distribution system three-phase unbalance'. *IEE Conf. Publ.*, **110**, 83–92.

10

Harmonic elimination

10.1 PURPOSE OF HARMONIC FILTERS

The primary object of a harmonic filter is to reduce the amplitude of one or more fixed frequency currents or voltages.

When the only purpose is to prevent a particular frequency from entering selected plant components or parts of a power system (e.g. in the case of ripple control signals) it is possible to use a series filter consisting of a parallel inductor and capacitor which presents a large impedance to the relevant frequency. Such a solution, however, cannot be extended to eliminate the harmonics from arising at the source, because the production of harmonics by non-linear plant components (like transformers and static convertors) is essential to their normal operation.

In the case of static convertors, the harmonic currents are normally prevented from entering the rest of the system by providing a shunt path of low impedance to the harmonic frequencies. Combined series and shunt filters could be designed to minimise harmonic currents and voltage in the a.c. system regardless of its impedance, but they are expensive.

10.2 DEFINITIONS

The shunt filter is said to be tuned to the frequency that makes its inductive and capacitive reactances equal.

The quality of a filter (Q) determines the sharpness of tuning and in this respect filters may be either of the high Q type or they may be of the low Q type. The high Q filter is sharply tuned to one of the lower harmonic frequencies (e.g. the fifth) and a typical value is between 30 and 60. The low Q filter, typically in the region of 0.5–5, has a low impedance over a wide range of frequencies. When used to eliminate the higher order harmonics (e.g. 17th up) it is also referred to as a high pass filter. Typical examples of high and low Q filter circuits and their impedance variation with frequency are illustrated in Figures 10.1 and 10.2.

In the case of a tuned filter Q is defined as the ratio of the inductance (or the capacitance) at resonance to resistance, i.e.

$$Q = X_0/R. \tag{10.2.1}$$

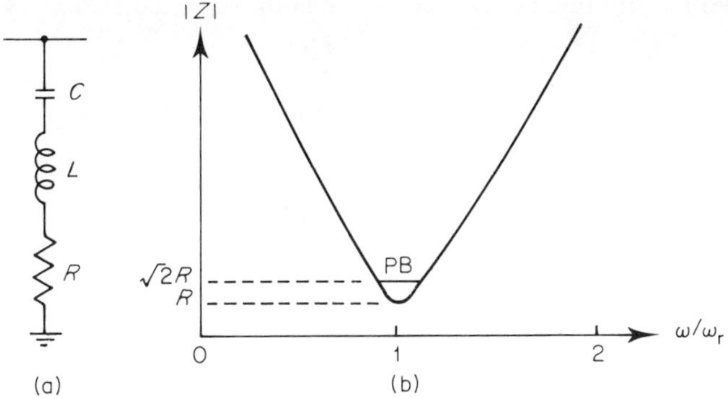

Figure 10.1. (a) Single-tuned shunt filter circuit and (b) single-tuned shunt filter impedance versus frequency

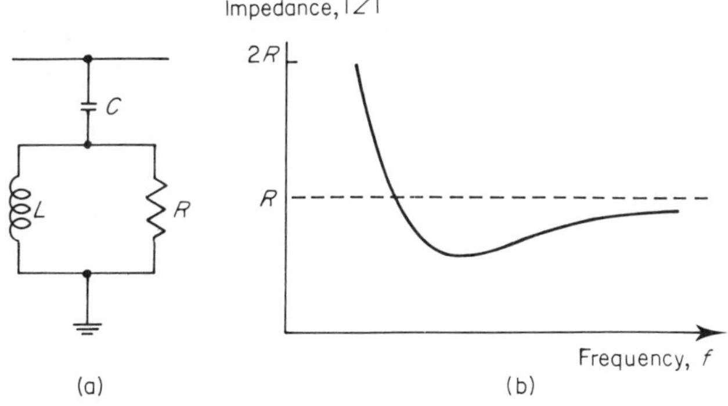

Figure 10.2. (a) Second order damped shunt filter circuit and (b) second order damped shunt filter impedance versus frequency

As shown in Figure 10.1(b) the filter pass band (PB) is defined as being bounded by the frequencies at which the filter reactance equals its resistance, i.e. the impedance angle is 45° and the impedance module $\sqrt{2}R$. The quality factor and pass band are related by the expression

$$Q = \omega_n/\mathrm{PB}, \tag{10.2.2}$$

where ω_n is the tuned angular frequency in radians per second.

The sharpness of tuning in high-pass damped filters is the reciprocal of that of tuned filters, i.e. $Q = R/X$.

The extent of filter detuning from the nominal tuned frequency is represented by a factor δ. This factor includes various effects, i.e. (i) variations in the fundamental (supply) frequency; (ii) variations in the filter capacitance and inductance caused by ageing and temperature; (iii) initial off-tuning caused by manufacturing tolerances and finite size of tuning steps.

The overall de-tuning, in per unit of the nominal tuned frequency, is

$$\delta = (\omega - \omega_n)/\omega_n. \tag{10.2.3}$$

Moreover, a change of L or C of say 2% causes the same detuning as a change of system frequency of 1%. Therefore δ is often expressed as

$$\delta = \frac{\Delta f}{f_n} + \frac{1}{2}\left(\frac{\Delta L}{L_n} + \frac{\Delta C}{C_n}\right). \tag{10.2.4}$$

10.3 FILTER DESIGN CRITERIA

The size of a filter is defined as the reactive power that the filter supplies at fundamental frequency. It is substantially equal to the fundamental reactive power supplied by the capacitors. The total size of all the branches of a filter is determined by the reactive power requirements of the harmonic source and by how much this requirement can be supplied by the a.c. network.

The ideal criterion of filter design is the elimination of all detrimental effects caused by waveform distortion, including telephone interference, which is the most difficult effect to eliminate completely. However, this ideal criterion is unrealistic both for technical and economical reasons. From the technical point of view, it is very difficult to estimate in advance the distribution of harmonics throughout the a.c. network. On the economic side, the reduction of telephone interference can normally be achieved more economically by taking some of the preventive measures in the telephone system and others in the power system.

A more practical criterion suggests reducing the problem to an acceptable level at the point of common coupling with other consumers, the problem being expressed in terms of harmonic current, harmonic voltage, or both. A criterion based on harmonic voltage is more convenient for filter design, because it is easier to guarantee staying within a reasonable voltage limit than to limit the current level as the a.c. network impedance changes.

In order to comply with the required harmonic limitations the design of filters involves the following steps:

(i) The harmonic current spectrum produced by the non-linear load is injected into a circuit consisting of filters in parallel with the a.c. system (Figure 10.3) at the relevant frequencies and the harmonic voltages are calculated.

(ii) The results of (i) are used to determine the specified parameters, i.e. voltage distortion D, TIF and IT factors.

(iii) The stresses in the filter components, i.e. capacitors, inductors and resistors are then calculated and with them their ratings and losses.

Three components require detailed consideration in filter design, i.e. current source, filter admittance and system admittance.

The current source content should be varied through the range of load and (in

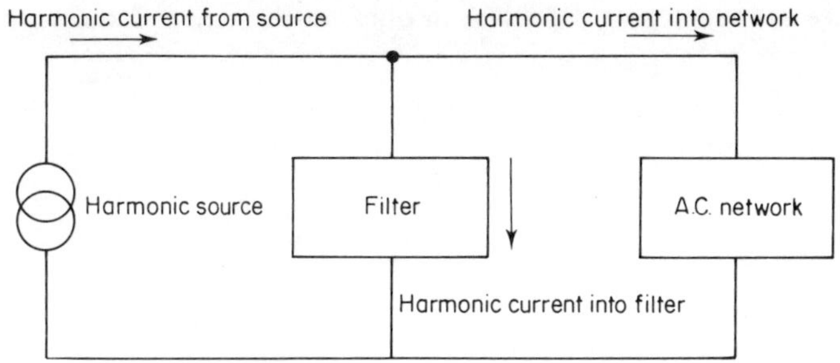

Figure 10.3. Circuit for the computation of voltage harmonic distortion

the case of static convertors) firing angle conditions. This subject has been discussed in Chapter 3. As far as filter and system admittances are concerned, it is essential to calculate the minimum total equivalent admittance at each harmonic frequency, which will result in maximum voltage distortion.

The loci of filter impedance (or admittance) are easy to obtain once a particular filter configuration has been decided. Source impedance loci are more difficult to determine with any degree of accuracy.

The obvious filter design is a single broad band-pass configuration capable of attenuating the whole spectrum of injected harmonics (e.g. from the fifth up in the case of a six-pulse convertor). However, the capacitance required to achieve such target is too large, and it is usually more economical to attenuate the lower harmonics by means of single arm tuned filters.

10.4 TUNED FILTERS

A single tuned filter is a series *RLC* circuit (as shown in Figure 10.1) tuned to the frequency of one harmonic (generally a lower characteristic harmonic). Its impedance is given by

$$Z_1 = R + j\left(\omega L - \frac{1}{\omega C}\right), \tag{10.4.1}$$

which at the resonant frequency (f_n) reduces to R. There are two basic design parameters to be considered prior to the selection of R, L and C. These are the quality factor (Q), and the relative frequency deviation (δ), already defined.

In order to express the filter impedance in terms of Q and δ the following relationships apply:

$$\omega = \omega_n(1 + \delta); \tag{10.4.2}$$

$$\omega_n = \frac{1}{\sqrt{LC}} \tag{10.4.3}$$

(tuned angular frequency in radians per second);

$$X_0 = \omega_n L = \frac{1}{\omega_n C} = \sqrt{\frac{L}{C}} \tag{10.4.4}$$

(reactance of inductor or capacitor in ohms at the tuned frequency);

$$Q = \frac{X_0}{R}; \tag{10.4.5}$$

$$C = \frac{1}{\omega_n X_0} = \frac{1}{\omega_n R Q}; \tag{10.4.6}$$

$$L = \frac{X_0}{\omega_n} = \frac{RQ}{\omega_n}. \tag{10.4.7}$$

Substituting equations (10.4.2), (10.4.6), (10.4.7) in equation (10.4.1) yields

$$Z_f = R\left(1 + jQ\delta\left(\frac{2+\delta}{1+\delta}\right)\right) \tag{10.4.8}$$

or, considering that δ is relatively small as compared with unity,

$$Z_f \simeq R(1 + j2\delta Q) = X_0(Q^{-1} + j2\delta) \tag{10.4.9}$$

and

$$|Z_f| \simeq R(1 + 4\delta^2 Q^2)^{1/2}. \tag{10.4.10}$$

It is generally more convenient to deal with admittances rather than impedances in filter design, i.e.

$$Y_f \simeq \frac{1}{R(1 + j2\delta Q)} = G_f + jB_f, \tag{10.4.11}$$

where

$$G_f = \frac{Q}{X_0(1 + 4\delta^2 Q^2)}, \tag{10.4.12}$$

$$B_f = -\frac{2\delta Q^2}{X_0(1 + 4\delta^2 Q^2)}. \tag{10.4.13}$$

The harmonic voltage at the filter busbar is

$$V_n = \frac{I_n}{Y_{nf} + Y_{sn}} = \frac{I_n}{Y_n} \tag{10.4.14}$$

and therefore to minimize the voltage distortion it is necessary to increase the overall admittance of the filter in parallel with the a.c. system.

In order to predict the largest V_n, the variables that are not accurately known have to be chosen pessimistically; these are the frequency deviation δ and the network admittance Y_{sn}. Since the harmonic voltage increases with δ, the largest expected deviation δ_m must be used in the analysis. Again, the worst realistic system condition (the lowest admittance) must be represented.

Within certain limits the designer can decide on the values of Q and filter size (VA ratings at fundamental frequency).

In terms of Q and δ equation (10.4.14) can be written as follows:

$$|V_n| = I_n\left\{\left[G_{sn} + \frac{1}{R(1+4Q^2\delta^2)}\right]^2 + \left[B_{sn} - \frac{2Q\delta}{R(1+4Q^2\delta^2)}\right]^2\right\}^{-1/2} \quad (10.4.15)$$

The case of a purely inductive a.c. network admittance, often used in filter design, is unduly pessimistic.

The impedance loci indicate that generally the harmonic impedances can be circumscribed in a part of the plant R, jX determined by two straight lines and a circle passing through the origin. The maximum phase angle of the network impedance can thus be limited to below 90° and generally decreases with increasing frequency (except in cable networks for high harmonic orders). The highest harmonic voltage is then obtained by using ϕ_{sn} with a sign opposite to that of δ.

Then equation (10.4.15) becomes

$$|V_n| = I_n\{(|Y_{sn}|\cos\phi_{sn} + G_f)^2 + (-|Y_{sn}|\sin\phi_{sn} + B_f)^2\}^{-1/2}, \quad (10.4.16)$$

taking ϕ_{sn} positive and δ negative.

Since $|Y_{sn}|$ is unrestricted, the admittance giving maximum $|V_n|$ is

$$|Y_{sn}| = \frac{\cos\phi_{sn}(2Q\delta\tan\phi_{sn} - 1)}{R(1+4Q^2\delta^2)}, \quad (10.4.17)$$

giving

$$|V_n| = I_n\omega_n L\left[\frac{1+4Q^2\delta^2}{Q(\sin\phi_{sn} + 2Q\delta\cos\phi_{sn})}\right]. \quad (10.4.18)$$

There is an optimum Q which results in the lowest harmonic voltage, i.e.

$$Q = \frac{1+\cos\phi_{sn}}{2\delta\sin\phi_{sn}} \quad (10.4.19)$$

for which

$$|V_n| = I_n\delta\omega_n L\left[\frac{4}{1+\cos\phi_{sn}}\right] = \frac{2I_nR}{\sin\phi_{sn}}. \quad (10.4.20)$$

Nevertheless, it should be noted that filters are not usually designed to give minimum harmonic voltage under these conditions. Normally a higher Q is selected in order to reduce losses.

A condition that also has to be considered in the design of filters, and which can restrict the operation of the convertors, is an outage of one or more filter branches. The remaining filter branches may then be overstressed as they have to take the total harmonic current generated by the convertor.

Graphic approach

A graphic explanation is given by Kimbark[(1)] which helps to understand the selection of optimum Q, i.e. the value that maximizes Y_n.

For a given maximum value of the frequency deviation factor δ_m, and using a fixed reactance X_0 and variable resistance R, the locus of the filter admittance, i.e.

$$Y_f = \frac{1}{R(1 + j2\delta Q)}$$

is a semicircle of diameter $1/(2\delta_m X_0)$ tangent to the G-axis at the origin, as shown by the dashed line in Figure 10.4. The same figure displays (shaded area) the system admittance domain (Y_{sn}), obtained by inverting the impedance locus, and the minimum admittance for each frequency lies on the boundary of the shaded area.

For a given Y_{nf} the shortest vector Y_n is perpendicular to and ends on the boundary. The vector diagram of Figure 10.4, drawn for a positive δ_m and negative $\phi = \phi_m$, produces the highest harmonic voltage. Moreover, the optimum value of Y_{nf} is that which terminates on the semicircle at a point where the boundary at angle $+\phi_m$ is tangent to the semicircle. This optimum case is illustrated in Figure 10.4, where at point D, Y_{nf} maximizes V_n and Y_{sn} minimizes it.

For such condition the filter admittance can be shown to be

$$|Y_{nf}| = \frac{\cos(\phi_m/2)}{2\delta_m X_0} \tag{10.4.21}$$

and

$$|Y_n| = |Y_{nf}|\cos(\phi_m/2) = \frac{1 + \cos\phi_m}{4\delta_m X_0}. \tag{10.4.22}$$

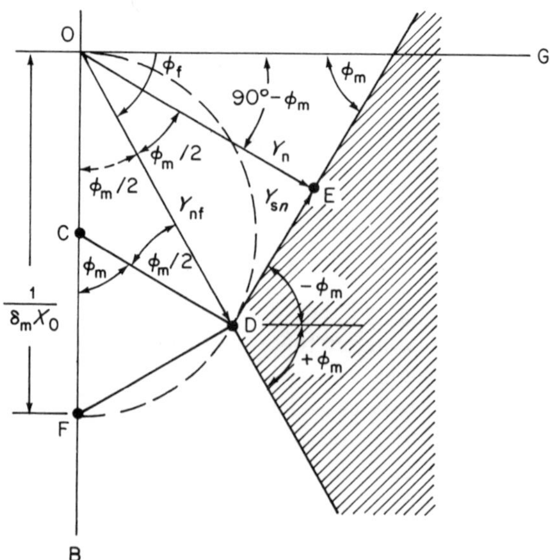

Figure 10.4. Construction for finding optimum Q and worst network admittance Y_{sn}, drawn for $\phi_m = 60°$. From Kimbark, *Direct Current Transmission*. Copyright © 1971 John Wiley & Sons, Inc. Reprinted by permission of John Wiley & Sons, Inc.

The quality factor of the chosen Y_{nf} is

$$Q = \frac{X_0}{R} = \frac{X_0}{X_f/(\tan \phi_f)} \tag{10.4.23}$$

but (from equation (10.4.9))

$$X_f = 2\delta_m X_0 \tag{10.4.24}$$

and (from Figure 10.4)

$$\tan \phi_f = \cot (\phi_m/2). \tag{10.4.25}$$

Therefore

$$Q = \frac{\cot (\phi_m/2)}{2\delta_m} = \frac{\cos \phi_m + 1}{2\delta_m \sin \phi_m}. \tag{10.4.26}$$

After the individually tuned filter Qs values have been determined, the entire filter configuration must be used to determine the network admittance Y_n that yields the minimum total admittance Y at each harmonic frequency.

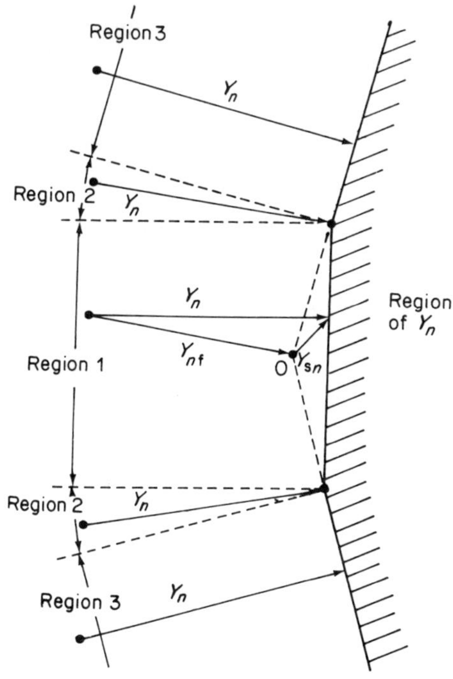

Figure 10.5. Determination of network admittance Y_{sn} for minimum resultant admittance Y_n corresponding to filter admittances Y_f lying in different regions. From Kimbark, *Direct Current Transmission*. Copyright © 1971 John Wiley & Sons, Inc. Reprinted by permission of John Wiley & Sons, Inc.

In practice, the minimal possible system admittances are also limited by a minimum conductance thus resulting in the admittance domain shown shaded in Figure 10.5.

At any harmonic frequency the equivalent admittance of the filter configuration consists of a vector that ends at point O and starts in one of three regions of the admittance plane as shown in Figure 10.5.

At frequencies for which tuned filters are provided, the origin of the filter admittance is likely to lie in region 3, i.e. the total filter admittance is relatively large. At other frequencies, however, the filter admittance origin may lie in region 1 or 2.

The most pessimistic values of the network admittance are those which result in the lowest total admittance; these are clearly defined in the graph: (i) in region 1 the resultant admittance vector Y_n ends on the vertical (i.e. minimum conductance) part of the boundary; (ii) in region 2, Y_n ends on the corner of the boundary; (iii) in region 3, Y_n is perpendicular to the nearer angular limit.

Double tuned filters[3]

The equivalent impedances of two single tuned filters (Figure 10.6(a)) near their resonance frequencies are practically the same as those of a double tuned filter configuration, illustrated in Figure 10.6(b), subject to the following relationships between their components

$$C_1 = C_a + C_b, \tag{10.4.27}$$

$$C_2 = \frac{C_a C_b (C_a + C_b)(L_a + L_b)^2}{(L_a C_a - L_b C_b)^2}, \tag{10.4.28}$$

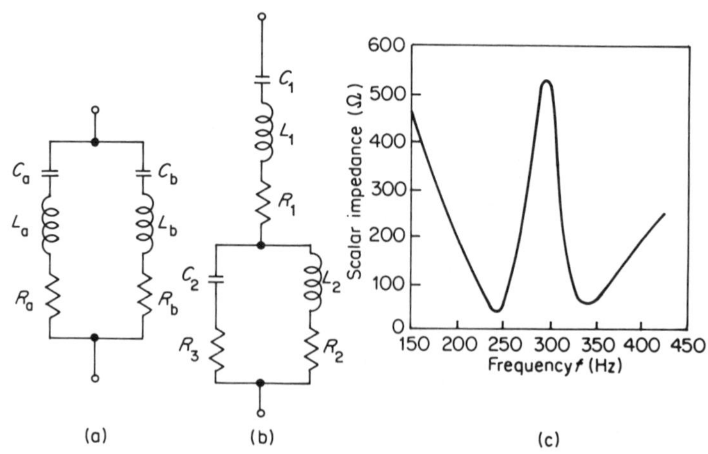

Figure 10.6. Transformation form (a) two single-tuned filters to (b) double-tuned filters. (c) Double-tuned at Echinghen for fifth and seventh harmonics: calculated scalar impedance versus frequency

$$L_1 = \frac{L_a L_b}{L_a + L_b}, \tag{10.4.29}$$

$$L_2 = \frac{(L_a C_a - L_b C_b)^2}{(C_a + C_b)^2 (L_a + L_b)}, \tag{10.4.30}$$

$$R_2 = R_a \left[\frac{a^2(1 - x^2)}{(1 + ax^2)^2(1 + x^2)} \right] + R_b \left[\frac{1 - x^2}{(1 + ax^2)^2(1 + x^2)} \right] + R_1 \left[\frac{(1 - x^2)(1 - ax^2)}{(1 + x^2)(1 + ax^2)} \right], \tag{10.4.31}$$

where

$$a = \frac{C_a}{C_b} \quad \text{and} \quad x = \sqrt{\frac{L_b C_b}{L_a C_a}}.$$

The above practical approximation is carried out by omitting resistor R_1, which is therefore determined by the minimum resistances of the inductor L_1. This has the advantage of reducing the power loss at fundamental frequency as compared with the single tuned filter configurations. The main advantage of the double-tuned filter is in high voltage applications, because of the reduction in the number of inductors to be subjected to full line impulse voltages.

By way of example the equivalent impedances of the double tuned filters used at the Echinghen terminal of the Cross-Channel link[(4)] are illustrated in Figure 10.6(c).

Triple and quadruple-tuned filters can also be designed but these are rarely justified because of the difficulty of adjustment.

Automatically tuned filters

In tuned filter design it is advantageous to reduce the maximum frequency deviation. This can be achieved by making the filters tunable by either automatically switching the capacitance or by varying the inductance. A range of $\pm 5\%$ is usually considered adequate. A control system, which measures the harmonic frequency reactive power in the filter, and which control the L or the C based on the sign and the magnitude of this reactive power, has been used in high voltage d.c. convertors. Automatically tuned filters offer the following advantages over fixed filters:

(i) The capacitor rating is lower.
(ii) The capacitor used can combine a high temperature coefficient of capacitance and a high reactive power rating per unit of volume and per unit of cost.
(iii) Because of the higher Q, the power loss is smaller.

Advantages (i) and (ii) reduce the cost of the capacitor which is the most expensive component of the filter. Advantage (iii) reduces the cost of the resistor and the cost of the system losses.

10.5 DAMPED FILTERS

The damped filter offers several advantages:

(i) Its performance and loading is less sensitive to temperature variation, frequency deviation, component manufacturing tolerances, loss of capacitor elements, etc.
(ii) It provides a low impedance for a wide spectrum of harmonics without the need for subdivision of parallel branches with increased switching and maintenance problems.
(iii) The use of tuned filters often results in parallel resonance between the filter and system admittances at a harmonic order below the lower tuned filter frequency, or in between tuned filter frequencies. In such cases the use of one or more damped filters is a more acceptable alternative.

The main disadvantages of the damped filter are as follows:

(iv) To achieve a similar level of filtering performance the damped filter needs to be designed for higher fundamental VA ratings, though in most cases a good performance can be met within the limits required for power factor correction.
(v) The losses in the resistor and reactor are generally higher.

Types of damped filters

Four types of damped filters are shown in Figure 10.7, i.e. first order, second order, third order and C-type.

(a) The first order filter is not normally used, as it requires a large capacitor and has excessive loss at the fundamental frequency.
(b) The second order type provides the best filtering performance, but has higher fundamental frequency losses as compared with the third order filters.

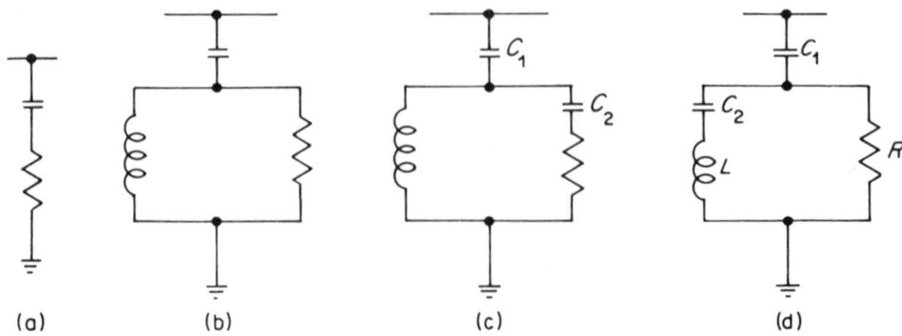

Figure 10.7. High-pass damped filters: (a) first order; (b) second order; (c) third order; (d) C-type

(c) Its main advantage over (b) is a substantial reduction in fundamental frequency loss, owing to the increased impedance at that frequency caused by the presence of the capacitor C_2. Moreover, the rating of C_2 is very small compared with C_1.

(d) The filtering performance of the newly introduced C-type[(2)] lies in between those of (b) and (c). Its main advantage is a considerable reduction in fundamental frequency loss since C_2 and L are series tuned at that frequency. This filter is more susceptible to fundamental frequency deviations and component value drifts.

Design of damped filters

When designing a damped filter the Q is chosen to give the best characteristic over the required frequency band and there is no optimal Q as with tuned filters.

The behaviour of damped filters has been described by Ainsworth[(3)] with the help of two parameters, i.e.

$$f_0 = \frac{1}{2\pi CR}, \tag{10.5.1}$$

$$m = \frac{L}{R^2 C}. \tag{10.5.2}$$

Typical values of m are between 0.5 and 2. For a given capacitance these parameters (and hence L and R) are decided to achieve an appropriately high admittance over the required frequency range.

The conductance and susceptance terms of a second order damped filter admittance are

$$G_f = \frac{m^2 x^4}{R_1[(1 - mx^2)^2 + m^2 x^2]}, \tag{10.5.3}$$

$$B_f = \frac{x}{R_1}\left[\frac{1 - mx^2 + m^2 x^2}{(1 - mx^2)^2 + m^2 x^2}\right], \tag{10.5.4}$$

where $x = f/f_0$.

The minimum total admittance (i.e. the filter Y_F plus the a.c. system Y_{sn}) can be shown to be

$$Y = B_f \cos \phi_m + G_f \sin \phi_m \tag{10.5.5}$$

with both terms in equation (10.5.5) being positive and x being less than the value that gives

$$|\cot \phi_f| = |G_f/B_f| = |\tan \phi_m|. \tag{10.5.6}$$

For greater values of x the minimum total admittance is that of the filter (i.e. with $Y_{sn} = 0$).

Figure 10.8 illustrates typical minimal admittances for a second order damped

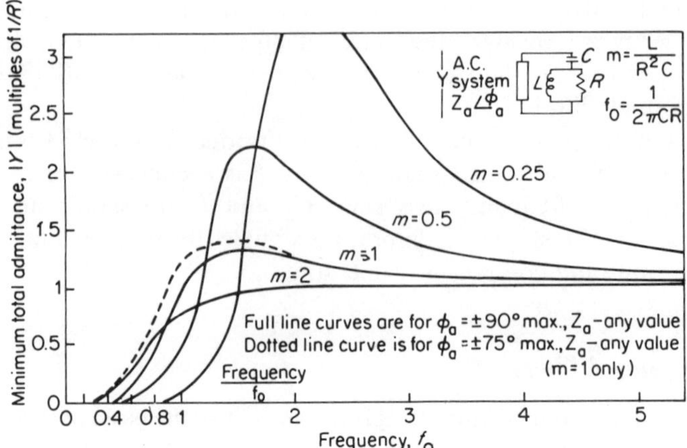

Figure 10.8. Admittance of second order low-pass filter

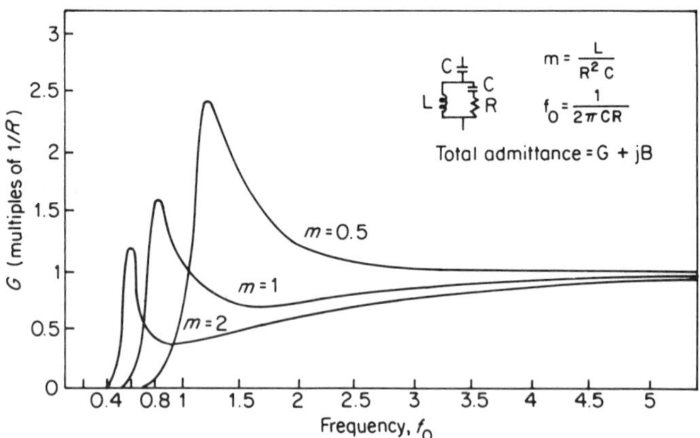

Figure 10.9. Conductance component for third order low-pass filter

filter in parallel with a lossless system (i.e. $\phi_m = \pm 90°$). For comparison the conductance component G_f of a third order damped filter, for the case of equal capacitors, is shown in Figure 10.9. These figures show that the third order filter peaks are much sharper than those of the second order.

10.6 TYPICAL FILTER CONFIGURATIONS

Static convertors of large ratings are normally designed for at least 12-pulse operation. Very often, however, to cope with maintenance and other partial temporary outages, six-pulse operation is permitted. Under such conditions they produce considerable fifth and seventh harmonics as well as the characteristic 12-pulse related orders. These are conventionally filtered by using a hybrid combination of tuned branches for the low orders, i.e. fifth, seventh, 11th and 13th, and a high-pass damped filter for the 17th and higher orders.

Let us illustrate the conventional design with a numerical example obtained from Ainsworth.[3]

A six-pulse convertor bridge is rated at 100 kV, 100 MW d.c., operating at $\alpha = 15°$. The bridge is connected to a 275 kV, 50 Hz a.c. system via a 275/83 kV convertor transformer with 15% leakage reactance. The secondary fundamental current is 780 A and that of the primary 236 A. The filters, to be connected to the primary side, consist of resonant arms for the fifth, seventh, 11th and 13th harmonics and a second order high-pass arm.

For a total filter size of 50 MVAr, and assuming that the capacitance is to be equally divided among the filter branches, each branch requires 0.417 μF. If the capacitor temperature coefficient is 0.05% per degree Celsius, the inductor temperature coefficient 0.01% per degree Celsius, ambient temperature ± 20 °C and frequency tolerance $\pm 1\%$, then from equation (10.2.4)

$$\delta = \tfrac{1}{100}\{1 + \tfrac{1}{2}[0.05 \times 20 + 0.01 \times 20]\} = 0.016.$$

Let the a.c. system impedance be of any magnitude but its phase angle restricted to $\phi_a < 75°$ at any frequency. The optimum Q (giving the lowest harmonic voltage) is then obtained from equation (10.4.26), i.e.

$$Q = \frac{1 + \cos 75°}{2(0.016)\sin 75°} = 41.$$

With Q and C known, the values of L and R of the resonant arms can then be determined.

The damped arm components are found from equations (10.5.1) and (10.5.2) by choosing $m = 1$ and $f_0 = 17 \times 50 = 850$ Hz. Since C has been fixed above (i.e.

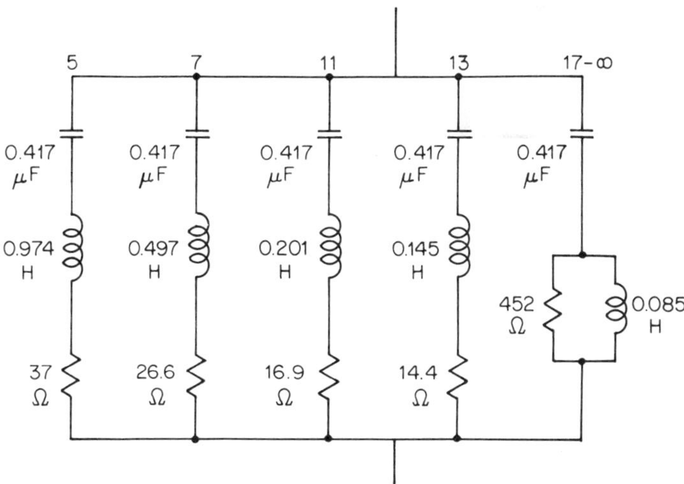

Figure 10.10. Example of a.c. filter design

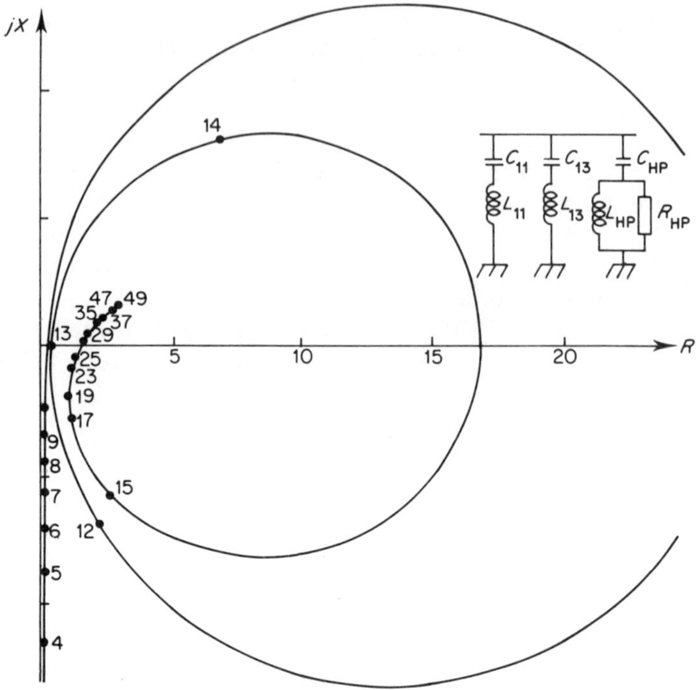

Figure 10.11. Filter configuration for 12-pulse operation and typical impedance locus

0.41 μF), the resulting values of inductor and resistor are 0.085 H and 452 Ω respectively.

The complete circuit design is then as shown in Figure 10.10.

Whenever a reasonable 12-pulse operation can be guaranteed under all operating conditions the fifth and seventh harmonic filters can be eliminated. An example of the configuration used in recent high voltage d.c. convertors and the corresponding impedance locus are shown in Figure 10.11. The locus exhibits resonant points at the 11th, 13th and 27th harmonics and reasonably low impedances to the fifth and seventh which will cope with the levels expected under slight unbalanced conditions.

10.7 BAND PASS FILTERING FOR 12-PULSE CONVERTORS

The use of a conventional filter design, for convertor stations with separate tuned filters of the series resonant type, for the 11th and 13th harmonics and a high pass filter for the higher orders, will usually give a more effective reduction of harmonics than required. The reason for this is that the minimum size of the filters is usually determined by the available economic size of capacitor units and the minimum amount of reactive power generation required by the convertors.

Table 10.1. Derivation of the maximum r.m.s. distortion for the filter configuration of four second-order plus four C-type filters at 2000 MW d.c. load

Harmonic order	2	3	5	7	11
Distortion due to pre-existing distortion on the a.c. system (%)	0.39	0.34	0.22	0.12	0.01
Distortion due to convertor harmonic current (%)	0.13	0.42	0.56	0.29	0.43
Distortion from third-harmonic current generated by the convertor due to negative-sequence on the a.c. system (%)	—	0.19	—	—	—
Distortion due to static VAR compensator harmonic current (%)	—	0.03	0.03	0.02	0.01
Distortion from third-harmonic current generated by the compensator due to unbalance on the a.c. system (%)	—	0.17	—	—	—
Total contribution due to each harmonic (%)	0.41	1.16	0.60	0.31	0.43

Therefore the filter design can be simplified, either by replacing the tuned filters for harmonics 11 and 13 by a single filter of the damped type, or replacing all filters by a single damped filter. In the first case, the damped filter replacing the two tuned filters should be tuned to about the 12th harmonic and a fairly high Q can be selected (20–50), while the damped filter for the higher harmonics has a much smaller Q (2–4). In the second case, the single damped filter is also tuned to about the 12th harmonic but a fairly low Q has to be chosen (2–6) to get a sufficiently low impedance at higher harmonics.

Moreover, the hybrid design discussed in the last section (Figure 10.11) exhibits increasing impedance at the lower harmonic frequencies.

With the large ratings of recent high-voltage d.c. projects there is an increased probability of low order harmonic resonance between the system impedance and the filter capacitance.

The resonance is of the series or parallel type, depending on whether the source of low order harmonic is the a.c. system or convertor respectively. As a result of system unbalance a significant third harmonic current is produced by the convertor. Moreover, the third harmonic produced is of positive sequence and will not be absorbed by the transformer delta winding.

An alternative filter design has been proposed[(5)] to eliminate the low order resonance, which consists of a '*C*'-type and a second-order damped filter as shown in Figure 10.12.

It is unduly pessimistic to consider the possibility of a number of harmonics in near resonance simultaneously. In the design of the 2000 MW Cross-Channel high voltage d.c. link the following combination of system impedances has been used:

(i) The harmonic order that produces the highest voltage distortion is assumed to be at or near resonance with the system.

(ii) Other harmonics in the range 2–25 are selected from tables containing information of the system impedances under all likely system and planned outage conditions.

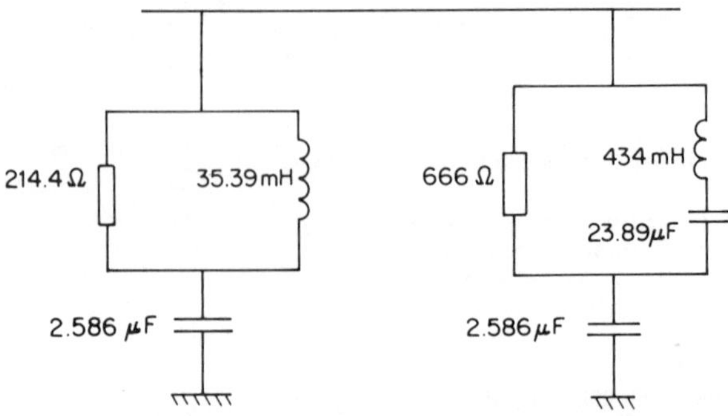

Figure 10.12. Combined second order and *C*-type damped filters

(iii) The remaining harmonics in the range 25–49 are assumed to lie within a wide radius of 750 Ω, centred at $R = 750\,\Omega$, limited by impedance angles of 73° capacitive and 85° inductive.

The performance characteristics predicted for the Cross-Channel scheme (at the Sellindge end) are illustrated in Table 10.1. Since in this case the highest distortion is caused by the third harmonic, this harmonic is taken as the arithmetic sum of the content produced by the various sources listed.

10.8 FILTER COMPONENT PROPERTIES

From knowledge of the fundamental and harmonic voltages at the relevant busbars the current and voltage ratings of the capacitors, inductors and resistors can be calculated, and with them the active and reactive powers and losses.

To prevent damage of these components their ratings must be based on the most severe conditions expected, i.e. the highest fundamental voltage, highest effective frequency deviation, harmonic currents from other sources and from possible resonances between the filter and a.c. system.

Capacitors

Capacitors are composed of standard units which are connected in series and/or parallel for obtaining the desired overall voltage and kVA rating.

The main properties of capacitors are as follows:[6] (i) temperature coefficient of capacitance, (ii) reactive power per unit volume, (iii) power loss, (iv) reliability, (v) cost.

A very low temperature coefficient of capacitance is desirable for tuned filters in order to avoid de-tuning caused by change of capacitance with ambient temperature or with self-heating of the capacitors. Note that this property is unimportant for damped filters or for power capacitors.

Capacitors obtain their high reactive power per unit volume by having low losses and by operating at very high voltage stresses. For this reason, prolonged operation at moderate over-voltage must be avoided to prevent thermal destruction of the dielectric; and even very brief operation at higher over-voltage must be avoided to prevent destructive ionization of the dielectrics.

The required reactive-power rating of the capacitor is the sum of the reactive powers at each of the frequencies to which it is subjected.

Inductors

Inductors used in filter circuits need to be designed bearing in mind the high frequencies involved, i.e. skin-effect and hysteresis losses must be included in the power loss calculation. Also the effect of the flux level in the iron, i.e. the detuning caused by magnetic non-linearity, must therefore be considered. This normally leads to the use of low flux densities when using iron cores. Alternatively, filter inductors are better designed with non-magnetic cores.

The Q at the predominant harmonic frequency may be selected for lowest cost and is usually between 50 and 150. However, lower Q values are normally required and these are derived by using a series resistor.

Inductor ratings depend mainly on the maximum r.m.s. current and on the insulation level required to withstand switching surges. Normally the R and L form the ground side of a tuned filter.

10.9 FILTER COSTS

An effective filter adequately suppresses harmonics at the least cost and supplies some reactive power, but perhaps not all that is required. The cost of losses incurred in the filters may be charged to reactive power supply and some to filtering, although there is no logical basis for the division.

The following assumptions are usually made in the cost analysis of filter components:

(i) In a typical installation, a capacitor bank consists of a 'matrix' of capacitor units, each having a nominal rating at the prescribed operating voltage and protected by an external fuse.

The cost of a capacitor bank is thus approximately constant up to the rating of the minimum matrix containing full units. For higher ratings, one or more units are added to each series group as required and a reasonably accurate cost per MVAr or SIZE can be arrived at. The situation is complicated further by the availability of standard units with different nominal ratings, e.g. 50, 100, 150 kVAr etc. and the incremental cost varies for different bands of capacitor bank SIZES. Although such factors would have to be included in the development of an accurate cost equation, here we are assuming that the capacitors' cost is proportional to their ratings.

(ii) Although the cost of filter inductors depends greatly on the method of construction (e.g. oil insulated/cooled units, natural air-cooled reactors of open construction, etc.), their cost does not vary greatly for units of different rating. The cost approximation used in the analysis, therefore, is of the form

$$\text{Inductor cost} = U_K + U_L \times (\text{total MVAr rating}),$$

where U_K is a constant cost component and U_L is the inductor incremental cost per MVAr rating.

(iii) The power rating of the resistor necessary for Q-adjustment in each filter branch will doubtless affect the cost to some extent. The nominal resistance of the unit is difficult to predict, however, in a general analysis, because it obviously depends on the natural Q-factor of the inductor. For this reason and also because the cost of an air-cooled resistor is small compared with that of the other components, a constant cost per resistor is allocated in the analysis. If an oil-cooled unit is used, the cost would be more significant but it would, in fact, become virtually independent of power rating.

(iv) Finally it is assumed that the resistance of the inductor, for the purposes of power loss estimation, is constant at all frequencies.

Single tuned filter

In a high-Q circuit it may be assumed that

$$V_c = V_L + V_s \tag{10.9.1}$$

where V_c, V_L and V_s represent the capacitor, inductor and supply voltages respectively.

The filter size is expressed as

$$\text{SIZE}, S = \frac{V_s^2}{X_c - X_L}, \tag{10.9.2}$$

where X_c and X_L are the fundamental frequency reactances of the capacitor and inductor.

But for a filter tuned to harmonic n,

$$X_0 = nX_L = X_c/n,$$

i.e.

$$X_L = X_c/n^2 \quad \text{and} \quad V_L = V_c/n^2.$$

Therefore

$$S = V_s^2/[X_c(1 - 1/n^2)] = (V_s^2/X_c)[n^2/(n^2 - 1)] \qquad \text{MVAr}. \tag{10.9.3}$$

Also

$$V_c - V_L = V_c(1 - 1/n^2) = V_s$$

i.e.

$$V_c = [n^2/(n^2 - 1)]V_s \qquad \text{kV}. \tag{10.9.4}$$

The loadings for each filter component are determined for cost evaluation as follows:

CAPACITOR

Fundamental loading:

$$V_c^2/X_c = (V_s^2/X_c)[n^2/(n^2-1)]^2$$
$$= S[n^2/(n^2-1)] \quad \text{MVAr.} \tag{10.9.5}$$

Harmonic loading:

$$I_n^2(X_c/n) = [(I_n^2 \cdot V_s^2)/(S \cdot n)][n^2/(n^2-1)] \quad \text{MVAr.} \tag{10.9.6}$$

Power loss:

$$K_{CL} \cdot (\text{total loading}) = K_{CL}[S + (I_n^2 V_s^2)/(S \cdot n)][n^2/(n^2-1)] \quad \text{kW}, \tag{10.9.7}$$

where
K_{CL} is the loss factor of the capacitors (in kW/MVA_r).

INDUCTOR

Fundamental loading:

$$V_L^2/X_L = (V_c/n^2)^2 \cdot (n^2/X_c) = V_c^2/n^2 X_c$$
$$= (S/n^2)[n^2/(n^2-1)] \quad \text{MVAr}. \tag{10.9.8}$$

Harmonic loading is the same as for the capacitor since the reactances are equal at harmonic frequency.

For cost purposes, it is convenient to consider the losses in the total effective resistance R, where

$$R = X_0/Q = X_c/nQ.$$

The fundamental current is

$$I_1 = S/V_s \quad \text{kA}$$

and the total power loss

$$(I_1^2 + I_n^2)R = (S^2/V_s^2)X_c/nQ + I_n^2 X_c/nQ$$
$$= [S^2/nQ](1/S)[n^2/(n^2-1)] + [I_n^2 V_s^2/nSQ][n^2/(n^2-1)]$$
$$= [S/nQ + I_n^2 V_s^2/nSQ][n^2/(n^2-1)] \times 10^3 \quad \text{kW}. \tag{10.9.9}$$

For comparison purposes, the cost of energy losses is expressed in terms of equivalent capital cost by use of a present value factor

$$P_v = [(1+i)^N - 1]/[i(1+i)^N], \tag{10.9.10}$$

where i is the interest rate and N is the budgeted filter life.

Thus the present value cost of energy losses is

$$P_v U_u F_u \times 365 \times 24 \times (\text{total power loss})$$
$$= 8760\ P_v U_u F_u \times (\text{total power loss}), \tag{10.9.11}$$

where U_u is the cost of energy loss per kilowatt-hour and F_u is the filter utilization factor, and the complete expression for the total cost is

$$\text{TCOST} = U_{\text{T}} + [n^2/(n^2-1)]\left\{U_{\text{c}}\left(S + \frac{V_{\text{s}}^2 I_n^2}{nS}\right) + U_{\text{L}}\left(\frac{S}{n^2} + \frac{V_{\text{s}}^2 I_n^2}{nS}\right)\right.$$

$$\left. + 8760\, P_{\text{v}} U_{\text{u}} F_{\text{u}}\left[K_{\text{CL}}\left(S + \frac{V_{\text{s}}^2 I_n^2}{nS}\right) + 10^3\left(\frac{S}{nQ} + \frac{I_n^2 V_{\text{s}}^2}{nSQ}\right)\right]\right\},$$

i.e.

$$\text{TCOST} = U_{\text{T}} + AS + \frac{B}{S}, \tag{10.9.12}$$

where U_{T} is the total constant cost of the filter branch, U_{c} is the capacitor incremental cost per MVAr, U_{L} is the inductor incremental cost per MVAr,

$$A = [n^2/(n^2-1)]\left[U_{\text{c}} + \frac{U_{\text{L}}}{n^2} + 8760 P_{\text{v}} U_{\text{u}} F_{\text{u}}\left(K_{\text{CL}} + \frac{10^3}{nQ}\right)\right] \tag{10.9.13}$$

and

$$B = [n^2/(n^2-1)][V_{\text{s}}^2 I_n^2/n]\left[U_{\text{c}} + U_{\text{L}} + 8760\, P_{\text{v}} U_{\text{u}} F_{\text{u}}\left(K_{\text{CL}} + \frac{10^3}{Q}\right)\right] \tag{10.9.14}$$

As SIZE S is varied, the minimum total cost occurs when

$$\text{d(TCOST)}/\text{d}S = 0,$$

i.e. when

$$S_{\text{MIN}} = \sqrt{\frac{B}{A}} \qquad \text{MVAr}. \tag{10.9.15}$$

Band-pass filter

The component loadings at fundamental and all harmonic frequencies may be determined as for the single-tuned filter, i.e.

$$S = (V_{\text{s}}^2/X_{\text{c}})[n_0^2/(n_0^2-1)] \qquad \text{MVAr}, \tag{10.9.16}$$

where n_0 is the ratio of the tuned frequency to the supply frequency.

CAPACITOR RATING

The fundamental loading is

$$S[n_0^2/(n_0^2-1)] \qquad \text{MVAr}, \tag{10.9.17}$$

The loading at harmonic n is

$$I_n^2(X_{\text{c}}/n) \tag{10.9.18}$$

and using equation (10.9.16) this becomes

$$\frac{1}{S}(I_n^2/n)V_{\text{s}}^2 n_0^2/(n_0^2-1). \tag{10.9.19}$$

Thus the total harmonic loading is

$$\left[\frac{1}{S}(V_s^2 n_0^2)/(n_0^2-1)\right]\sum_{n=n_{\min}}^{n_{\max}}(I_n^2/n) \qquad \text{MVAr}. \tag{10.9.20}$$

INDUCTOR RATING

Referring to Figure 10.2(a), for a Q value of 1.5, say,

$$R = 1.5X_0 = 1.5n_0X_L.$$

Thus if the filter is tuned to a frequency close to the 17th harmonic,

$$R \approx 25X_L.$$

Since $I_c = I_L + jI_R$ it follows that, at fundamental frequency,

$$I_c \approx I_L$$

and the fundamental loading is

$$\begin{aligned} I_L^2X_L &= I_c^2X_c/n_0^2 \\ &= (S/V_s)^2\,[V_s^2/n_0^2S]\,[n_0^2/(n_0^2-1)] \\ &= (S/n_0^2)[n_0^2/2-1)] \qquad \text{MVAr}. \end{aligned} \tag{10.9.21}$$

At harmonic n,

$$\begin{aligned} (I_L)_n &= I_nR/(R+jX_L) \\ &= I_nQ/[Q+(jn/n_0)] \end{aligned} \tag{10.9.22}$$

and

$$|(I_L)_n| = I_nQ/[Q^2+(n/n_0)^2]^{1/2}. \tag{10.9.23}$$

The inductive reactance at harmonic n is

$$\begin{aligned} (X_L)_n &= X_0(n/n_0) = (n/n_0)(X_c/n_0) \\ &= (n/n_0^2)(V_s^2/S)[n_0^2/(n_0^2-1)]. \end{aligned}$$

Thus the loading at harmonic n is

$$(I_L)_n^2(X_L)_n = \frac{1}{S}Q^2V_s^2[n_0^2/(n_0^2-1)]\,[nI_n^2/(Q^2n_0^2+n^2)] \qquad \text{MVAr}, \tag{10.9.24}$$

and the total harmonic loading is

$$\frac{1}{S}Q^2V_s^2[n_0^2/(n_0^2-1)]\sum_{n=n_{\min}}^{n_{\max}}\left[\frac{nI_n^2}{Q^2n_0^2+n^2}\right] \qquad \text{MVAr}, \tag{10.9.25}$$

POWER LOSSES

(i) Power loss in the capacitor is

$$K_{CL}\times(\text{total rating in kilowatts}). \tag{10.9.26}$$

(ii) The inductor series resistance at fundamental frequency is

$$R_L = X_0/Q_L = (n_0/Q_L)X_L, \tag{10.9.27}$$

where Q_L is the quality factor of the inductor, and the corresponding power loss is

$$\begin{aligned} I_L^2 R_L &= (n_0/Q_L)(\text{MVAr loading}) \\ &= [S/(n_0 Q_L)][n_0^2/(n_0^2-1)] \qquad \text{MW.} \end{aligned} \tag{10.9.28}$$

Similarly the harmonic power loss is

$$\sum (I_L)_n^2 (R_L)_n = \frac{1}{S}(Q^2 V_s^2 n_0/Q_L)[n_0^2/(n_0^2-1)] \sum_{n=n_{min}}^{n_{max}} \frac{I_n^2}{Q^2 n_0^2 + n^2} \quad \text{MW.} \tag{10.9.29}$$

(iii) The power loss in the shunt resistor R may also be expressed as a fraction of the inductor loading. At the fundamental frequency,

$$R = QX_0 = Qn_0 X_L,$$

$$|I_R| = \frac{|I_L|X_L}{R} = \frac{I_L X_L}{Qn_0 X_L} = \frac{I_L}{Qn_0}, \tag{10.9.31}$$

and the power loss is

$$\begin{aligned} I_R^2 R &= (1/Qn_0) I_L^2 X_L \\ &= (1/Qn_0)(\text{MVAr loading}) \\ &= [S/(Qn_0^3)][n_0^2/(n_0^2-1)] \times 10^3 \qquad \text{kW.} \end{aligned} \tag{10.9.32}$$

At harmonic n,

$$|(I_R)_n| = |(I_L)_n|(X_L/R) \tag{10.9.33}$$

and the power loss is

$$\sum (I_R)_n^2 (R)_n = \frac{1}{S}(QV_s^2/n_0)[n_0^2/(n_0^2-1)] \sum_{n=n_{min}}^{n_{max}} \left[\frac{n^2 I_n^2}{Q^2 n_0^2 + n^2}\right] \times 10^3 \quad \text{kW.} \tag{10.9.34}$$

TOTAL COST

Applying the present-value factor to energy costs and collecting terms in S and in $1/S$ as for the single-tuned filter, it can easily be shown that once again the total cost is given by

$$\text{TCOST} = U_T + AS + \frac{B}{S}, \tag{10.9.35}$$

where

$$A = \left[U_c + \frac{U_L}{n_0^2} + 8760 P_v U_u F_u \left(K_{CL} + \frac{10^3}{Q_L n_0} + \frac{10^3}{Qn_0^3}\right)\right](n_0^2/(n_0^2-1)] \tag{10.9.36}$$

and

$$B = [n_0^2/(n_0^2 - 1)]\,V_s^2 \sum_{n=n_{\min}}^{n_{\max}} I_n^2 \left[\frac{U_c}{n} + \frac{Q^2 U_L n}{Q^2 n_0^2 + n^2} \right.$$

$$\left. + 8760\, P_v U_u F_u \left(\frac{K_{CL}}{n} + \frac{Q^2 n_0 \times 10^3}{Q_L (Q^2 n_0^2 + n^2)} + \frac{Q n^2 \times 10^3}{n_0 (Q^2 n_0^2 + n^2)} \right) \right]. \quad (10.9.37)$$

As before, TCOST is a minimum when

$$S = S_{MIN} = \sqrt{\frac{B}{A}} \qquad \text{MVAr.} \quad (10.9.38)$$

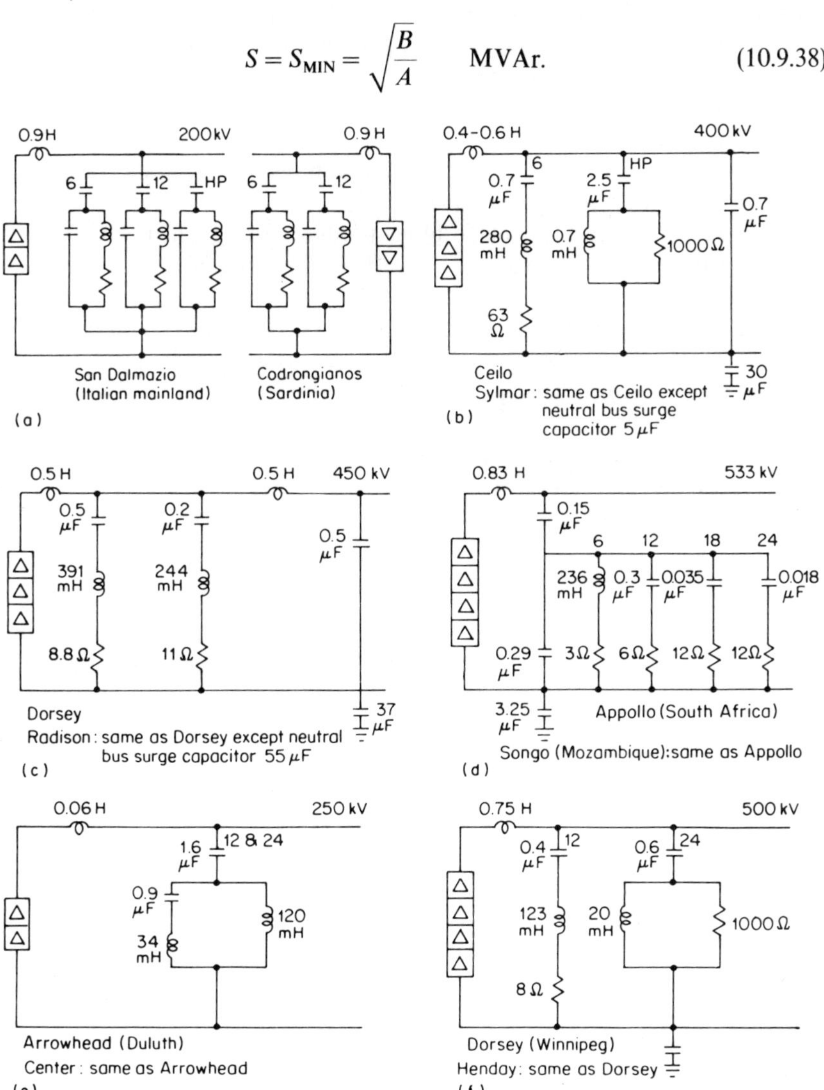

Fig. 10.13. D.C. filter circuits of various hvdc schemes (from reference 7).

10.10 D.C. SIDE FILTERS

Although the rectifier voltage ripple of static convertors generates harmonic currents, these are rarely filtered out because they do not have a direct effect on other processes or consumers.

High voltage d.c. transmission is a special case, where overhead lines are used, because of the interference problems. The amplitudes of the harmonic currents, discussed in Chapter 3, depend on known elements such as the delay and extinction angles, the overlap angle, the impedances of d.c. circuit (i.e. smoothing reactors, damping circuits, surge capacitors and the line itself).

Three different criteria have been used to define the performance of d.c. filters in d.c. transmission schemes: (i) maximum voltage TIF on the d.c. high voltage bus; (ii) maximum permissible noise to ground in telephone lines close to the high voltage d.c. line; (iii) maximum induced noise intensity in a parallel test line 1 km away from the high voltage d.c. line.

Typical types and location of d.c. filters used in existing schemes are illustrated in Figure 10.13, and the subject is discussed thoroughly by Harrison and Krishnayya.[(7)]

Component ratings are considerably different to those for an a.c. filter, since the harmonic current is reduced to a relatively small value by the large d.c. smoothing reactor; consequently the capacitor cost is almost entirely dependent on its capacitance and the d.c. voltage. The capacitor has the greatest cost and is chosen first; the inductor is then fixed for a given frequency. The selection of the quality factors is made as for an a.c. filter (e.g. by using equation (10.4.26)).

10.11 ALTERNATIVE IDEAS FOR HARMONIC ELIMINATION

Because of the complexity and cost of filters, there have been several attempts to achieve harmonic control by other means. These are: (i) elimination by magnetic flux compensation, (ii) elimination by harmonic injection, and (iii) elimination by d.c. ripple injection.

Magnetic flux compensation

This method of harmonic elimination is basically introduced in Figure 10.14. A current transformer is used to detect the harmonic components coming from the non-linear load and these are fed, through an amplifier, into the tertiary windings of a transformer in such a manner as to cause cancellation of the harmonic currents concerned. The main area of concern with this system involves the coupling of the output of the amplifier to the tertiary winding, in such a way that the fundamental current flow does not damage the amplifier. A quaternary winding and filter are used, as shown in Figure 10.14, to reduce the fundamental current in the amplifier output.

One advantage with this scheme is its ability to take account of uncharacteristic harmonics such as the third and ninth. A disadvantage with this scheme is its

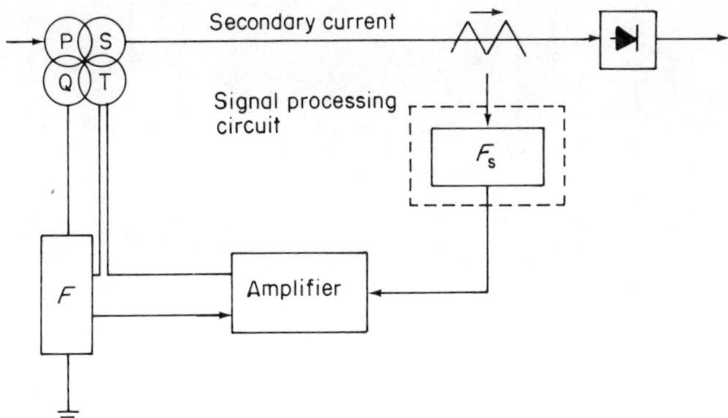

Figure 10.14. The basic configuration of the harmonic current elimination method by flux compensation

inability effectively to remove the lower order harmonics without the need for a very high power feedback amplifier.

The system has been developed to an advanced experimental stage by Sasaki and others[8] and it is now up to technology to develop a suitable amplifier for the economic implementation of such a scheme.

Harmonic injection

Another means by which harmonics can be eliminated is by adding a harmonic current to the rectangular wave form produced, as shown in Figure 10.15. Such a scheme, as originally proposed by Bird[9] and later generalized by Ametani,[10] utilizes the triplen harmonics and an external triplen harmonic source. This principle (illustrated in Figure 10.16) can be used to reduce some of the harmonics at a given operating point.

The advantage of these schemes over filtering is that the system impedance is not part of the design criteria. One disadvantage is the need to maintain a triplen

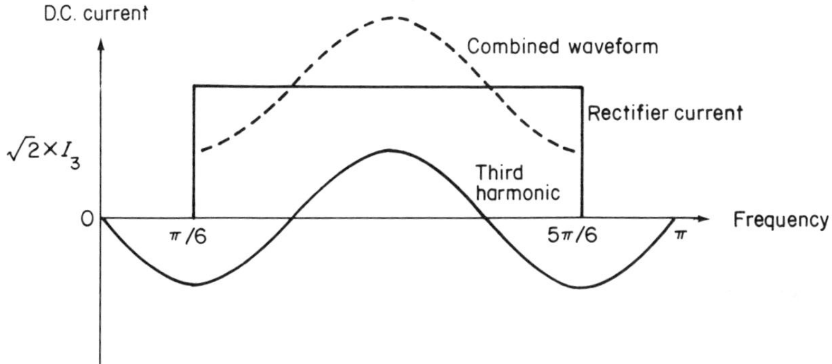

Figure 10.15. Adding a third harmonic current to rectifier waveform

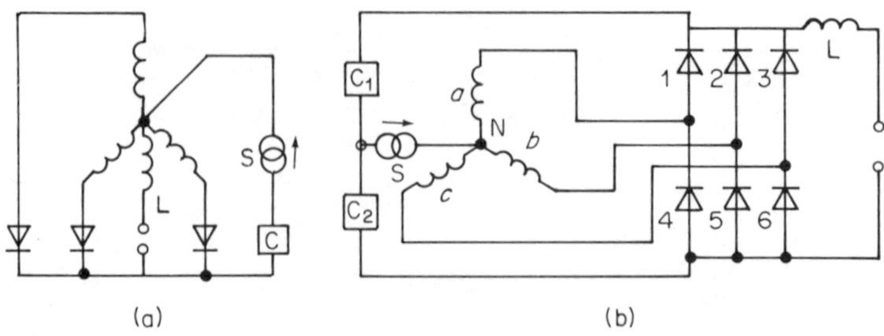

Figure 10.16. Classical harmonic injection: (a) three-phase half-wave rectifier; (b) three-phase bridge rectifier. L, smoothing reactor; N, neutral point of d.c. windings; S, injected current source; C, cutoff circuit of direct current

harmonic source in synchronism with the system. A further disadvantage is the power dissipation requirement of the triplen harmonic generator, often up to 10% of the d.c. power of the rectifier. However, they suffer from the following disadvantages: (i) need of a triplen harmonic current generator and its synchronization to the supply main frequency; (ii) difficulty in adjusting the amplitude of the sinusoidal injected current to suit each particular operating condition; (iii) difficulty in adjusting the phase of the sinusoidal injected current to suit each particular operating condition; (iv) inability to nullify more than one harmonic order at any operating point; (v) poor efficiency due to ineffective dissipation of the triplen harmonic power injected; (iv) the passive system proposed by Bird, although solving (i) and (ii) above, is only applicable to rectifiers operating with 0° delay.

D.C. ripple injection [11,12]

Static convertor schemes produce a ripple voltage at their d.c. output. With six-pulse rectification the ripple has a period of $1/6T$, where $T = 1/f$.

However, with respect to the star point of the convertor side transformer windings, each d.c. pole has non-sinusoidal ripple voltage of period $1/3T$, i.e. a triple frequency voltage. This voltage has the same phase relationship on each d.c. pole and is referred to as the common mode d.c. ripple voltage.

The principle of d.c. ripple reinjection, illustrated in Figure 10.17, is applicable to naturally commutated convertors with 120° conducting valves. The transformers must be star-connected on the rectifier side and must have either a delta primary or delta tertiary winding.

The primary winding of a single-phase transformer is connected to the common mode d.c. ripple voltage. This transformer provides the commutating voltage for a single-phase full-wave rectifier (or feedback convertor) connected to the secondary winding. The output of the feedback convertor is connected in series with the d.c. output of any of the six-pulse convertor configurations. The frequency of the harmonic injection is determined by the supply frequency and therefore the problem of synchronising the harmonic source with the mains frequency does not arise.

The problem of injected current phase adjustment is solved by using

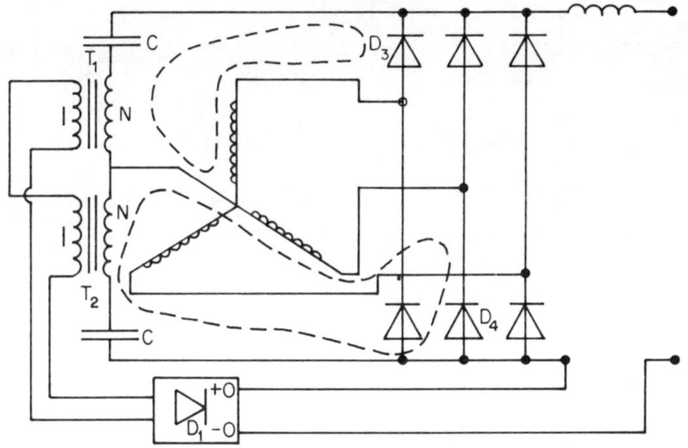

Figure 10.17. Bridge rectifier with ripple re-injection. T_1, T_2, feedback rectifier transformers; N:1, turns ratio; C, blocking capacitor; D_1, feedback rectifier. Broken lines indicate the paths of injected current while D_3 and D_4 are on

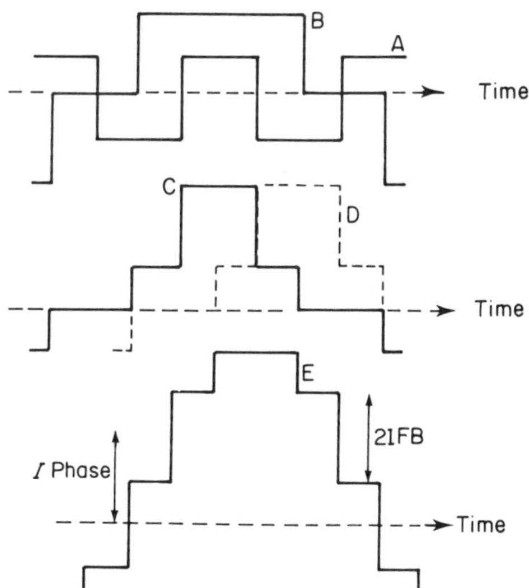

Figure 10.18. Inductive d.c. load: trace A, triple frequency injected current; trace B, rectifier current before modification; trace C, modified phase current, rectifier winding; trace D, second phase displaced 120°; trace E, resultant phase current on delta primary

controlled-rectifier feedback. The firing angle control of the feedback convertor is thus locked to the main rectifier control, i.e. if the thyristors of the feedback convertor are fired 30° after the corresponding main convertor thyristors then the waveforms illustrated in Figure 10.18 result.

A Fourier analysis of waveform (E) shows that a particular ratio of the injected to rectifier current ratio all harmonics of order harmonic order $6n \pm 1$ (where

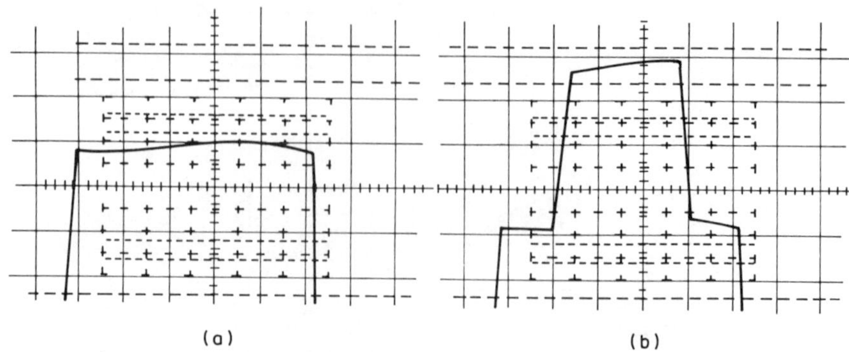

Figure 10.19. A.C. current waveforms: (a) without ripple re-injection; (b) with ripple re-injection

$n = 1, 3, 5, \ldots$) are zero, while the other harmonic orders (i.e. for $n = 2, 4, 6, \ldots$) retain the same relationship with the fundamental as before. The result is that the original six-pulse convertor configuration has been converted into a 12-pulse convertor system from the point of view of a.c. and d.c. system harmonics.

Typical experimental waveforms of the current waveforms are illustrated in Figure 10.19(a) without and in Figure 10.19(b) with ripple reinjection.

10.12 REFERENCES

1. Kimbark, E. W. (1971). *Direct Current Transmission*, John Wiley, New York, Chap. 8.
2. Stanley, C. H., Price, J. J., and Brewer, G. L. (1977). 'Design and performance of a.c. filters for 12-pulse h.v.d.c. schemes'. *IEE Conf. Publ.*, **154**, 158–161.
3. Ainsworth, J. D. (1965). 'Filters, damping circuits and reactive voltamperes in h.v.d.c. convertors'. In *High Voltage Direct Current Convertors and Systems* (B. J. Cory, ed.). Macdonald, London, pp. 137–174.
4. Harris, K. A. (1965). 'Some design aspects of the Cross-Channel power link'. In *High Voltage Direct Current Convertors and Systems* (B. J. Cory, ed.), Macdonald, London, pp. 196–198.
5. Abramovich, B. J., *et. al.* (1982). 'Harmonic filters for the Sellindge convertor stations'. *GEC J. Sci. Tech.*, **48**, 35–38.
6. Powell, R. O. M. (1966). 'The design of capacitor components of large high voltage a.c. filter networks'. *IEE Conf. Publ.*, **22**, 275–276.
7. Harrison, R. E., and Krishnayya, P. C. S. (1978). 'System considerations in the application of d.c. filters for h.v.d.c. transmission'. CIGRE paper 14–09. Paris.
8. Sasaki, H., and Machida, T. (1971). 'A new method to eliminate a.c. harmonic currents by magnetic flux compensation. Considerations on basic design'. *Trans. IEEE*, **PAS-90**, 2009–2019.
9. Bird, B. M., Marsh, J. F., and McLellan, P. R. (1969). 'Harmonic reduction in multiplex convertors by triple-frequency current injection'. *Proc. IEE*, **116**, 1730–1734.
10. Ametani, A. (1972). 'Generalized method of harmonic reduction in a.c.–d.c. converter by harmonic current injection'. *Proc. IEE*, **119**, 857–864.
11. Baird, J. F., and Arrillaga, J. (1980). 'Harmonic reduction in d.c.-ripple reinjection'. *Proc. IEE, C*, **127**, 294–303.
12. Arrillaga, J., Joosten, A. P. B., and Baird, J. F. (1983). 'Increasing the pulse number of a.c.–d.c. convertors by current reinjection techniques'. *Trans. IEEE*, **PAS-102**, 2649–2655.

Index